Autologous Blood Concentrates, Second Edition

Library of Congress Cataloging-in-Publication Data

Names: Garg, Arun K., 1960- author.
Title: Autologous blood concentrates / Arun K. Garg.
Description: Second edition. | Batavia, IL : Quintessence Publishing Co, Inc, [2021] | Includes bibliographical references and index. | Summary: "Presents the science and biology of PRP and other autologous blood concentrates before demonstrating its use in implant surgery, soft and hard tissue healing, facial cosmetics, and other clinical applications"-- Provided by publisher.
Identifiers: LCCN 2021005706 (print) | LCCN 2021005707 (ebook) | ISBN 9781647240837 (hardcover) | ISBN 9781647240844 (epub)
Subjects: MESH: Platelet-Rich Plasma | Blood Platelets--physiology | Blood Transfusion, Autologous | Plasma--physiology | Bone Regeneration--physiology | Oral Surgical Procedures
Classification: LCC QP97 (print) | LCC QP97 (ebook) | NLM WH 400 | DDC 612.1/17--dc23
LC record available at https://lccn.loc.gov/2021005706
LC ebook record available at https://lccn.loc.gov/2021005707

A CIP record for this book is available from the British Library.
ISBN: 9781867150837

QUINTESSENCE PUBLISHING USA

Quintessence Publishing Co, Inc
411 N Raddant Road
Batavia, IL 60510
www.quintpub.com

5 4 3 2 1

Editor: Leah Huffman
Design: Sue Zubek
Production: Angelina Schmelter

Printed in Croatia

Autologous Blood Concentrates

Second Edition

Arun K. Garg, DMD

Founder and President
Implant Seminars
Miami, Florida

Former Professor of Surgery
Division of Oral and Maxillofacial Surgery
Miller School of Medicine
University of Miami
Miami, Florida

QUINTESSENCE PUBLISHING

Berlin | Chicago | Tokyo
Barcelona | London | Milan | Mexico City | Moscow | Paris | Prague | Seoul | Warsaw
Beijing | Istanbul | Sao Paulo | Zagreb

Contents

Preface

In the nearly 25 years since Bob Marx and I developed the original formula for PRP, I have followed its gradual evolution from controversial idea to vital wound healing agent with keen interest—not unlike an anxious parent. In 2005, our co-authored book—*Dental and Craniofacial Applications of PRP* (Quintessence)—introduced the concept of platelet-rich plasma (PRP) to the world and provided scientific and clinical proof of its efficacy. Together, we spent the next decade training other clinicians on the proper use of PRP.

Everything changed in 2010, when it became public knowledge that PRP was the secret to Tiger Woods's speedy recovery from a torn ACL. Commercial interests quickly co-opted the conversation, drowning out the voices of those who, like Marx and me, did not want to see this low-cost biotechnology exploited by profit-hungry manufacturers of centrifuge devices. There was also an enormous amount of misinformation being promulgated by certain medical/dental experts who recognized the enormous therapeutic potential of autologous growth factors and seized the opportunity to establish a name for themselves in the scientific community. The medical literature became saturated with articles introducing new terminology to describe slightly modified growth factor compositions, often without much (if any) additional clinical benefit. The result was an alarming lack of standardization in protocols, a nomenclature best described as an alphabet soup of acronyms, and an overwhelming sense of confusion among clinicians. In 2015, when I published the first edition of this book, my primary motivation was to "set the record straight."

So the first edition of this book was my effort to refocus the conversation about platelet-derived therapies in order to make PRP accessible again to the practicing clinician. In titling the book, I made the major concession of using a generic term—autologous blood concentrates—as a way to signal my desire to focus on the science and not the politics. I wanted the book to reach clinicians regardless of which machine or nomenclature they were most familiar with.

In the world of regenerative biotechnology, 6 years is a very long time, and much has happened since the first edition of this book was published. This new second edition has been thoroughly revised, updated, and expanded to reflect current understanding, applications, and protocols of PRP for clinicians who have been using or wish to start using PRP in their practice. The centerpiece of this edition is a completely new chapter that details the step-by-step formulas and processes for preparing eight configurations of PRP and the specific indications for using each one. How and when to apply the various configurations in clinical dentistry—for soft tissue preservation, hard tissue preservation and regeneration, and facial rejuvenation procedures—is the subject of subsequent chapters. Because the use of PRP requires the clinician or an assistant to perform a venipuncture, the final chapter is a comprehensive guide to the principles and practice of phlebotomy.

I continue to engage in clinical PRP research, both in my private practices and through my charitable foundation, and will remain a passionate advocate for its use for the benefit of patients everywhere.

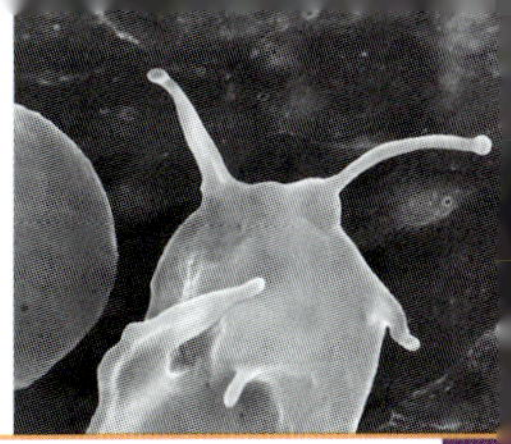

1

Autologous Blood Concentrates: The Science of Natural Wound Healing

In many ways, traditional surgery and the medical arts have always tried to remove barriers to natural wound healing. Removal of these barriers proved that through replicable conditions and cases, standardized protocols could be created and followed to enhance wound healing. Over time, replication of results led to even more standardized techniques and procedures. For example, wound debridement and administering antibiotics demonstrably helped prevent infection, and stabilizing wounds and placing tissues in closer physical proximity promoted healing. These particular kinds of standardized, replicable surgical techniques can be labeled assistive or nonobstructive.[1] However, beginning in the last quarter of the 20th century, a truly "proactive" phase in surgical medicine began with the discovery that macrophages, reacting with oxygen, release growth factors that promote wound healing.[2-6] An assortment of cellular/tissue and oxygen-related therapies followed,[7-16] culminating only about two decades ago in the use of growth factors produced from concentrated autologous blood platelets to promote wound healing.[17-22] The result is medical science's present focus on platelet and other biologic/regenerative therapies as critical means for promoting, initiating, and sustaining wound healing.[23]

In the mid-1980s, platelets were understood essentially as cells that helped to stop bleeding. Over the next 20 years, the discovery of the various growth factors released by platelets gave birth to regenerative medical therapies, most of which are still in their infancy.[24-35] How the growth factors and functional matrix delivered by autologous blood concentrates induce wound healing is widely understood. The focus of current research is replicating and standardizing the preparation and administration of the autologous-derived product to best suit the donor-patient. Though a variety of preparation techniques, products, and nomenclatures have been tried, the good news is that no significant difference in the osteogenesis of growth factors has been evidenced.[36-40]

Nevertheless, firmly establishing the science of platelet-rich plasma and other platelet-derived products requires an investigation of platelet biology, the release of growth factors, and the practical application for soft tissue healing and bone tissue regeneration. So far, the scientific journey of autologous blood concentrates has been remarkably expansive. The future of this journey promises to be more focused, even single-minded, toward its scientific destination—even more standardized products and procedures based on replicable results.

Platelet Biology

The first autologous blood concentrate was introduced in the literature as *autologous fibrin adhesive* and later changed to *platelet-rich plasma* (PRP). That term became standard, first in the oral surgery literature and then in all medical and dental surgical specialties. While many other terms have been used to describe autologous blood concentrates—particularly in niche markets as a way to sell specific centrifuges and/or test tubes—PRP will be used throughout this book.

As an actor in the performance of regenerative medicine,[41-47] PRP provides two of the three essential components for allowing a wound to heal in place: growth factors and a scaffolding stage (Fig 1-1). The third ingredient for

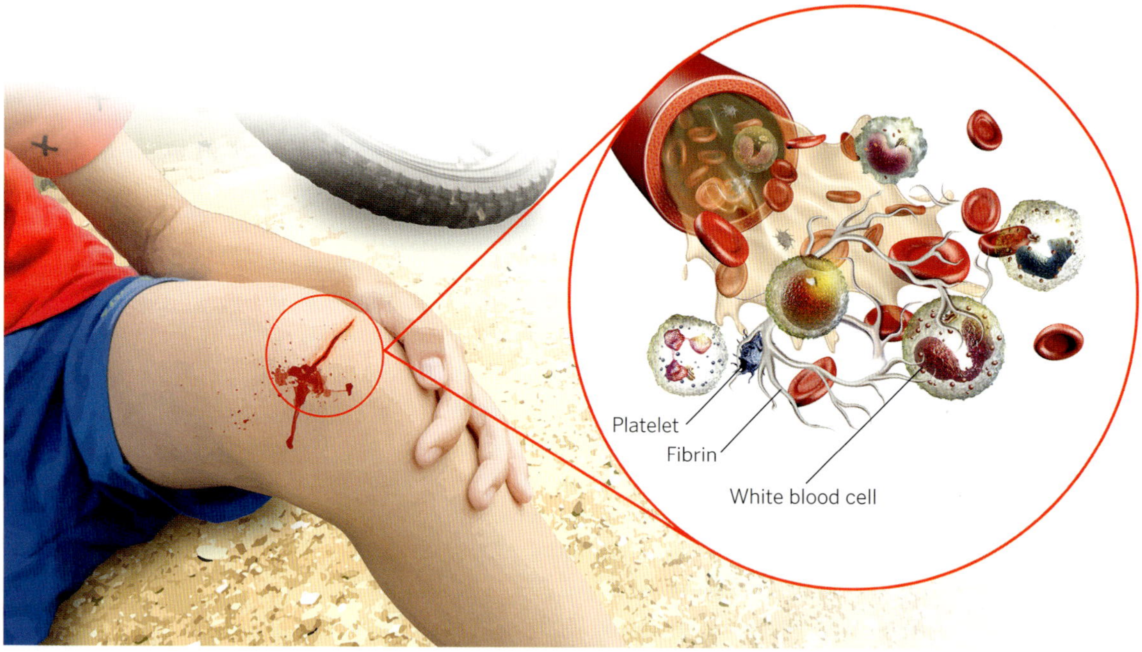

FIG 1-1 Components of blood that are concentrated in PRP.

in situ tissue regeneration is the cells. PRP is a patient's own blood concentrate, modified in a relatively quick, efficient, safe, and simple procedure, to obtain a dense concentration of platelets. The autologous nature of PRP precludes disease transmission to the patient or other adverse reactions. To provide wound healing benefit to the patient, PRP generally must have at least four to seven times the normal concentration of platelets, or roughly 1 million platelets per microliter.[48] A blood clot in a wound consists mostly of red blood cells and much smaller percentages of platelets and white blood cells (Fig 1-2). Applying PRP to such a wound essentially replaces red blood cells with growth factor–producing platelets and a fibrin network, thus (at least in theory) greatly enhancing the healing of wounds and migration of cells, as well as the regeneration of bone and soft tissue.

When bone marrow megakaryocytes undergo cytoplasmic fragmentation, the anuclear platelet cells enter the circulatory system. The relatively tiny platelet is about one-fourth the size of a red blood cell (approximately 8 μm in diameter) and six to seven times smaller than a lymphocyte; however, the platelet's membrane extends pseudopodially via invaginations, which provide an expansive, dynamic, and vigorous surface area for the cell membrane during activation.[1,49] The plasticity and resilience of the platelet's pseudopodic membrane enable its vascular-sealing qualities, along with its ability to form a thrombus and fibrin clot, as well as clot retraction when its hemostatic labors are complete[50] (Fig 1-3). Generally, the larger and younger the platelet, the greater its hemostatic qualities and the greater the quantities of growth factors contained within it.[23,51,52]

The short lives of platelets (240 hours or less) are very actively spent synthesizing and secreting growth factors as part of the blood-clotting process. The platelet contains lysosomes, ribosomes, mitochondria, and an assortment of intercellular proteins that help form its shape as well as its mobility. The platelet cell also contains storage organelles that consist of lysosomal granules (for storing enzymes for digestion), dense granules (for storing and secreting adenosine diphosphate [ADP]), and alpha granules (for summoning and activating other platelets via nascent growth factors) (Fig 1-4).[53-55]

The growth factors stored in the alpha granules include platelet-derived growth factor (PDGF) isomers labeled AA, BB, and AB (referred to as polypeptide "dimers" because

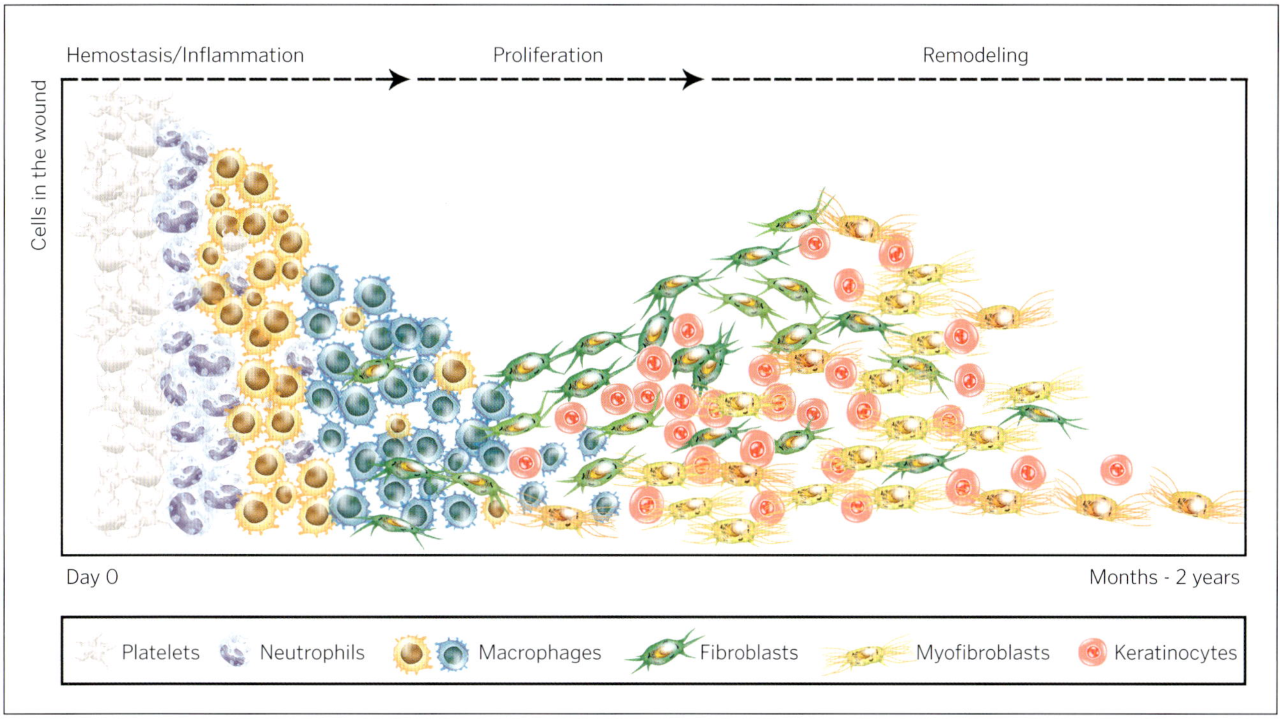

FIG 1-2 The classic wound healing cascade.

of their two active sites, which are actually antiparallel monomers), transforming growth factor (TGF) isomers beta 1 and 2; vascular endothelial growth factor (VEGF); and epithelial growth factor (EGF). Growth factors not contained in platelets include insulinlike growth factors (IGF) 1 and 2 and bone morphogenetic protein (BMP). The blood-clotting process activates the alpha granules in platelets to secrete growth factors, both when the platelets circulate normally in the blood and when the platelets are concentrated in PRP. Alpha granules move toward the membrane and bind themselves to its surface, causing histone and carbohydrate side chains to combine with, and to activate, growth factor proteins.[1]

In addition to their basic hemostatic roles, platelets have been found to play a number of nonhemostatic functions.[56,57] Tissue repair and inflammation are two of several functions that researchers are currently exploring.[58,59] The alpha granules in platelets are the producers/directors of the diverse roles medical science is learning platelets can play beyond the traditional role of hemostasis. Ironically, in fact, the granules contain substances that normally work in opposition to each other. The platelet's ability (when properly signaled) to release different substances specifically

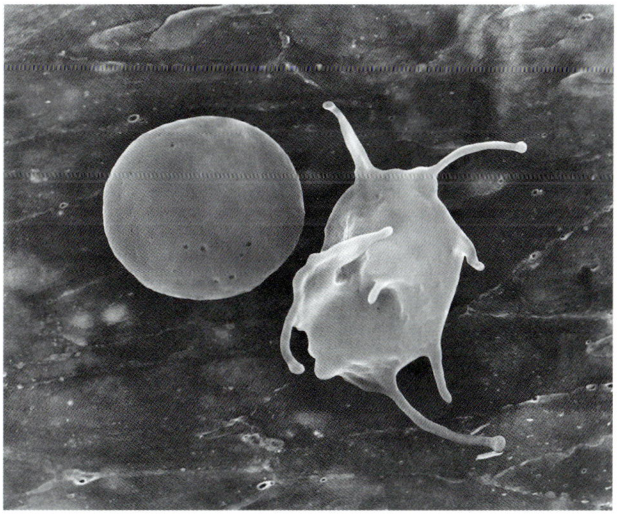

FIG 1-3 Scanning electron micrograph (SEM) appearance of a platelet before *(left)* and after *(right)* activation.

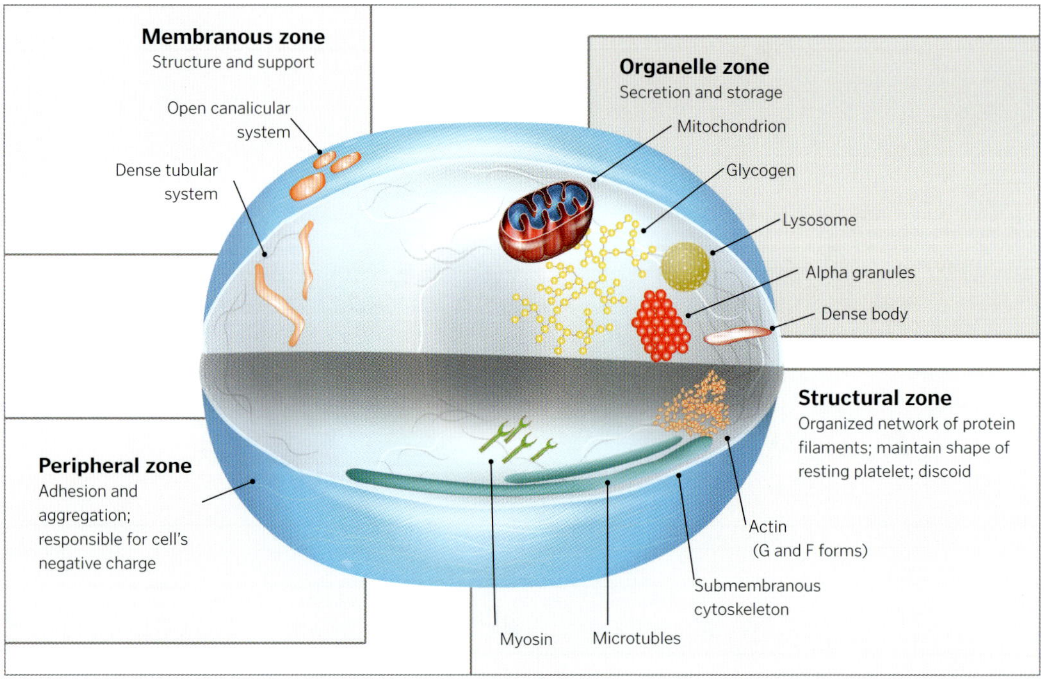

FIG 1-4 Shape and functions of platelet cells.

required by other cells, molecules, and tissues explains why it is the focus of so many different types of current medical research, including tissue inflammation and regeneration, neurology, autoimmunity, hemostasis, wound healing, atherosclerosis, and a diverse range of others.[60-66]

PRP Growth Factors

There are a number of different types of growth factors contained in platelets.[67-69] These growth factors are polypeptides, accounting not only for tissue and organ morphogenesis (from shortly after human conception to adulthood) but also for cell differentiation and proliferation. Their crucial activity in cell healing makes them especially important to current research in tissue engineering and regenerative medicine (Fig 1-5). The growth factors that have received the most attention from medical researchers and practitioners include PDGF-AA, PDGF-BB, and PDGF-AB; TGF-β1, TGF-β2, and TGF-β3; fibroblastic growth factor (FGF); IGF; EGF; and VEGF.

PDGF

PDGFs, found in several types of human cells but mostly in platelets, were the first type of growth factor discovered in alpha granules.[27] PDGF adheres only to target cells with receptive surface membranes. When a wound is treated with concentrated platelets, the release of PDGF triggers activity in fibroblasts, neutrophils, and macrophages, stimulating the latter to additionally release growth factors that help to heal injured tissues.[70-72] Specifically, the target cells' transmembrane receptors are activated by the platelet-released growth factors, and the receptors' intracytoplasmic qualities activate the signal transducer proteins, one of which migrates to the target cell's nucleus. There, the protein initiates a particular, regulated gene sequence—which may, for example, include the production of osteoid.

Alternatively, the sequence might lead to the synthesis of collagen. The regulated nature of the sequence precludes an "overdose" of concentrated growth factors. Growth factors' amutagenic qualities testify to their ability as natural proteins to initiate and participate in the regular genetic activities associated with the controlled mechanisms for wound healing.[1,73-76] As isomers of a single protein, and functioning as mitogens, PDGFs—the most common growth factors—perform different but often complementary tasks (Fig 1-6). This prompts certain cells to replicate, specifically mesenchymal stem cells, osteoblasts (producing osteoid), endothelial cells (secreting basal lamina), and fibroblasts (producing collagen).[77-83]

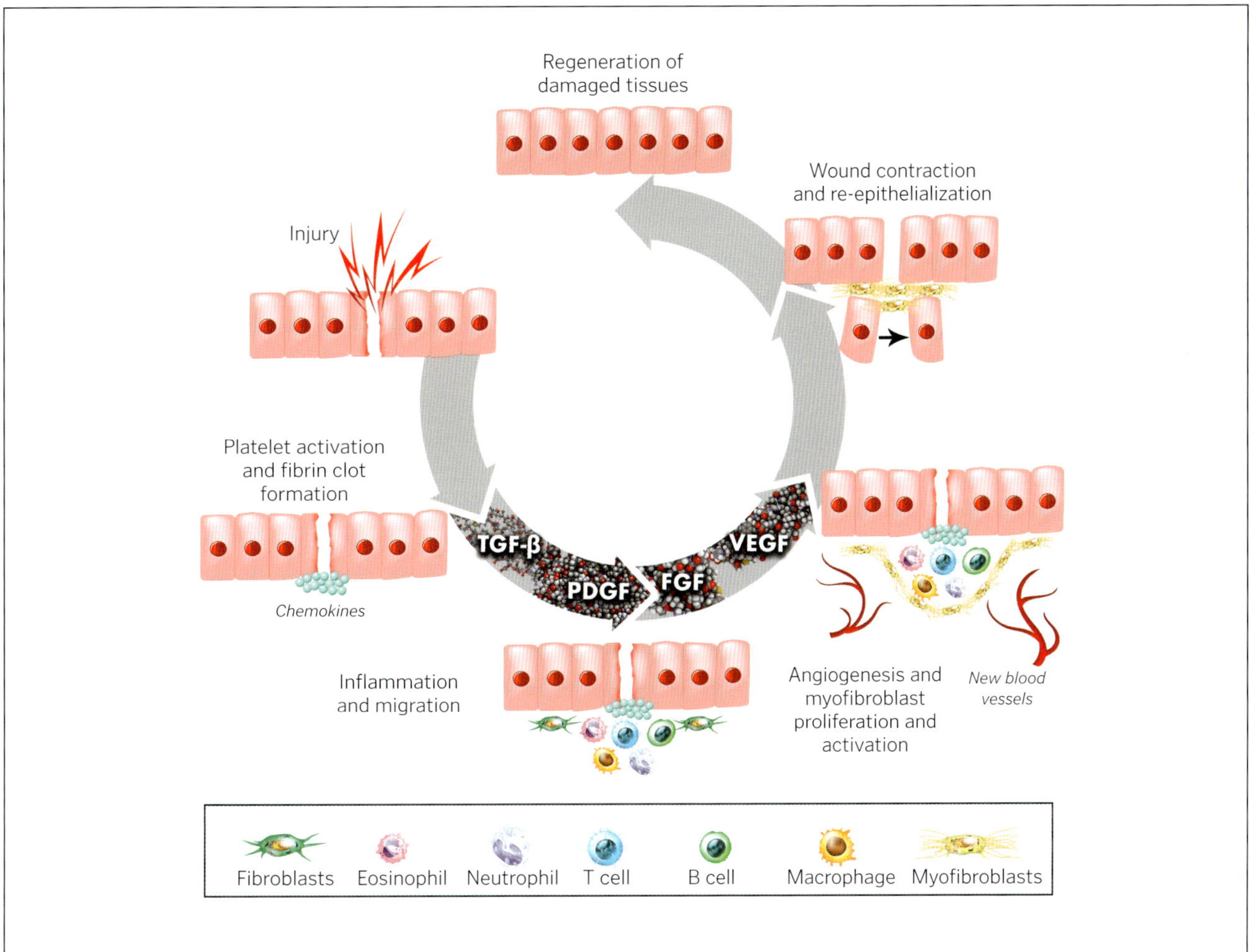

FIG 1-5 Wound healing and tissue regeneration cycle.

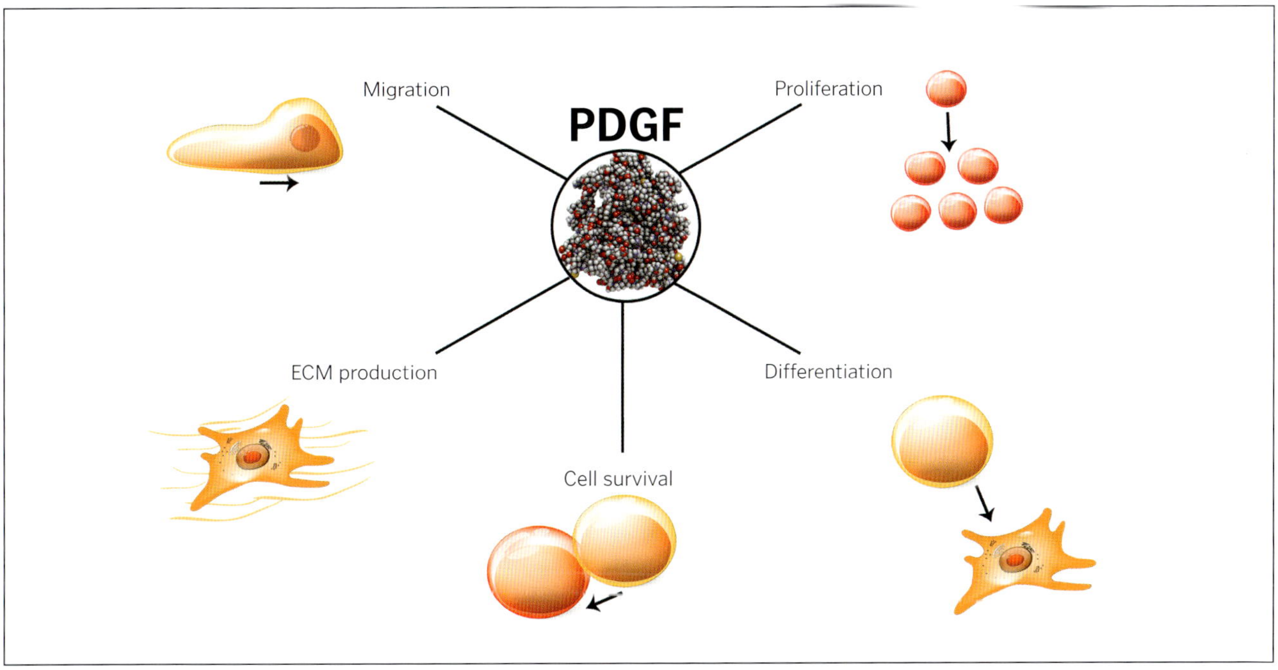

FIG 1-6 Roles of PDGF.

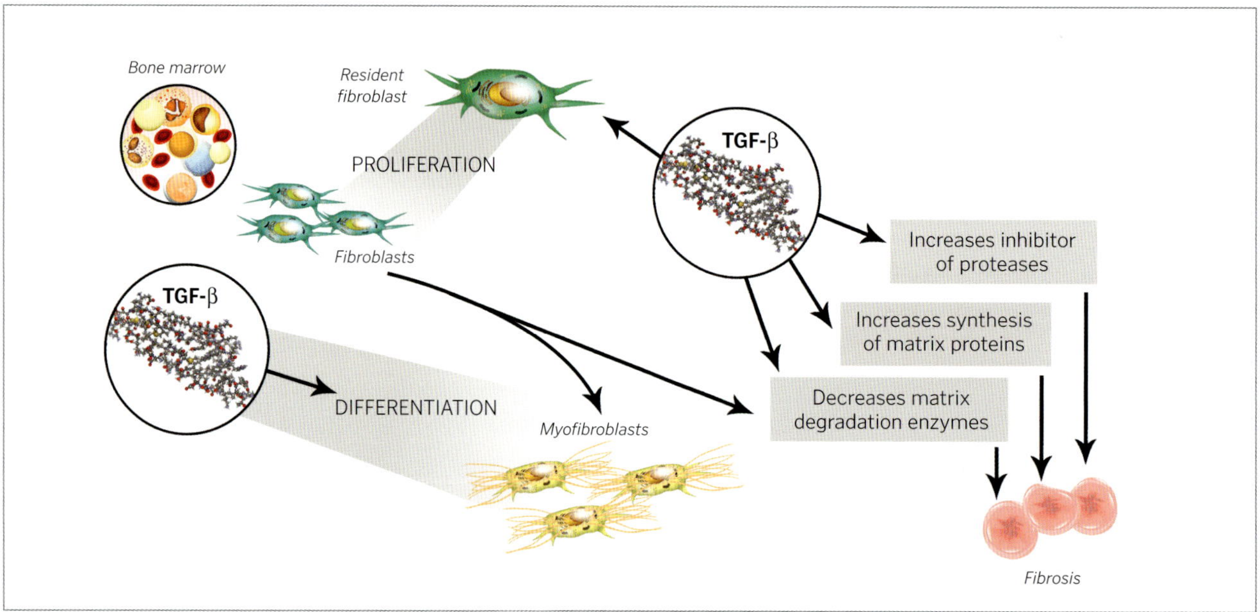

FIG 1-7 TGF-β stimulates the proliferation and migration of fibroblasts and the process of epithelial-mesenchymal transition, but it also stimulates fibrotic responses, which can lead to end-stage organ failure in the heart, kidney, etc.

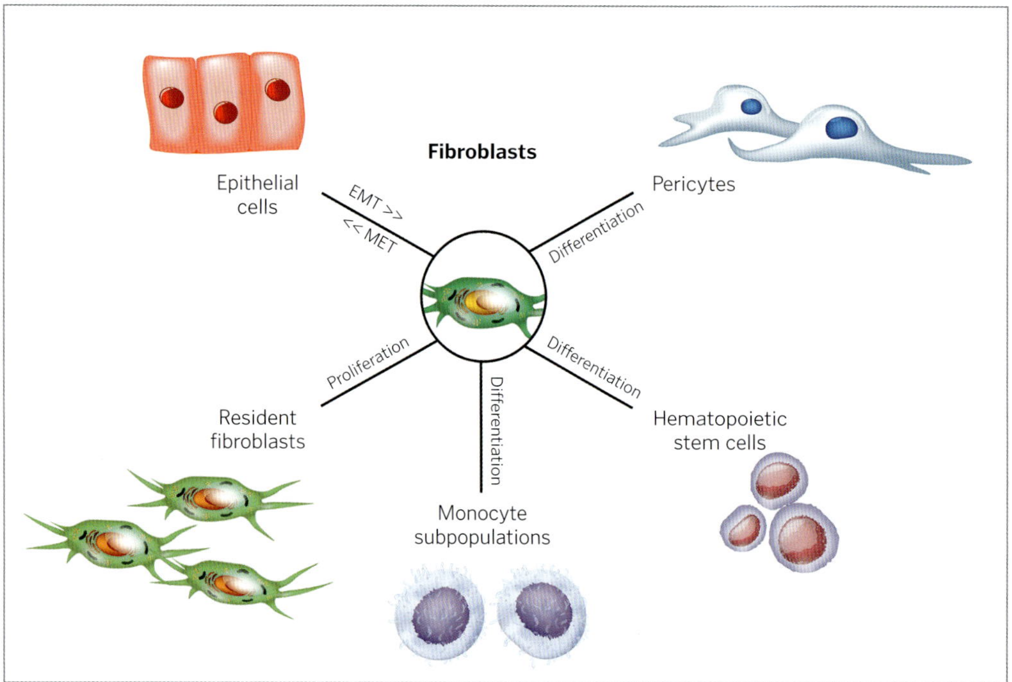

FIG 1-8 Fibroblasts have different origins at different developmental stages (EMT, epithelial-mesenchymal transition; MET, mesenchymal-epithelial transition).

TGF-β1, TGF-β2, and TGF-β3

There are dozens of TGFs (including BMPs), and three of them (TGF-β1, TGF-β2, and TGF-β3) are protein growth factors that behave not only as mitogens (for cell replication) but also as morphogens (for cell differentiation).[84,85] As cell preservers, TGF-β factors serve essential functions in wound healing, from the fetus to the adult.[86–89] Like PDGFs, TGFs exert their influence in both healthy and pathologic cell activities, making their therapeutic use extremely problematic and challenging in that they often behave in opposite and even contradictory ways in both soft and hard tissues, including their roles in proliferation, migration, and differentiation[90–93] (Fig 1-7).

FGF

FGF is an important mitogen inducer of fibroblast and endothelial cell proliferation. It also stimulates angiogenesis and plays a vital role in the repair of skeletal muscles and tendons. New blood vessel formation is due in large part to the activities of FGF, and it stimulates the migration of the macrophage as well as epithelium for epidermis formation[40,94,95] (Fig 1-8).

IGF

IGFs are peptide hormones that have been found to promote cell growth in in vitro experiments. IGFs were also found to reduce levels of blood glucose in various tissues, hence their name. Their ability to stimulate glandular activities in humans differentiates them from other PRP growth factors. For example, IGF-1 mainly adheres to and stimulates receptors in cells of pituitary growth hormones. Most cells increase in size and number as a direct result of the synthesis and secretion of IGF-1 by tissues stimulated by growth hormones.

Growth factor production, along with growth hormone production and concentrations, wax and wane as part of the natural maturation process of humans before, during, and after pubescence, particularly for IGF-1. Liver production accounts for most IGF concentrations. IGFs have their most potent growth-stimulating effect on themselves and on nearby cells, and they (both IGF-1 and IGF-2) aid in bone cell mitosis. As an osteoblast secretion, IGF assists in osteogenesis and bone ossification via proliferation and differentiation of cells[96-98] (Fig 1-9).

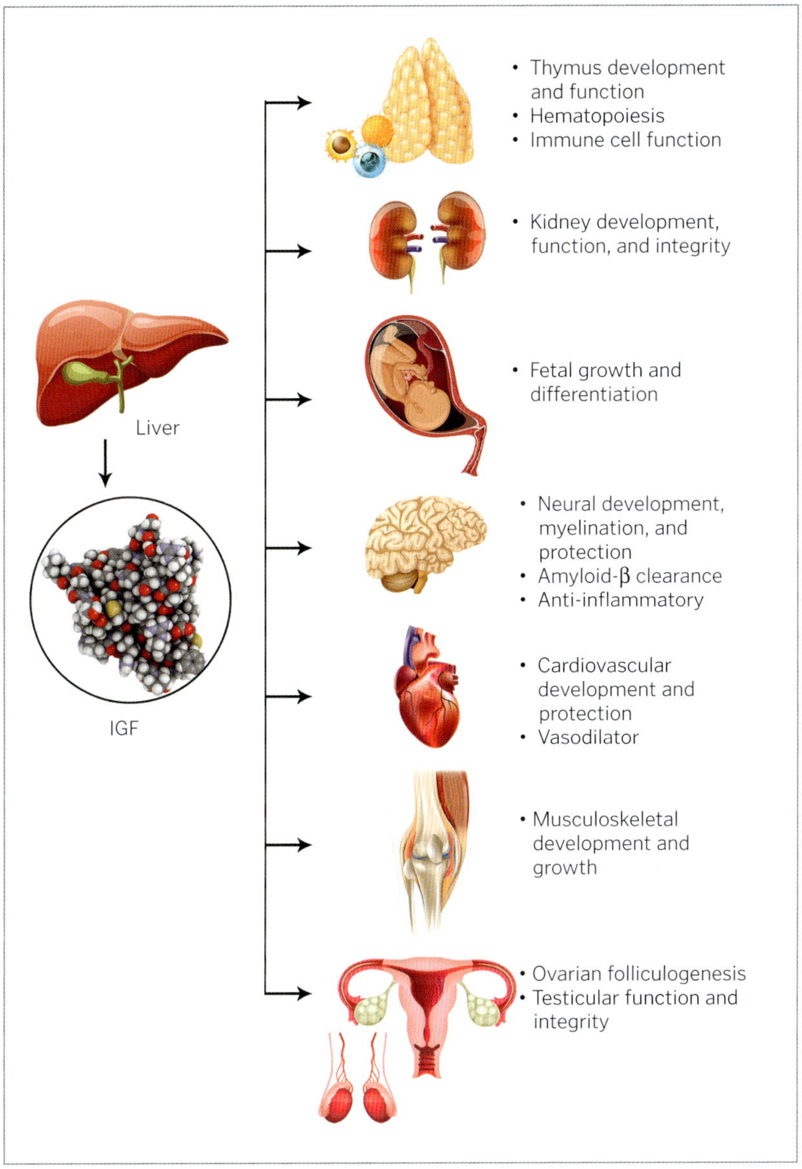

FIG 1-9 Some of the roles of IGFs.

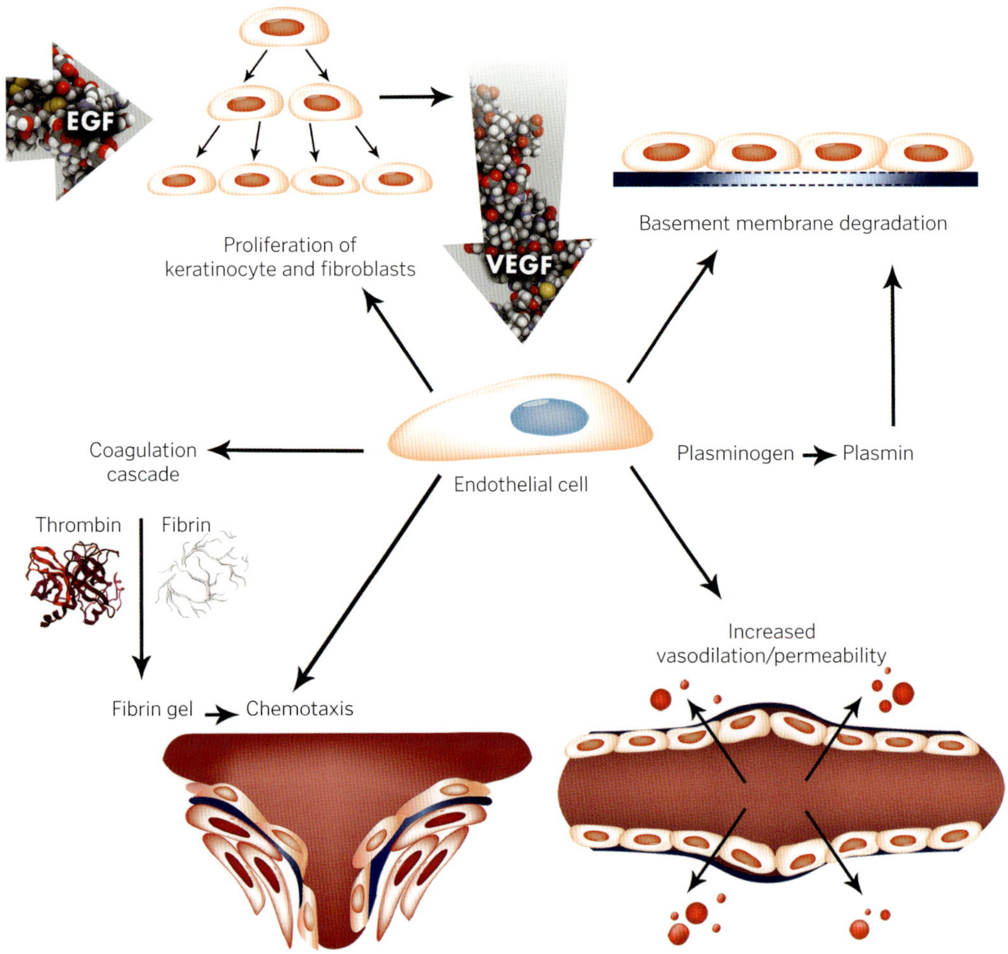

FIG 1-10 VEGF stimulates the endothelial cell, enhancing multiple phases of the angiogenic cascade.

EGF

The protein-based EGF causes the basal cells of the skin and mucous membranes to replicate, migrate, and form essential elements of these membranes.[99] EGF likely positively affects the generation of tissues as well as wound healing because of the way it controls the proliferation, growth, and migration of epithelial cells while strengthening the formation of new blood vessels. A therapeutic application of EGF has been used in preventing and treating dermatitis resulting from overexposure to x-rays or radium.

VEGF

As its name suggests, the protein-based VEGF helps to develop blood vessels by interacting with endothelial cells, stimulating the synthesis of basal lamina, and recruiting pericytes[69,100] (Fig 1-10). Like several other growth factors, VEGF is known for and studied scientifically mainly due to its activity in pathologic states rather than healthy ones. The role it plays in the formation of new blood vessels of cancerous tumors has provided much of what we know of its abilities to stimulate growth in other cells. Tumors, just like healthy cells, can synthesize growth factor proteins—including VEGF—that promote the formation of new blood vessels from existing ones (angiogenesis). Instead of angiogenesis taking place for normal body growth or tissue repair, in this pathologic process it enables the spread of cancerous cells. In this case, VEGF activity leads to the development of capillaries within the tumor because of its stimulation of endothelial cells, which are the raw materials needed for angiogenesis. Endothelial cell division leads to the growth and migration of the tumor cells, developing a cascade of shared growth factors between their cells. Possible cancer therapies, therefore, include ways to introduce proteins into the tumor that slow or stop angiogenesis.

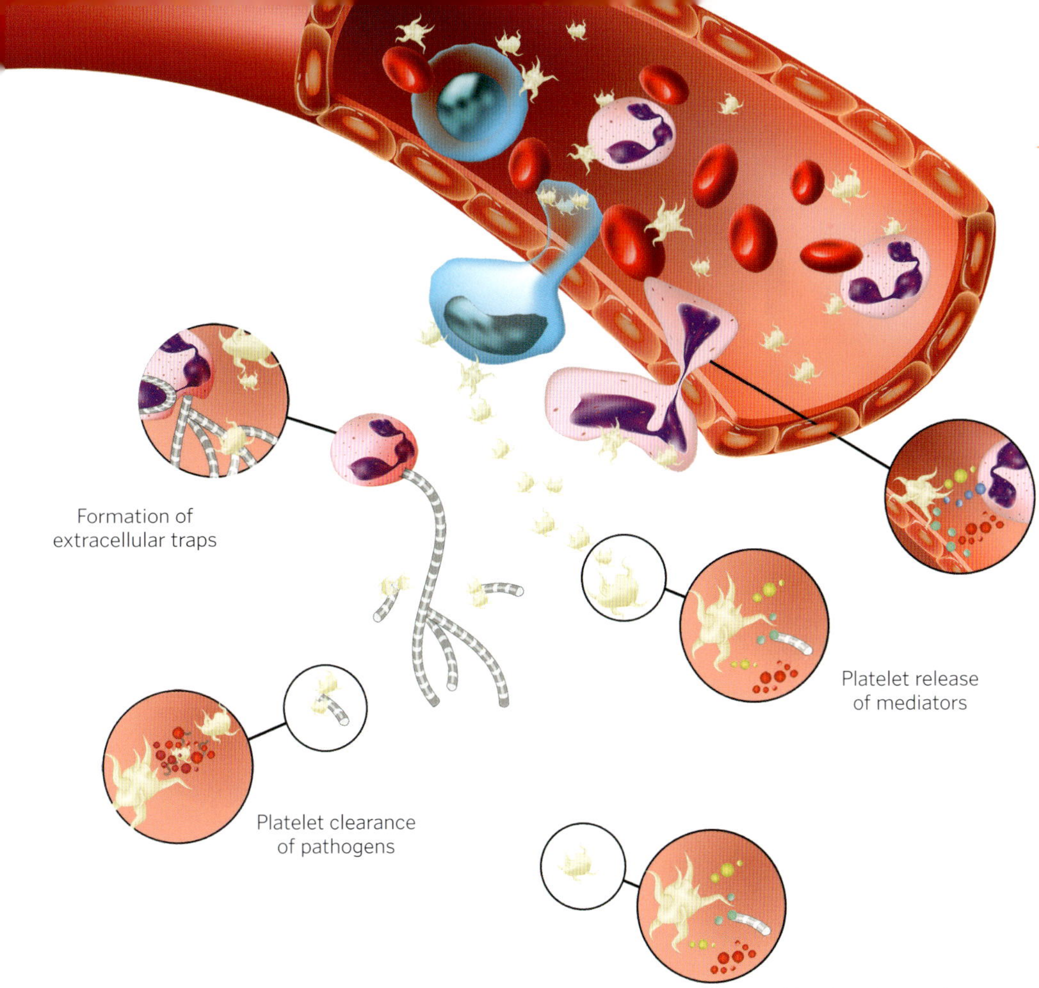

FIG 1-11 Effects of platelet-leukocyte aggregates on differentiation, activation, and cytokine release.

PRP and Soft Tissue Healing

PRP has shown great promise clinically and histologically, especially for healing soft tissue in a standard wound of a donor site for a split-thickness skin graft.[101,102] Additionally, healing therapies for burns may benefit significantly from PRP application. The platelets release growth factors and cell-signaling cytokines, such as interleukin and interferon, that act to regulate inflammation and infection in the immune system[103] (Fig 1-11).

When compared to non-PRP–assisted clotting, PRP-assisted clotting is remarkable for its rapidity of healing in the basal cells at the edge of the wound, where EGF induces epithelial proliferation; subsequent migration to the granulation tissue helps the clot's cell adhesion molecules. Unlike an unassisted clot, the PRP clot reveals the bundles of fibroblasts and collagen, evidence of an expanding epithelium, and more mature healing. This comparatively accelerated maturity is also evidenced by increasingly reduced vascularity and fibroblastic cellularity over time, as well as quicker fleshlike appearance in 2 to 6 months. Reduced pain in the first 7 days of the wound, and reduced scarring over time, are also notable differences effected by the PRP clot. These benefits are also demonstrated in healing of other soft tissue wounds, including mucosal flaps, dermal fat grafts, and similar wounds.[1]

Bone and soft tissue healing can be strengthened in a variety of surgical procedures when a concentrated mixture of autologous platelets is placed at the wound site. The relative ease of methods for obtaining PRP makes it an attractive regenerative adjunct therapy for many surgical treatments. Promoters of PRP tout its ability not only to help restore damaged bone and soft tissue but also to enhance wound healing, lower the patient's pain and discomfort after the surgical procedure, and reduce the rates of infection and loss of blood.[104-106] Much of the recent

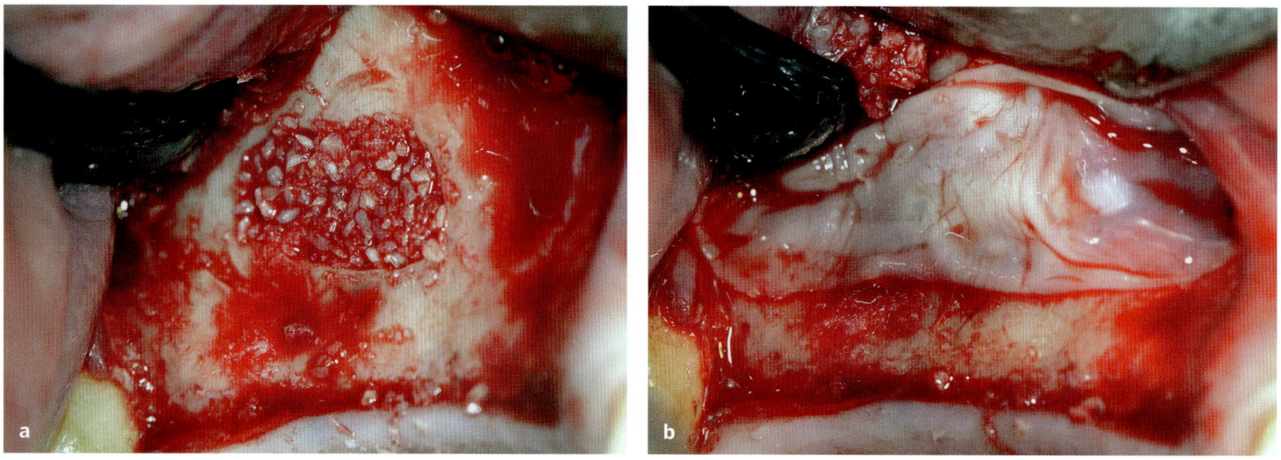

FIG 1-12 Application of PRP in a sinus elevation procedure. *(a)* Once an adequate volume of graft material hydrated with PRP liquid has been placed, the window is ready for a PRP membrane. *(b)* The PRP membrane is placed over the lateral window, and then the flap is sutured back in place.

literature on PRP has been devoted to its wide range of applications in tissue healing and repair, including maxillofacial, periodontal, oral (Fig 1-12), and plastic surgery; heart and spine surgery; and chronic ulcers of the skin and soft tissue.[107–110] Some studies have suggested that in addition to enhancing wound healing, PRP provides antimicrobial qualities to inhibit postoperative infections in oral surgery.[111] PRP has been used in such soft tissue therapies as ligament, muscle, and tendon repair[38,112]; rotator cuff tears[113–115]; skin ulcers[116,117]; acne scarring[118]; and limb amputation.[119–121]

PRP and Bone Regeneration

PRP accelerates and expands cells' wound healing response and acts biochemically to set the rate and amount of regeneration in bone. Within 10 days, its activity is complete, but this short action has long-lasting effects. For example, the alpha granules in platelets degranulate within several minutes of clot formation, and within 1 hour 90% of their growth factors are released, stimulating osteoprogenitor, endothelial, and mesenchymal stem cells. A graft-surrounding matrix is formed by fibrin, fibronectin, and vitronectin. PDGFs have a mitogenic effect on osteoblast, endothelial, and mesenchymal stem cells. The latter are also acted upon mitogenetically and angiogenetically by TGF-β isomers, which induce osteoblastic differentiation as well. While capillary ingrowth is promoted by VEGF, the lack of epithelial cells renders EGF inert (Fig 1-13a). Within about 72 hours, osteoprogenitor cell mitosis begins and capillary buds appear (Fig 1-13b). In the entire first phase of bone graft healing (about 2½ to 3 weeks), the graft is penetrated by capillaries, and osteoprogenitor cells have greatly proliferated (Fig 1-13c). During this phase, cell instability and infection are common, with the potential for lysing and arresting the development of wound healing. Obviously, prevention of infection and contamination are essential, as is graft stability.[1]

The hypoxic and acidic atmosphere of the wound itself attracts the circulating macrophage and blood monocyte (soon a wound macrophage), both of which assist bone regeneration via the secretion of more growth factors. The clot now contains fibrin, fibronectin, and vitronectin, acting as a matrix for the ingrowth of blood vessels as well as the proliferation and migration of cells. Between 3 and 6 weeks, the proliferation and differentiation of osteoprogenitor cells in the matrix produce osteoid (Fig 1-14), which signals the next (second) phase of healing, when graft and bone join and when adventitial cells develop to support the vascular ingrowth (Fig 1-15). Hypoxia diminishes due to the oxygen provided by the increased blood flow, preventing hyperplasia. By week 6, osteoclasts resorb the osteoid, releasing BMPs and IGF factors 1 and 2, causing the differentiation of nearby osteoblasts and mesenchymal stem cells for maturing bone replacement (Fig 1-16). Mineralized dense bone thus becomes the normal formation now, in the third phase

FIG 1-13 *(a)* The biochemical environment of an autogenous bone graft. *(b)* As early as 3 days after graft placement, significant cell divisions and penetration of capillary buds into the graft can be seen. *(c)* By 17 to 20 days, complete capillary penetration and profusion of the graft has taken place, and osteoid production has been initiated. (Reprinted with permission from Marx and Garg.[1])

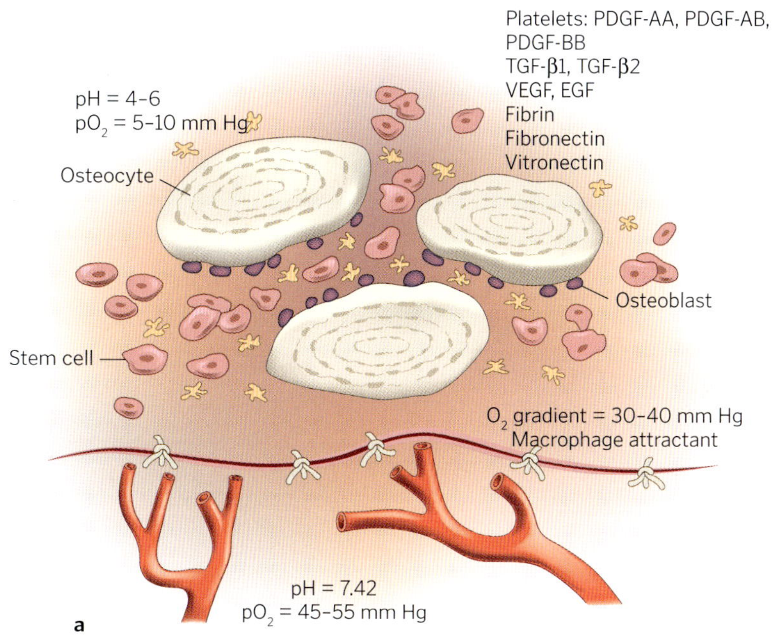

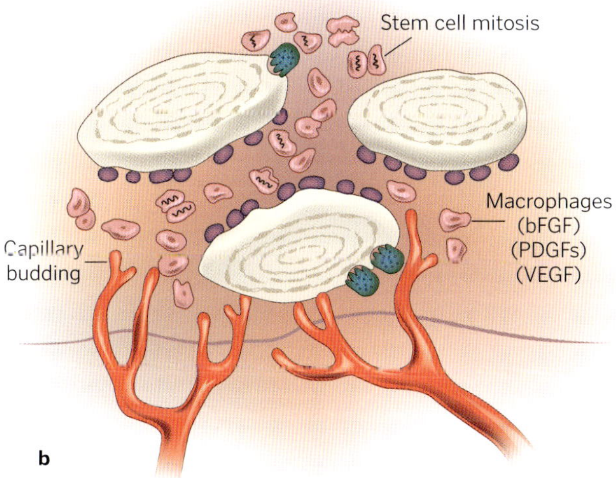

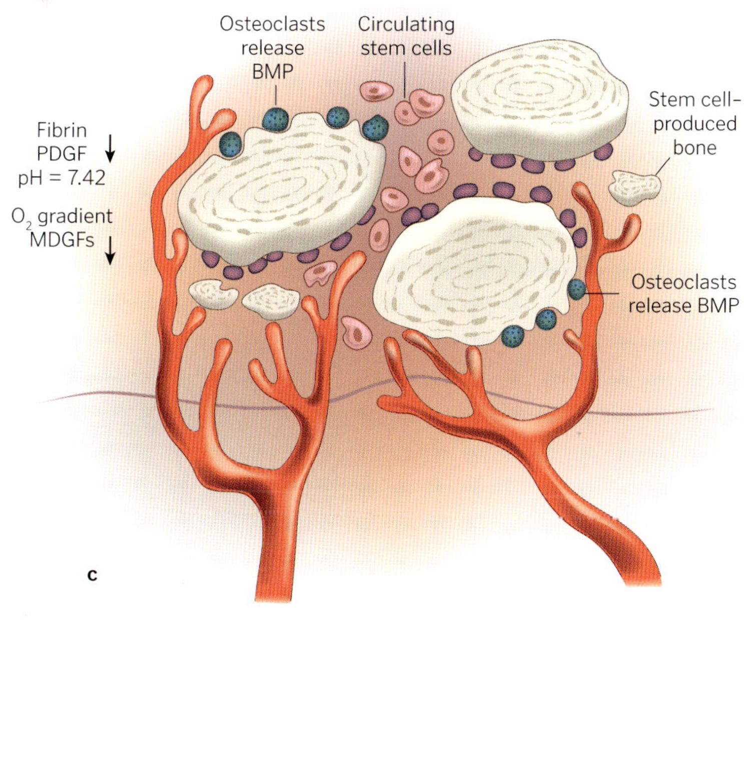

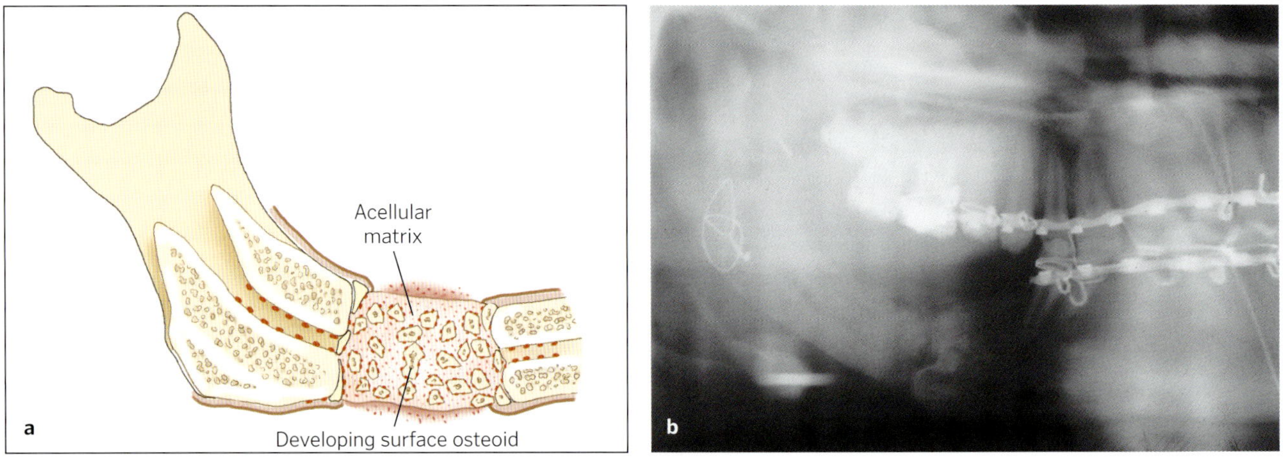

FIG 1-14 *(a)* Acellular matrix along with surface osteoid developing on the endosteal surfaces of the transplanted bone and the resection edges of the host bone in a 3-week autogenous bone graft. *(b)* Corresponding radiograph shows a not-yet-mineralized graft with a "cloudy" appearance indicative of a graft that is not yet consolidated. The radiolucent line between the graft and host bone is the result of a dying-back resorption of the host bone from periosteal reflection. (Reprinted with permission from Marx and Garg.[1])

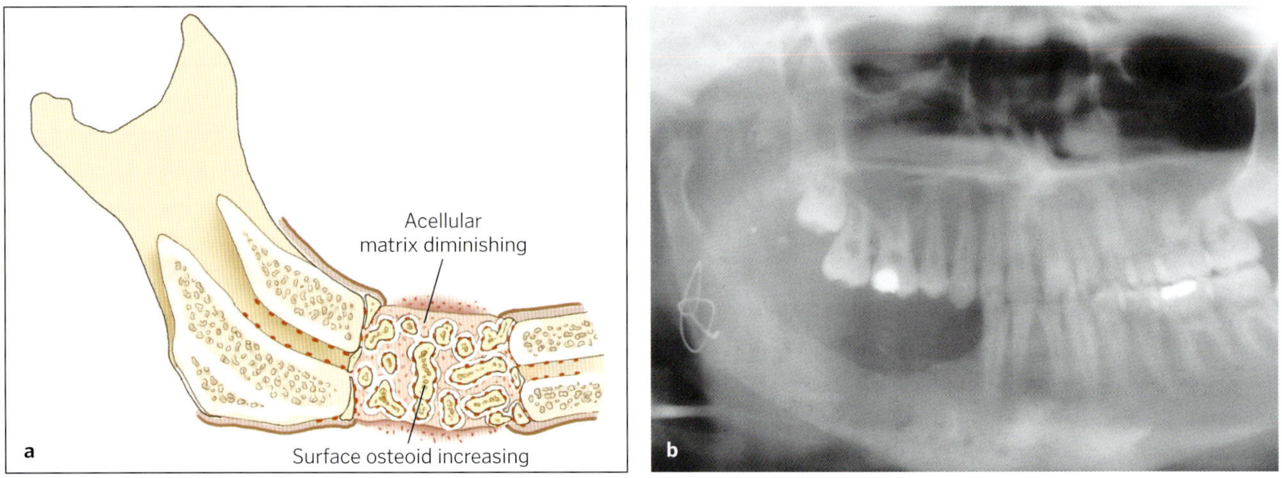

FIG 1-15 *(a)* By fusing graft particles together and to the host bone, the graft has produced sufficient osteoid to consolidate by 6 weeks. *(b)* Corresponding radiograph shows condensation of the cloudy graft appearance, indicative of osteoid production and graft organization. The radiolucent line between the graft and host bone has nearly disappeared as a result of osteoconduction between the graft and host bone edge. (Reprinted with permission from Marx and Garg.[1])

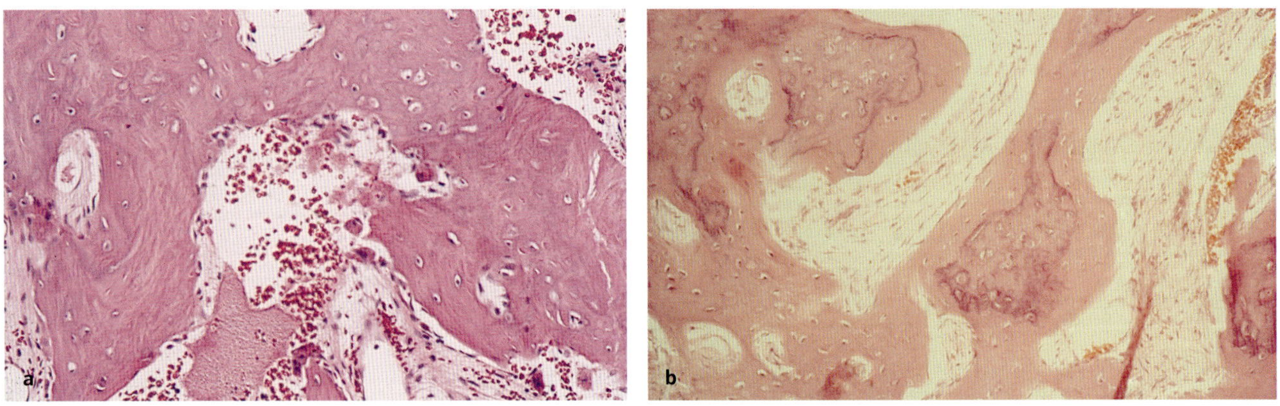

FIG 1-16 *(a and b)* At about 6 weeks, the graft begins a major resorption-remodeling cycle in which osteoclasts resorb the disorganized immature bone and release BMP and insulinlike growth factors, thus inducing formation of new bone that will mature during function. (Reprinted with permission from Marx and Garg.[1])

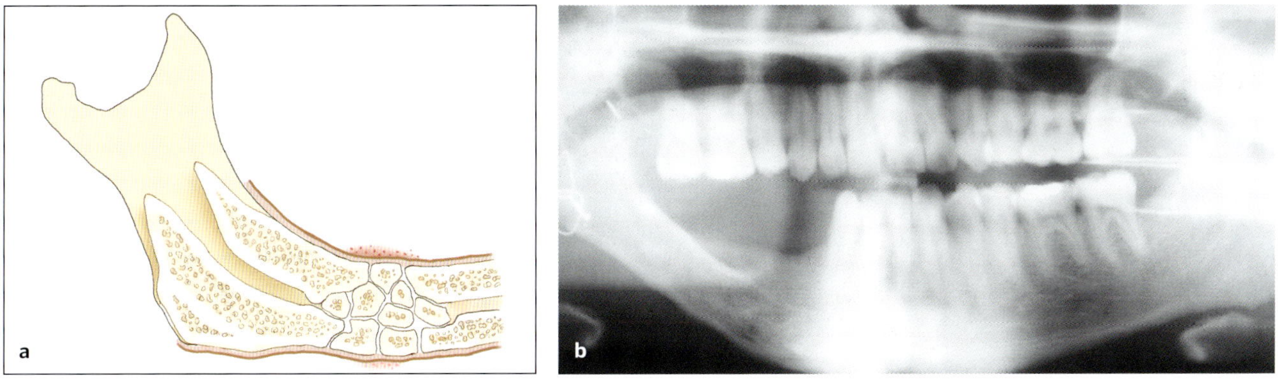

FIG 1-17 *(a)* After 6 weeks, the graft will be consolidated and fused to the host bone. It then enters the lifelong resorption-remodeling cycle of the remainder of the skeleton. *(b)* Radiographically, bone maturation is characterized by the development of a normal trabecular pattern and an increased density. Here, an inferior border outline, an external oblique ridge, and a coronoid process attest to the remodeling of bone under function. (Reprinted with permission from Marx and Garg.[1])

of bone regeneration, as the graft-fused bone life cycle parallels the regular turnover rate of bone replacement in the body (Fig 1-17).

Platelet growth factors not only induce bone cell regeneration but also double the normal increase in mineral density in bone, with faster-forming and more quickly maturing bone, including significantly increased trabecular bone values[48] (Fig 1-18). Bone-related therapies for PRP include oral and cranial surgery,[48,122–124] spinal fusion,[125–128] osteogenesis distraction,[129–131] foot and ankle fractures,[132–134] bone grafting,[135,136] oral implants,[137–143] and diabetic fractures.[78,144,145]

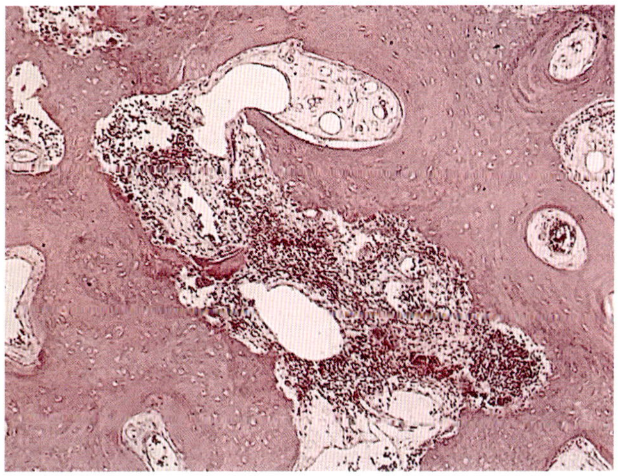

FIG 1-18 Histomorphometry of an autogenous bone graft without PRP at 4 months shows that the graft has a 60% trabecular bone density, consists mostly of immature bone, and is undergoing active resorption-remodeling. (Reprinted with permission from Marx and Garg.[1])

What Is PRP?

Efforts to standardize protocols for the clinical application of autologous platelet-derived product have often been accompanied by nonstandardized nomenclature for that product. Many researchers believe the term *platelet-rich plasma* is too general and incomplete, leading to confusion not only in preparation and application but also in cataloging in the scientific databases, thus hampering the flow of information in the scientific community. Other names that have been applied over the years include *platelet-rich fibrin* (PRF), *autologous platelet-rich plasma* (aPRP), *platelet gel* (PG), *autologous platelet concentrates* (APCs), *preparation rich in growth factors* (PRGF), *platelet-derived growth factor* (PDGF), *platelet-leukocyte gel* (PLG), *concentrated growth factors* (CGFs), and numerous others (Fig 1-19).

FIG 1-19 The terminology for PRP is a confusing alphabet soup of acronyms. Chapter 4 presents the names and formulations for eight configurations of PRP and the step-by-step instructions for preparing them.

A simplified terminology-classification system for platelet concentrates has been widely adopted. It consists of eight categories based on its consistency and clinical application. The researchers and clinicians who devised this classification system believe it underscores the profound potential of these products and the need for all stakeholders to learn and demonstrate their awareness of the products' complexity. The composition and application of these other forms of PRP are detailed in the chapters of this book.

Historically, the majority of studies of the clinical efficacy of PRP did not provide sufficient information to allow their PRP preparation protocol to be reproduced. This not only added to the confusion over terminology but also made it impossible to compare the PRP products that were being tested. Several recent papers have addressed this confusion and shared ideas about how to standardize the protocol to make PRP preparation simple and straightforward, and most importantly, to obtain a consistent and effective platelet yield.[146-148] In 2017, Chahla et al published a systematic review of the clinical orthopedic literature on PRP and a call for more standardized reporting of PRP preparation protocols and composition.[146] Although 105 studies met the inclusion criteria for analysis, only 11 of them (10%) provided sufficient detail of their preparation protocol to allow it to be reproduced. Additionally, only 17 studies (16%) provided quantitative metrics on the composition of the PRP product they delivered to patients.

Ongoing Controversies

Since the mid-1990s, clinicians have used an assortment of PRP preparations in several different therapies, ranging from oral and maxillofacial surgery and implantology to orthopedic surgery, burn treatment and the treatment of other wounds difficult to heal, soft tissue cosmetic surgery, soft tissue disease and injuries, and tissue engineering. Invariably, the literature that documents these therapies calls not only for high-level studies to better measure the efficacy of the variety of PRP therapies but also for standardizing the preparation and application of platelet-derived products.[1-4]

Any discussion of PRP application must be preceded by a review of platelet sequestration principles. Only then can a discussion of progress toward consensus on PRP protocols take place. Such an approach reveals that the clinical preparation and application of PRP across medical disciplines is a dynamic, evolving process, and that we disrupt this dynamism at our medical/scientific peril. Too-hasty attempts to reach consensus could inhibit researchers' efforts to discover the full potential of PRP and other biologic therapies.

Naming and classifying platelet derivatives

Nearly 25 years after the earliest reports appeared, the literature on PRP for both human and veterinary medicine still lacks consensus on standardized protocols for preparing, naming,

Table 1-1 Factors that influence the quality of platelet concentrates

Preparation step	Critical factor	Details
Blood collection	Needle Tubing of butterfly needle Syringe/tube Lag	Gauge, length, material, surface modification Diameter, length, material Materials, surface modification Distance between blood-collection space and centrifuge
Centrifuge	Tube Rotator Centrifugal condition	Shape, material, surface modification Swing or angle Force, duration
Other handling	Pipetting Coagulation	Technique, material $CaCl_2$, thrombin, glassware

classifying, and applying platelet concentrates. PRP appears to remain the most likely term, given that plasma resuspension of any platelet concentrate is required before application.[1] A high-density fibrin concentrate can facilitate cell migration and the release of cytokines. However, the number of leukocytes affects the concentrate's wound healing ability, leading to the often contradictory results of different studies.

Standardizing PRP sequestration and application protocols

A rush to standardize PRP protocols could prevent researchers from critical discoveries found only through overcoming clinical obstacles that may defy established norms. For example, a definition of PRP in 2001 proposed an optimal clinical healing concentration of 1,000,000 platelets/µL in a 5-mL volume of plasma for bone.[149] But in 2008, a platelet gel study concluded that a concentration of about 1.5 × 10^6 plt/mL appeared to be optimal for proliferation, migration, and invasion of endothelial cells, showing that higher concentrations of growth factors can adversely affect wound healing for soft tissues.[150]

Despite the general similarity in the protocols for preparing PRP, a number of variables affect whole blood centrifugation for platelet concentration and volume: platelet size, anatomical differences of patients, hematocrit variability, the amount and location of autologous blood drawn, the centrifugal forces and number/duration of spins, and temperature variants (including a refrigerated centrifuge; Table 1-1).[151–157]

Compounding the problems of standardization are variations in centrifugation terminology (for example, rotation-per-minute versus g-force), centrifuge rotor radius, patient age and sex,[158] activating or not activating PRP before application,[159,160] using noncommercial PRP kits, needle bore size, and types of anticoagulants or lack thereof.

The variety of methods for delivering PRP to a wound site demonstrates the evolving and sometimes competing techniques for applying PRP and other biologics in wound healing; however, reliable clinical results often require replicable delivery methods in addition to standard production methods. For example, hydrogels, sponges, and nanofiber scaffold fabrication can be used for treating bone defects with PRP.[159]

So clinicians must be wary about too-rigid standardization of the principles and technologies of platelet sequestration, the nomenclature and classification of platelet-rich products, and the application of those products for wound healing. The dynamic and often contradictory nature of PRP derivations and clinical applications across the medical spectrum may in fact represent opportunities for greater scientific and medical understanding and advancement.

Conclusion

The dental and medical community has traveled a great distance since the mid-1980s when platelets were understood essentially as cells promoting hemostasis. The discovery of growth factors released by platelets introduced regenerative medical therapies that have become the present focus of a great deal of speculation and experimentation in the medical profession.

The succeeding chapters cover both the present and future science of autologous blood concentrates by casting more light than heat on the ongoing conversations concerning standardization and replication of techniques and procedures. It is the author's hope that these chapters will contribute valuable insights concerning biologic/regenerative therapies, helping to remove much of the skepticism regarding the applications of PRP and related autologous products used in a growing number of medical fields, but particularly the fields of oral and maxillofacial surgery, periodontics, dental implants, and facial cosmetic surgery.

References

1. Marx RE, Garg A. Dental and Craniofacial Applications of Platelet-Rich Plasma. Chicago: Quintessence, 2005.
2. Knighton DR, Silver IA, Hunt TK. Regulation of wound-healing angiogenesis-effect of oxygen gradients and inspired oxygen concentration. Surgery 1981;90:262–270.
3. Knighton DR, Hunt TK, Scheuenstuhl H, Halliday BJ, Werb Z, Banda MJ. Oxygen tension regulates the expression of angiogenesis factor by macrophages. Science 1983;221:1283–1285.
4. Marx RE, Johnson RP. Studies in the radiobiology of osteoradionecrosis and their clinical significance. Oral Surg Oral Med Oral Pathol 1987;64:379–390.
5. Hunt TK. The physiology of wound healing. Ann Emerg Med 1988;17:1265–1273.
6. Marx RE, Ehler WJ, Tayapongsak P, Pierce LW. Relationship of oxygen dose to angiogenesis induction in irradiated tissue. Am J Surg 1990;160:519–524.
7. Davis JC, Hunt TK (eds). Problem Wounds—The Role of Oxygen. New York: Elsevier, 1988.
8. Marx RE, Smith BR. An improved technique for development of the pectoralis major myocutaneous flap. J Oral Maxillofac Surg 1990;48:1168–1180.
9. Cordeiro PG, Disa JJ, Hidalgo DA, Hu QY. Reconstruction of the mandible with osseous free flaps: A 10-year experience with 150 consecutive patients. Plast Reconstr Surg 1999;104:1314–1320.
10. Crowther M, Brown NJ, Bishop ET, Lewis CE. Microenvironmental influence on macrophage regulation of angiogenesis in wounds and malignant tumors. J Leukoc Biol 2001;70:478–490.
11. Moeller BJ, Cao Y, Vujaskovic Z, Li CY, Haroon ZA, Dewhirst MW. The relationship between hypoxia and angiogenesis. Semin Radiat Oncol 2004;14:215–221.
12. Ramakrishnan VR, Yao W, Campana JP. Improved skin paddle survival in pectoralis major myocutaneous flap reconstruction of head and neck defects. Arch Facial Plast Surg 2009;11:306–310.
13. Hunter S, Langemo DK, Anderson J, Hanson D, Thompson P. Hyperbaric oxygen therapy for chronic wounds. Adv Skin Wound Care 2010;23:116–119.
14. Larsson A, Uusijärvi J, Eksborg S, Lindholm P. Tissue oxygenation measured with near infrared spectroscopy during normobaric and hyperbaric oxygen breathing in healthy subjects. Eur J Appl Physiol 2010;109:757–761.
15. Vanni CM, Pinto FR, de Matos LL, de Matos MG, Kanda JL. The subclavicular versus the supraclavicular route for pectoralis major myocutaneous flap: A cadaveric anatomic study. Eur Arch Otorhinolaryngol 2010;267:1141–1146.
16. Schreml S, Szeimies RM, Prantl L, Karrer S, Landthaler M, Babilas P. Oxygen in acute and chronic wound healing. Br J Dermatol 2010;163:257–268.
17. Knighton DR, Hunt TK, Thakral KK, Goodson WH 3rd. Role of platelets and fibrin in the healing sequence: An in vivo study of angiogenesis and collagen synthesis. Ann Surg 1982;196:379–388.
18. Hunt TK, Knighton DR, Thakral KK, Goodson WH 3rd, Andrews WS. Studies on inflammation and wound healing: Angiogenesis and collagen synthesis stimulated in vivo by resident and activated wound macrophages. Surgery 1984;96:48–54.
19. Dvorak HF, Harvey VS, Estrella P, Brown LF, McDonagh J, Dvorak AM. Fibrin containing gels induce angiogenesis. Implications for tumor stroma generation and wound healing. Lab Invest 1987;57:673–686.
20. Ofosu FA. The blood platelet as a model for regulating blood coagulation on cell surfaces and its consequences. Biochemistry (Mosc) 2002;67:47–55.
21. Nurden AT, Nurden P, Sanchez M, Andia I, Anitua E. Platelets and wound healing. Front Biosci 2008;13:3532–3548.
22. Roy S, Driggs J, Elgharably H, et al. Platelet-rich fibrin matrix improves wound angiogenesis via inducing endothelial cell proliferation. Wound Repair Regen 2011;19:753–766.
23. Malhotra A, Pelletier MH, Yu Y, Walsh WR. Can platelet-rich plasma (PRP) improve bone healing? A comparison between the theory and experimental outcomes. Arch Orthop Trauma Surg 2013;133:153–165.
24. Ross R, Glomset J, Kariya B, Harker L. A platelet-dependent serum factor that stimulates the proliferation of arterial smooth muscle cells in vitro. Proc Natl Acad Sci U S A 1974;71:1207–1210.
25. Witte LD, Kaplan KL, Nossel HL, Lages BA, Weiss HJ, Goodman DS. Studies of the release from human platelets of the growth factor for cultured human arterial smooth muscle cells. Circ Res 1978;42:402–409.
26. Ross R, Nist C, Kariya B, Rivest MJ, Raines E, Callis J. Physiological quiescence in plasma-derived serum: Influence of platelet-derived growth factor on cell growth in culture. J Cell Physiol 1978;97(3 pt 2 suppl 1):497–508.
27. Kaplan DR, Chao FC, Stiles CD, Antoniades HN, Scher CD. Platelet alpha granules contain a growth factor for fibroblasts. Blood 1979;53:1043–1052.
28. Assoian RK, Komoriya A, Meyers CA, Miller DM, Sporn MB. Transforming growth factor-beta in human platelets. Identification of a major storage site, purification, and characterization. J Biol Chem 1983;258:7155–7160.
29. Derynck R. Transforming growth factor-alpha: Structure and biological activities. J Cell Biochem 1986;32:293–304.
30. Karey KP, Marquardt H, Sirbasku DA. Human platelet-derived mitogens. I. Identification of insulinlike growth factors I and II by purification and N alpha amino acid sequence analysis. Blood 1989;74:1084–1092.

31. Karey KP, Sirbasku DA. Human platelet-derived mitogens. II. Subcellular localization of insulinlike growth factor I to the alpha-granule and release in response to thrombin. Blood 1989;74:1093–1100.
32. Brunner G, Nguyen H, Gabrilove J, Rifkin DB, Wilson EL. Basic fibroblast growth factor expression in human bone marrow and peripheral blood cells. Blood 1993;81:631–638.
33. Dooley DC, Oppenlander BK, Spurgin P, et al. Basic fibroblast growth factor and epidermal growth factor downmodulate the growth of hematopoietic cells in long-term stromal cultures. J Cell Physiol 1995;165:386–397.
34. Banks RE, Forbes MA, Kinsey SE, et al. Release of the angiogenic cytokine vascular endothelial growth factor (VEGF) from platelets: Significance for VEGF measurements and cancer biology. Br J Cancer 1998;77:956–964.
35. Hormbrey E, Gillespie P, Turner K, et al. A critical review of vascular endothelial growth factor (VEGF) analysis in peripheral blood: Is the current literature meaningful? Clin Exp Metastasis 2002;19:651–663.
36. Sánchez M, Anitua E, Andia I. Poor standardization in platelet-rich therapies hampers advancement. Arthroscopy 2010;26:725–726.
37. Taylor DW, Petrera M, Hendry M, Theodoropoulos JS. A systematic review of the use of platelet-rich plasma in sports medicine as a new treatment for tendon and ligament injuries. Clin J Sport Med 2011;21:344–352.
38. Moraes VY, Lenza M, Tamaoki MJ, Faloppa F, Belloti JC. Platelet-rich therapies for musculoskeletal soft tissue injuries. Cochrane Database Syst Rev 2014;4:CD010071.
39. Kim TH, Kim SH, Sándor GK, Kim YD. Comparison of platelet-rich plasma (PRP), platelet-rich fibrin (PRF), and concentrated growth factor (CGF) in rabbit-skull defect healing. Arch Oral Biol 2014;59:550–558.
40. Sandrey MA. Autologous growth factor injections in chronic tendinopathy. J Athl Train 2014;49:428–430.
41. Stellos K, Langer H, Daub K, et al. Platelet-derived stromal cell-derived factor-1 regulates adhesion and promotes differentiation of human CD34+ cells to endothelial progenitor cells. Circulation 2008;117:206–215.
42. Okabe K, Yamada Y, Ito K, Kohgo T, Yoshimi R, Ueda M. Injectable soft-tissue augmentation by tissue engineering and regenerative medicine with human mesenchymal stromal cells, platelet-rich plasma and hyaluronic acid scaffolds. Cytotherapy 2009;11:307–316.
43. Kon E, Filardo G, Di Martino A, Marcacci M. Platelet-rich plasma (PRP) to treat sports injuries: Evidence to support its use. Knee Surg Sports Traumatol Arthrosc 2011;19:516–527.
44. Wu CC, Chen WH, Zao B, et al. Regenerative potentials of platelet-rich plasma enhanced by collagen in retrieving pro-inflammatory cytokine-inhibited chondrogenesis. Biomaterials 2011;32:5847–5854.
45. Bava ED, Barber FA. Platelet-rich plasma products in sports medicine. Phys Sportsmed 2011;39:94–99.
46. Torricelli P, Fini M, Filardo G, et al. Regenerative medicine for the treatment of musculoskeletal overuse injuries in competition horses. Int Orthop 2011;35:1569–1576.
47. Smyth NA, Murawski CD, Fortier LA, Cole BJ, Kennedy JG. Platelet-rich plasma in the pathologic processes of cartilage: Review of basic science evidence. Arthroscopy 2013;29:1399–1409.
48. Marx RE, Carlson ER, Eichstaedt RM, Schimmele SR, Strauss JE, Georgeff KR. Platelet-rich plasma: Growth factor enhancement for bone grafts. Oral Surg Oral Med Oral Pathol Oral Radiol Endod 1998;85:638–646.
49. Ono A, Westein E, Hsiao S,et al. Identification of a fibrin-independent platelet contractile mechanism regulating primary hemostasis and thrombus growth. Blood 2008;112:90–99.
50. Cosemans JM, Angelillo-Scherrer A, Mattheij NJ, Heemskerk JW. The effects of arterial flow on platelet activation, thrombus growth, and stabilization. Cardiovasc Res 2013;99:342–352.
51. Karpatkin S. Heterogeneity of human platelets. VI. Correlation of platelet function with platelet volume. Blood 1978;51:307–316.
52. Hartley PS. Platelet senescence and death. Clin Lab 2007;53:157–166.
53. White JG. Electron microscopy methods for studying platelet structure and function. Methods Mol Biol 2004;272:47–63.
54. Coppinger JA, Maguire PB. Insights into the platelet releasate. Curr Pharm Des 2007;13:2640–2646.
55. Bijak M, Saluk J, Ponczek MB, Nowak P, Wachowicz B. The synthesis of proteins in unnucleated blood platelets [in Polish]. Postepy Hig Med Dosw (Online) 2013;67:672–679.
56. Weyrich A, Cipollone F, Mezzetti A, Zimmerman G. Platelets in atherothrombosis: New and evolving roles. Curr Pharm Des 2007;13:1685–1691.
57. Linden MD, Jackson DE. Platelets: Pleiotropic roles in atherogenesis and atherothrombosis. Int J Biochem Cell Biol 2010;42:1762–1766.
58. Schattner M. Platelets and galectins. Ann Transl Med 2014;2:85.
59. Romaniuk MA, Rabinovich GA, Schattner M. Galectins in the regulation of platelet biology. Methods Mol Biol 2015;1207:269–283.
60. Garraud O, Cognasse F. Platelet toll-like receptor expression: The link between "danger" ligands and inflammation. Inflamm Allergy Drug Targets 2010;9:322–333.
61. Nurden AT. Platelets, inflammation and tissue regeneration. Thromb Haemost 2011;105(suppl 1):S13–S33.
62. Goubau C, Buyse GM, Di Michele M, Van Geet C, Freson K. Regulated granule trafficking in platelets and neurons: A common molecular machinery. Eur J Paediatr Neurol 2013;17:117–125.
63. Habets KL, Huizinga TW, Toes RE. Platelets and autoimmunity. Eur J Clin Invest 2013;43:746–757.
64. Koseoglu S, Flaumenhaft R. Advances in platelet granule biology. Curr Opin Hematol 2013;20:464–471.
65. Golebiewska EM, Poole AW. Platelet secretion: From haemostasis to wound healing and beyond. Blood Rev 2015;29:153–162.
66. Duchêne J, von Hundelshausen P. Platelet-derived chemokines in atherosclerosis. Hamostaseologie 2015;35:137–141.
67. van den Dolder J, Mooren R, Vloon AP, Stoelinga PJ, Jansen JA. Platelet-rich plasma: Quantification of growth factor levels and the effect on growth and differentiation of rat bone marrow cells. Tissue Eng 2006;12:3067–3073.
68. Nikolidakis D, Jansen JA. The biology of platelet-rich plasma and its application in oral surgery: Literature review. Tissue Eng Part B Rev 2008;14:249–258.
69. Lubkowska A, Dolegowska B, Banfi G. Growth factor content in PRP and their applicability in medicine. J Biol Regul Homeost Agents 2012;26(2 suppl 1):3S–22S.
70. Uutela M, Wirzenius M, Paavonen K, et al. PDGF-D induces macrophage recruitment, increased interstitial pressure, and blood vessel maturation during angiogenesis. Blood 2004;104:3198–3204.
71. Everts PA, Knape JT, Weibrich G, et al. Platelet-rich plasma and platelet gel: A review. J Extra Corpor Technol 2006;38:174–187.
72. Appelmann I, Liersch R, Kessler T, Mesters RM, Berdel WE. Angiogenesis inhibition in cancer therapy: Platelet-derived growth factor (PDGF) and vascular endothelial growth factor (VEGF) and their receptors. Biological functions and role in malignancy. Recent Results Cancer Res 2010;180:51–81.
73. Caplan AI. Mesenchymal stem cells and gene therapy. Clin Orthop Relat Res 2000;(379 suppl):S67–S70.
74. Tocci A, Forte L. Mesenchymal stem cell: Use and perspectives. Hematol J 2003;4:92–96.
75. Reiser J, Zhang XY, Hemenway CS, Mondal D, Pradhan L, La Russa VF. Potential of mesenchymal stem cells in gene therapy approaches for inherited and acquired diseases. Expert Opin Biol Ther 2005;5:1571–1584.

76. Ye JS, Su XS, Stoltz JF, de Isla N, Zhang L. Signaling pathways involved in the process of mesenchymal stem cells differentiating into hepatocytes. Cell Prolif 2015;48:157–165.
77. Ball SG, Shuttleworth CA, Kielty CM. Mesenchymal stem cells and neovascularization: Role of platelet-derived growth factor receptors. J Cell Mol Med 2007;11:1012–1030.
78. Graham S, Leonidou A, Lester M, Heliotis M, Mantalaris A, Tsiridis E. Investigating the role of PDGF as a potential drug therapy in bone formation and fracture healing. Expert Opin Investig Drugs 2009;18:1633–1654.
79. Hellberg C, Ostman A, Heldin CH. PDGF and vessel maturation. Recent Results Cancer Res 2010;180:103–114.
80. Javed F, Al-Askar M, Al-Rasheed A, Al-Hezaimi K. Significance of the platelet-derived growth factor in periodontal tissue regeneration. Arch Oral Biol 2011;56:1476–1484.
81. Donovan J, Abraham D, Norman J. Platelet-derived growth factor signaling in mesenchymal cells. Front Biosci (Landmark Ed) 2013;18:106–119.
82. Iwayama T, Olson LE. Involvement of PDGF in fibrosis and scleroderma: Recent insights from animal models and potential therapeutic opportunities. Curr Rheumatol Rep 2013;15:304.
83. Shah P, Keppler L, Rutkowski J. A review of platelet derived growth factor playing pivotal role in bone regeneration. J Oral Implantol 2014;40:330–340.
84. Finnson KW, McLean S, Di Guglielmo GM, Philip A. Dynamics of transforming growth factor beta signaling in wound healing and scarring. Adv Wound Care (New Rochelle) 2013;2:195–214.
85. Pakyari M, Farrokhi A, Maharlooei MK, Ghahary A. Critical role of transforming growth factor beta in different phases of wound healing. Adv Wound Care (New Rochelle) 2013;2:215–224.
86. Ferguson MW, O'Kane S. Scar-free healing: From embryonic mechanisms to adult therapeutic intervention. Philos Trans R Soc Lond B Biol Sci 2004;359:839–850.
87. Bachem MG, Zhou Z, Zhou S, Siech M. Role of stellate cells in pancreatic fibrogenesis associated with acute and chronic pancreatitis. J Gastroenterol Hepatol 2006;21(suppl 3):S92–S96.
88. Occleston NL, Laverty HG, O'Kane S, Ferguson MW. Prevention and reduction of scarring in the skin by transforming growth factor beta 3 (TGFbeta3): From laboratory discovery to clinical pharmaceutical. J Biomater Sci Polym Ed 2008;19:1047–1063.
89. Lo DD, Zimmermann AS, Nauta A, Longaker MT, Lorenz HP. Scarless fetal skin wound healing update. Birth Defects Res C Embryo Today 2012;96:237–247.
90. Barrientos S, Stojadinovic O, Golinko MS, Brem H, Tomic-Canic M. Growth factors and cytokines in wound healing. Wound Repair Regen 2008;16:585–601.
91. Yang B, Zhou L, Peng B, Sun Z, Dai Y, Zheng J. In vitro comparative evaluation of recombinant growth factors for tissue engineering of bladder in patients with neurogenic bladder. J Surg Res 2014;186:63–72.
92. Song F, Tang J, Geng R, et al. Comparison of the efficacy of bone marrow mononuclear cells and bone mesenchymal stem cells in the treatment of osteoarthritis in a sheep model. Int J Clin Exp Pathol 2014;7:1415–1426.
93. Mariani E, Pulsatelli L, Facchini A. Signaling pathways in cartilage repair. Int J Mol Sci 2014;15:8667–8698.
94. Mammoto T, Jiang A, Jiang E, Mammoto A. Platelet rich plasma extract promotes angiogenesis through the angiopoietin1-Tie2 pathway. Microvasc Res 2013;89:15–24.
95. Jeong DU, Lee CR, Lee JH, et al. Clinical applications of platelet-rich plasma in patellar tendinopathy. Biomed Res Int 2014;2014:249498.
96. Franceschi RT. Biological approaches to bone regeneration by gene therapy. J Dent Res 2005;84:1093–1103.
97. Wang Y, Nishida S, Sakata T, et al. Insulin-like growth factor-I is essential for embryonic bone development. Endocrinology 2006;147:4753–4761.
98. Sroga GE, Wu PC, Vashishth D. Insulin-like growth factor 1, glycation and bone fragility: Implications for fracture resistance of bone. PLoS One 2015;10:e0117046.
99. Omori E, Inagaki M, Mishina Y, Matsumoto K, Ninomiya-Tsuji J. Epithelial transforming growth factor β-activated kinase 1 (TAK1) is activated through two independent mechanisms and regulates reactive oxygen species. Proc Natl Acad Sci U S A 2012;109:3365–3370.
100. Stratman AN, Davis GE. Endothelial cell-pericyte interactions stimulate basement membrane matrix assembly: Influence on vascular tube remodeling, maturation, and stabilization. Microsc Microanal 2012;18:68–80.
101. Marx RE. Platelet-rich plasma: Evidence to support its use. J Oral Maxillofac Surg 2004;62:489–496.
102. Hallman M, Thor A. Bone substitutes and growth factors as an alternative/complement to autogenous bone for grafting in implant dentistry. Periodontol 2000 2008;47:172–192.
103. Akhundov K, Pietramaggiori G, Waselle L, et al. Development of a cost-effective method for platelet-rich plasma (PRP) preparation for topical wound healing. Ann Burns Fire Disasters 2012;25:207–213.
104. Pietrzak WS, Eppley BL. Platelet rich plasma: Biology and new technology. J Craniofac Surg 2005;16:1043–1054.
105. Eppley BL, Pietrzak WS, Blanton M. Platelet-rich plasma: A review of biology and applications in plastic surgery. Plast Reconstr Surg 2006;118:147e–159e.
106. Rodrigues SV, Acharya AB, Thakur SL. Platelet-rich plasma. A review. N Y State Dent J 2012;78:26–30.
107. Molloy T, Wang Y, Murrell G. The roles of growth factors in tendon and ligament healing. Sports Med 2003;33:381–394.
108. Alsousou J, Thompson M, Hulley P, Noble A, Willett K. The biology of platelet-rich plasma and its application in trauma and orthopaedic surgery: A review of the literature. J Bone Joint Surg Br 2009;91:987–996.
109. Carless PA, Rubens FD, Anthony DM, O'Connell D, Henry DA. Platelet-rich-plasmapheresis for minimising peri-operative allogeneic blood transfusion. Cochrane Database Syst Rev 2011;(3):CD004172.
110. Vu TD, Pal SN, Ti LK, et al. An autologous platelet-rich plasma hydrogel compound restores left ventricular structure, function and ameliorates adverse remodeling in a minimally invasive large animal myocardial restoration model: A translational approach: Vu and Pal "Myocardial Repair: PRP, Hydrogel and Supplements." Biomaterials 2015;45:27–35.
111. Drago L, Bortolin M, Vassena C, Taschieri S, Del Fabbro M. Antimicrobial activity of pure platelet-rich plasma against microorganisms isolated from oral cavity. BMC Microbiol 2013;13:47.
112. Sánchez M, Anitua E, Azofra J, Andía I, Padilla S, Mujika I. Comparison of surgically repaired Achilles tendon tears using platelet-rich fibrin matrices. Am J Sports Med 2007;35:245–251.
113. Randelli PS, Arrigoni P, Cabitza P, Volpi P, Maffulli N. Autologous platelet rich plasma for arthroscopic rotator cuff repair. A pilot study. Disabil Rehabil 2008;30:1584–1589.
114. Chahal J, Van Thiel GS, Mall N, et al. The role of platelet-rich plasma in arthroscopic rotator cuff repair: A systematic review with quantitative synthesis. Arthroscopy 2012;28:1718–1727.
115. Charousset C, Zaoui A, Bellaïche L, Piterman M. Does autologous leukocyte-platelet-rich plasma improve tendon healing in arthroscopic repair of large or massive rotator cuff tears? Arthroscopy 2014;30:428–435.
116. Bennett SP, Griffiths GD, Schor AM, Leese GP, Schor SL. Growth factors in the treatment of diabetic foot ulcers. Br J Surg 2003;90:133–146.

117. Kirsner RS, Warriner R, Michela M, Stasik L, Freeman K. Advanced biological therapies for diabetic foot ulcers. Arch Dermatol 2010;146:857-862.
118. Gawdat HI, Hegazy RA, Fawzy MM, Fathy M. Autologous platelet rich plasma: Topical versus intradermal after fractional ablative carbon dioxide laser treatment of atrophic acne scars. Dermatol Surg 2014;40:152-161 [erratum 2014;40:601].
119. Assenza M, Valesini L, Monacelli G, et al. Traumatic complex wounds, multidisciplinary approach: Our experience in a case series. Clin Ter 2010;161:e95-e99.
120. Serra R, Buffone G, Dominijanni A, Molinari V, Montemurro R, de Franciscis S. Application of platelet-rich gel to enhance healing of transmetatarsal amputations in diabetic dysvascular patients. Int Wound J 2013;10:612-615.
121. Yamada T, Onishi K, Utsunomiya M, Nakamura M. Our treatment strategy for critical limb ischemia. Int J Vasc Med 2013;2013:437471.
122. Toscano N, Holtzclaw D. Surgical considerations in the use of platelet-rich plasma. Compend Contin Educ Dent 2008;29:182-185.
123. Smith SE, Roukis TS. Bone and wound healing augmentation with platelet-rich plasma. Clin Podiatr Med Surg 2009;26:559-588.
124. Maney P, Amornporncharoen M, Palaiologou A. Applications of plasma rich in growth factors (PRGF) in dental surgery: A review. J West Soc Periodontol Periodontal Abstr 2013;61:99-104.
125. Lowery GL, Kulkarni S, Pennisi AE. Use of autologous growth factors in lumbar spinal fusion. Bone 1999;25(2 suppl):47S-50S.
126. Jenis LG, Banco RJ, Kwon B. A prospective study of autologous growth factors (AGF) in lumbar interbody fusion. Spine J 2006;6:14-20.
127. Dimar JR 2nd, Glassman SD, Burkus JK, Pryor PW, Hardacker JW, Carreon LY. Clinical and radiographic analysis of an optimized rhBMP-2 formulation as an autograft replacement in posterolateral lumbar spine arthrodesis. J Bone Joint Surg Am 2009;91:1377-1386.
128. Zhang H, Wang F, Ding L, et al. A meta-analysis of lumbar spinal fusion surgery using bone morphogenetic proteins and autologous iliac crest bone graft. PLoS One 2014;9:e97049.
129. Kitoh H, Kitakoji T, Tsuchiya H, Katoh M, Ishiguro N. Distraction osteogenesis of the lower extremity in patients with achondroplasia/hypochondroplasia treated with transplantation of culture-expanded bone marrow cells and platelet-rich plasma. J Pediatr Orthop 2007;27:629-634.
130. Kitoh H, Kawasumi M, Kaneko H, Ishiguro N. Differential effects of culture-expanded bone marrow cells on the regeneration of bone between the femoral and the tibial lengthenings. J Pediatr Orthop 2009;29:643-649.
131. Kim SJ, Pierce W, Sabharwal S. The etiology of short stature affects the clinical outcome of lower limb lengthening using external fixation. A systematic review of 18 trials involving 547 patients. Acta Orthop 2014;85:181-186.
132. Weibrich G, Kleis WK, Hitzler WE, Hafner G. Comparison of the platelet concentrate collection system with the plasma-rich-in-growth-factors kit to produce platelet-rich plasma: A technical report. Int J Oral Maxillofac Implants 2005;20:118-123.
133. Gandhi A, Bibbo C, Pinzur M, Lin SS. The role of platelet-rich plasma in foot and ankle surgery. Foot Ankle Clin 2005;10:621-637.
134. Bibbo C, Hatfield PS. Platelet-rich plasma concentrate to augment bone fusion. Foot Ankle Clin 2010;15:641-649.
135. Bielecki T, Gazdzik TS, Szczepanski T. Benefit of percutaneous injection of autologous platelet-leukocyte-rich gel in patients with delayed union and nonunion. Eur Surg Res 2008;40:289-296.
136. Nathan ST, Fisher BE, Roberts CS, Giannoudis PV. The management of nonunion and delayed union of patella fractures: A systematic review of the literature. Int Orthop 2011;35:791-795.
137. Del Corso M, Vervelle A, Simonpieri A, et al. Current knowledge and perspectives for the use of platelet-rich plasma (PRP) and platelet-rich fibrin (PRF) in oral and maxillofacial surgery part 1: Periodontal and dentoalveolar surgery. Curr Pharm Biotechnol 2012;13:1207-1230.
138. Simonpieri A, Del Corso M, Vervelle A, et al. Current knowledge and perspectives for the use of platelet-rich plasma (PRP) and platelet-rich fibrin (PRF) in oral and maxillofacial surgery part 2: Bone graft, implant and reconstructive surgery. Curr Pharm Biotechnol 2012;13:1231-1256.
139. Roffi A, Filardo G, Kon E, Marcacci M. Does PRP enhance bone integration with grafts, graft substitutes, or implants? A systematic review. BMC Musculoskelet Disord 2013;14:330.
140. Georgakopoulos I, Tsantis S, Georgakopoulos P, et al. The impact of platelet rich plasma (PRP) in osseointegration of oral implants in dental panoramic radiography: Texture based evaluation. Clin Cases Miner Bone Metab 2014;11:59-66.
141. Davis VL, Abukabda AB, Radio NM, et al. Platelet-rich preparations to improve healing. Part I: Workable options for every size practice. J Oral Implantol 2014;40:500-510.
142. Schwartz-Arad D, Ofec R, Eliyahu G, Ruban A, Sterer N. Long-term follow-up of dental implants placed in autologous onlay bone graft. Clin Implant Dent Relat Res 2016;18:449-461.
143. Ntounis A, Geurs N, Vassilopoulos P, Reddy M. Clinical assessment of bone quality of human extraction sockets after conversion with growth factors. Int J Oral Maxillofac Implants 2015;30:196-201.
144. Grant WP, Jerlin EA, Pietrzak WS, Tam HS. The utilization of autologous growth factors for the facilitation of fusion in complex neuropathic fractures in the diabetic population. Clin Podiatr Med Surg 2005;22:561-584.
145. Sood A, Cunningham C, Lin S. The BB Wistar rat as a diabetic model for fracture healing. ISRN Endocrinol 2013;2013:349604.
146. Chahla J, Cimque ME, Piuzzi NS, et al. A call for standardization in platelet-rich plasmapreparation protocols and composition reporting. A systematic review of the clinical orthopaedic literature. J Bone Joint Surg Am 2017;99:1769-1779.
147. Mautner K, Malanga GA, Smith J, et al. A call for a standard classification system for future biologic research: The rationale for new PRP nomenclature. Phys Med Rehabil 2015;7(4 suppl):S53-S59.
148. Dohan Ehrenfest DM, Rasmusson L, Albrektsson T. Classification of platelet concentrates: From pure platelet-rich plasma (P-PRP) to leucocyte- and platelet-rich fibrin (L-PRF). Trends Biotechnol 2009;27:158-167.
149. Marx RE. Platelet-rich plasma (PRP): What is PRP and what is not PRP? Implant Dent 2001;10:225-228.
150. Rughetti A, Giusti I, D'Ascenzo S, et al. Platelet gel-released supernatant modulates the angiogenic capability of human endothelial cells. Blood Transfus 2008;6:12-17.
151. Dhurat R, Sukesh M. Principles and methods of preparation of platelet-rich plasma: A review and author's perspective. J Cutan Aesthet Surg 2014;7:189-197.
152. Anitua E, Aguirre JJ, Algorta J, et al. Effectiveness of autologous preparation rich in growth factors for the treatment of chronic cutaneous ulcers. J Biomed Mater Res B Appl Biomater 2008;84:415-421.
153. Mazzocca AD, McCarthy MB, Chowaniec DM, et al. Platelet-rich plasma differs according to preparation method and human variability. J Bone Joint Surg Am 2012;94:308-316.
154. Araki J, Jona M, Eto H, et al. Optimized preparation method of platelet-concentrated plasma and noncoagulating platelet-derived factor concentrates: maximization of platelet concentration and removal of fibrinogen. Tissue Eng Part C Methods 2012;18:176-185.
155. Bausset O, Giraudo L, Veran J, et al. Formulation and storage of platelet-rich plasma homemade product. Biores Open Access 2012;1:115-123.

156. Amable PR, Carias RB, Teixeira MV, et al. Platelet-rich plasma preparation for regenerative medicine: optimization and quantification of cytokines and growth factors. Stem Cell Res Ther 2013;4:67.
157. Kececi Y, Ozsu S, Bilgir O. A cost-effective method for obtaining standard platelet-rich plasma. Wounds 2014;26:232–238.
158. Evanson JR, Guyton MK, Oliver DL, et al. Gender and age differences in growth factor concentrations from platelet-rich plasma in adults. Mil Med 2014;179:799–805.
159. Bowlin GL, Sell SA. Platelet-rich plasma in bone regeneration: Engineering the delivery for improved clinical efficacy. Biomed Res Int 2014;2014:392398.
160. Han B, Woodell-May J, Ponticiello M, Yang Z, Nimni M. The effect of thrombin activation of platelet-rich plasma on demineralized bone matrix osteoinductivity. J Bone Joint Surg Am 2009;91:1459–1470.

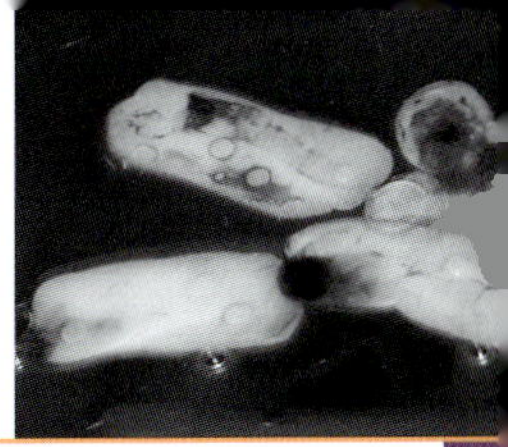

2

Medical and Surgical Applications of Autologous Blood Concentrates

The literature on PRP demonstrates the diversity of its clinical applications. Separate chapters in this text are devoted to clinical uses of PRP in plastic surgery and in oral, periodontal, and maxillofacial surgeries. Separate chapters are required because, for example in plastic surgery, there is a lengthy history of topical use of a variety of forms of platelet concentrates for skin wound healing, and there is a rapidly growing popularity of PRP use for esthetic and cosmetic purposes.[1,2] There is a similarly lengthy history of various forms of PRP being used in a variety of periodontal and dentoalveolar procedures as well as in bone grafting, implant surgery, and reconstructive surgery.[3,4]

Since its first surgical application in the late 1980s for autologous transfusion support in open-heart procedures, PRP has been used in a variety of clinical settings because of its ability to enhance tissue repair and healing at wound sites.[5] In April 2015, the National Center for Biotechnology Information listed more than 280 books, over 7,700 scientific and medical abstractions/citations, and more than 21,400 full-text journal articles on PRP, in a wide variety of fields, including cardiothoracic surgery, cosmetics, dentistry, maxillofacial surgery, neurosurgery, ophthalmology, orthopedics, otolaryngology, sports medicine, urology, and wound healing.

The application of PRP at wound sites is only natural because wound healing starts with the formation of a blood clot; the degranulation of platelets and the release of platelet growth factors regulate the wound healing process. The mechanisms of tissue repair that are aided by platelet-derived growth factors include cell migration and proliferation, the formation of new blood vessels from existing ones, extracellular matrix formation, and the remodeling of cells (Fig 2-1).

Over the last 30 years, the clinical use of PRP through various preparations and for a variety of therapies has grown in importance due mainly to the platelet growth factors released by PRP. These growth factor proteins exert tremendous influence on blood coagulation/clotting, immunities, angiogenesis, and wound healing. The concentration of platelets through cell sequestration to levels over 300 times their normal strength (with resultant concentrations in growth factors) allows clinicians to apply the concentrates to a number of wound healing therapies, including skin ulcers, oral and maxillofacial surgery and implantology, orthopedic surgery, burn treatment and the treatment of other wounds difficult to heal, soft tissue disease and injuries, and tissue engineering.[6]

This chapter focuses on the clinical application of PRP in the historically significant and wide-reaching medical areas of cardiothoracic surgery and orthopedics, respectively, as well as a number of miscellaneous clinical applications, including bone surgery, chronic wound healing, general surgery, neurosurgery, ophthalmology, otolaryngology, podiatry, and urology.

Cardiothoracic Surgery

Since the late 1980s, cardiac physicians have used cell-saving machines and platelet-rich plasmapheresis (allogeneic red blood cells and platelet concentrates) as methods to control nonsurgical bleeding after cardiopulmonary bypass (CPB) as well as postsurgical platelet

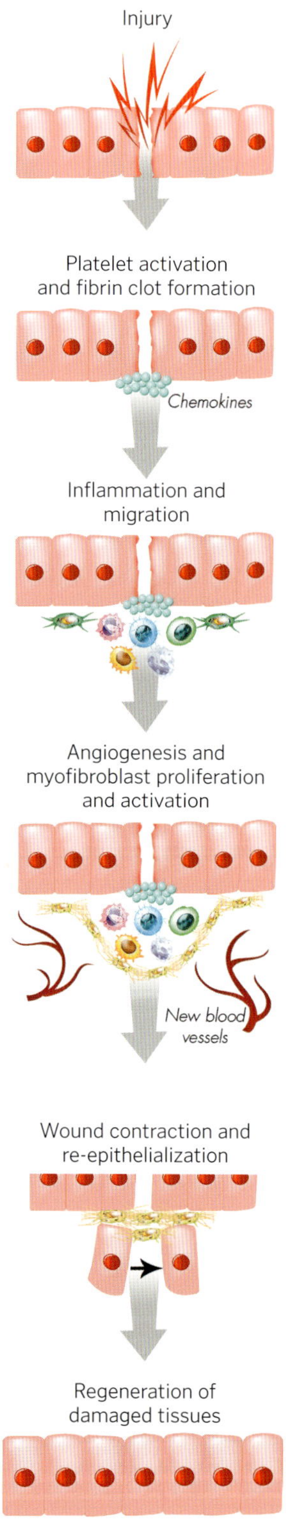

FIG 2-1 Wound healing mechanisms of tissue repair that are enhanced by the application of PRP.

Box 2-1 Effects of platelet-rich plasmapheresis and PRP transfusion on cardiopulmonary bypass surgery outcomes[8]

- Reduces postoperative blood loss
- Reduces postoperative fresh frozen plasma transfusion
- Reduces postoperative red blood cell transfusion
- Reduces postoperative allogeneic red blood cell transfusion

dysfunction. Autologous blood components—including PRP, platelet-poor plasma (PPP), and red blood cell concentrate—are produced presurgically for postsurgical infusion to combat platelet dysfunction and bleeding.[7] PRP has been used in cardiovascular surgery to maintain hemodynamic stability in patients following CPB.

A 2019 meta-analysis of randomized controlled clinical trials (RCTs) concluded that conducting platelet-rich plasmapheresis before CPB and transfusing PRP after reversal of heparin could reduce postoperative blood loss and hence the need for blood transfusions (Box 2-1).[8] The study examined 15 RCTs with a total of 1,002 patients. Half of the patients (501) received PRP and the other half served as controls. PRP reduced the total volume of postoperative blood loss, reduced postoperative fresh frozen plasma transfusion, reduced postoperative red blood cell transfusion, and reduced the proportion of patients requiring postoperative allogeneic red blood cell transfusion. However, a high degree of undetermined heterogeneity among the analyzed trials suggests that more precise RCTs are needed to confirm these conclusions.

PRP in the form of platelet gel has been used to help control bleeding and to prevent infection on CPB wound tissue of the sternum via syringe or spray-tip catheter application prior to closure (Fig 2-2). Platelet gel has also been injected into a patient's leg vein harvest site. Efforts to predict preoperatively the effectiveness of platelet infusion postsurgery are ongoing, including calibrated automated thrombography assays for estimations of blood loss.[9] PRP has also been combined successfully with antioxidant and anti-inflammatory agents in an intramyocardial hydrogel injection to treat injured heart muscles after an acute myocardial infarction in a randomized study that used a pig model.[10]

Postoperative wound dehiscence and infection are a common and extremely serious complication of cardiac surgery, affecting up to 8% of patients undergoing median sternotomy. They are associated with long-term antibiotic use, prolonged hospital stays, multiple operative procedures, and high costs. Over a 7-year period, 2,000 patients undergoing open cardiac operations requiring sternotomy were enrolled in a study evaluating the use of PRP to minimize the incidence of deep sternal wound infections (DSWI).[11] Half of the patients (1,000) received standard of care treatment for sternal closure, and the other

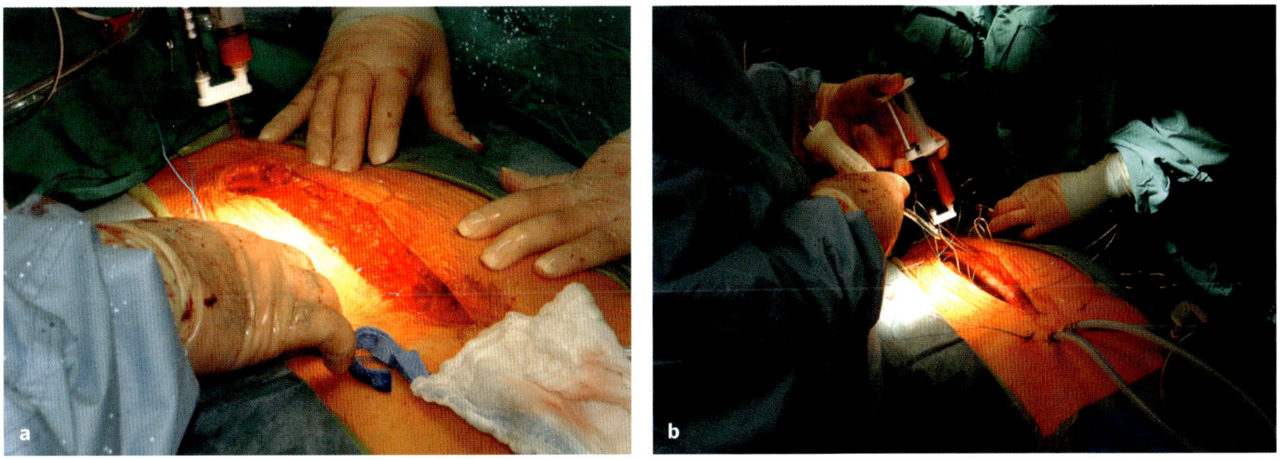

FIG 2-2 *(a and b)* Application of PRP gel with a spray-tip catheter to prevent infection in the sternum wound prior to closure in CPB surgery.

FIG 2-3 Injection of PRP into the subacromial space of the shoulder decreases the rate of retears in patients with rotator cuff injuries.

half received standard of care plus the application of PRP to the sternum at the time of closure. The outcome analysis showed that in the PRP group the incidence of DSWI decreased from 2.0% to 0.6%, sternal wound infections decreased from 8.0% to 2.0%, and the rate of readmission decreased from 4.0% to 0.8%. Moreover, PRP reduced the total costs associated with deep and superficial wound complications from $1.25 million to $593,000. Although this was not an RCT, the study nevertheless demonstrated the value of PRP for cardiac surgery patients and the health care system given its large sample size.

Sports Medicine

Rotator cuff injury (Fig 2-3) is a common musculoskeletal condition involving torn tendons and muscles. Pain and shoulder immobility can severely impact a person's quality of life and usually requires arthroscopic surgery. A systematic review and meta-analysis of RCTs was performed in 2019 to investigate whether application of PRP during arthroscopic repair of rotator cuff tears decreases retear rates and improves clinical outcomes. A total of 880 patients were enrolled across 13 RCTs, including 439 in

Box 2-2 Effects of application of PRP on retear rates following rotator cuff repair surgery[12]

- Positive effect on retear rates
- Reduced/improved constant pain scores

the PRP groups and 441 in the control groups (ie, no PRP during surgery), and the final follow-up was 6 to 16 months posttreatment. Of the 13 studies, 12 reported the rate of retear at the final follow-up. The pooled data showed that retears occurred in 63 of 392 patients (16%) in the PRP group and 90 of 381 patients (24%) in the control group. The authors concluded that PRP had a significantly positive effect on postoperative retear rates and on functional outcome measures, including constant shoulder scores, constant pain scores, UCLA shoulder scores, and visual analog scale (VAS) pain scores (Box 2-2).[12]

Pain relief and improved function in humans, along with improved cartilage repair in animals, are the focus of a 2013 sports medicine article, whose authors note the effect of PRP on stem cells and chondrocytes, as well as PRP's ability to improve cartilage regeneration, particularly in cases of osteoarthritis.[13] A 2015 study lists PRP therapy along with steroid injections and anesthetics as current treatment for skeletal muscle injuries, which are among the most common sports injuries, resulting in reduction of practice and play for athletes. The authors note that future therapies could include transforming growth factor β (TGF-β) and angiotensin II as well as muscle-derived stem cells and that effective treatment is predicated on accurate diagnoses of these injuries, including athlete history, physical exams, ultrasound, and magnetic resonance imaging (MRI).[14]

A 2014 Level III case-control study using clinical and postoperative MRI evaluation of patients undergoing arthroscopic rotator cuff repair for large or massive rotator tears treated with leukocyte-PRP (L-PRP) revealed that while functional outcomes were similar for two patient cohorts in the 2-year follow-up, and the L-PRP did not affect the quality of tendon healing, significantly smaller iterative tears resulted in the L-PRP patients.[15]

A 2015 Level III systematic review of Level I through III studies of PRP used for treating anterior cruciate ligament ruptures revealed that PRP might add a synergic element to faster healing, though clinical evidence remained unclear. PRP appeared not to affect tunnel healing. The review included assessment of graft-to-bone healing, graft maturation, and/or clinical outcomes for RCTs or prospective cohort studies. Eleven studies were included in the review, including 516 patients (266 reconstructions with PRP and 250 without PRP).[16]

A 2012 study of the sports applications of PRP referred to PRP as a "new biologic tool in sports medicine" yet concluded that "more vigorous study [is needed] in order to better understand how to apply it most effectively." The study notes the growing interest in PRP around the world for its sports applications and refers to the soundness of the basic science and preclinical data for PRP use in a number of different sports injuries and disorders. However, it further notes the limited peer-reviewed and human literature on PRP use. Despite these limitations, the study notes the faithful use of PRP by a number of world-class athletes as well as recreational athletes for sports-related injuries.[17]

Osteoarthritis

The widespread chronic disability associated with osteoarthritis has researchers longing to find a less invasive treatment method than those currently offered. PRP injections may in fact be that working alternative, anti-inflammatory method for treating osteoarthritis. An aging population and a (likely) corresponding rise in debilitating osteoarthritis are driving researchers to find alternatives to replacing articular cartilage, namely, using regenerative therapies on the tissues affected. For the healing of tendons in particular, PRP is a promising treatment alternative.

The destruction of joints from degeneration of articular cartilage is the primary symptom of cartilage defects and osteoarthritis. Researchers hope to find a treatment strategy for restoring functional hyaline cartilage as conventional treatments are unable to provide such restoration or its equivalent for function. Platelet-rich therapy offers some hope for a tissue-engineering solution to the effects of osteoarthritis via autocartilage cells or mesenchymal stem cells.

Osteoarthritis in the knee has been treated by intra-articular injections of PRP (Fig 2-4). A 2017 systematic review and meta-analysis of RCTs was performed to evaluate the temporal effect of PRP on knee pain and physical function. Fourteen RCTs that compared the efficacy and safety of PRP infiltration with other intra-articular

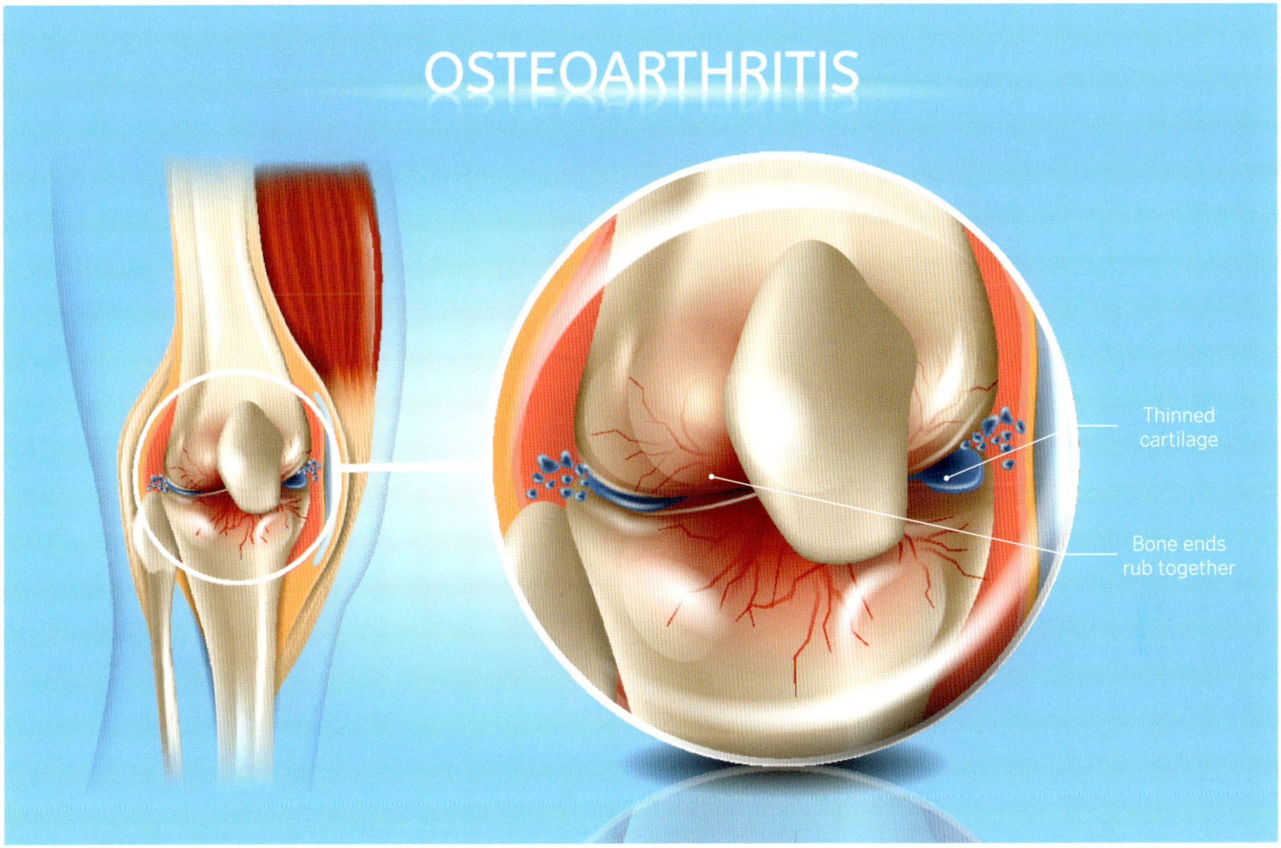

FIG 2-4 Intra-articular injection of PRP has been shown to significantly reduce pain and improve function in patients with osteoarthritis of the knee.

injections—namely, hyaluronic acid, ozone, corticosteroids, and a saline placebo—were included. A total of 1,423 participants were enrolled, and the follow-up ranged from 12 weeks to 12 months. For the PRP groups, the sample size ranged from 12 to 96 patients, and the control groups ranged from 11 to 96 patients. Analysis of the data consistently showed that intra-articular PRP injections significantly reduced knee pain, improved knee function, and were not associated with an increased risk of adverse effects. The superior effects of PRP over controls were observed at 3, 6, and 12 months posttreatment (Box 2-3).[18]

A 2014 comprehensive, systematic literature review of the effectiveness of PRP injections for osteoarthritis included RCTs and nonrandomized controlled trials that were subsequently graded for risk of bias and a level of evidence.[19] Results in 10 trials revealed that intra-articular PRP injections resulted in more reduction in pain than placebo after 6 months. Similar favorable results indicated that intra-articular PRP injections resulted in less pain in patients compared to hyaluronic acid therapy.

Box 2-3 Effects of intra-articular injection of PRP on osteoarthritis pain and function compared with hyaluronic acid, ozone, corticosteroids, and saline[18]

- Significantly improved knee pain
- Significantly improved knee function
- Produced no adverse effects

Another review of the effects of PRP in osteoarthritis patients, this one focused on the treatment of hip osteoarthritis, was conducted in 2018. Just four RCTs (with a total of 303 participants) met the inclusion criteria for this meta-analysis, which compared the effects of PRP vs hyaluronic acid, with pain as the outcome measure. According to the authors, the data showed that intra-articular injection of PRP was associated with a significant reduction in VAS pain at 2 months compared with hyaluronic acid, but it was no better than intra-articular injection of hyaluronic acid at 6 and 12 months.[20]

FIG 2-5 A 2017 meta-analysis of RCTs on therapeutic injection of PRP for recalcitrant tendinopathy (including tennis elbow) concluded that it is statistically significantly more efficacious than control injections (consisting of saline, local anesthetic, or corticosteroid) in reducing pain.

Tendinopathy

Tendinopathy is the number one reason patients seek medical care; it generally affects the elbow, rotator cuff, Achilles tendon, and patellar tendon. Patients who receive an early diagnosis often respond well to nonsurgical treatments, including behavior modification, gentle static stretching exercises, nonsteroidal anti-inflammatory drugs (NSAIDs), and eccentric loading. Those with recalcitrant tendinopathy receive therapy designed to reduce inflammation, such as local cooling, NSAIDs, and corticosteroids, all of which have limited efficacy.

The etiology of chronic tendinopathy remains controversial but is thought to result from an imbalance between the demands that are placed on a tendon and its ability to remodel. Because growth factors play an important role in the maintenance of normal tissue structure and healing of tissue lesions, many researchers believe that PRP holds promise as a regenerative therapy for the treatment of chronic tendinopathy (Fig 2-5).

A search of the literature yields a large body of evidence on the use of PRP as a therapy for chronic tendinopathy, and many of these studies report positive results. However, because study designs and PRP preparation methods vary widely, a systematic review and meta-analysis of RCTs was performed in 2017 to determine the efficacy of PRP versus a control to reduce pain associated with chronic symptomatic tendinopathy. A total of 18 studies, with a median sample size of 35 patients and a minimum of 3 months' follow-up, were included. Twelve of these studies involved patients with lateral epicondylopathy (also known as *tennis elbow*), and the other six treated patients with rotator cuff, Achilles tendon, and patellar tendon pathology.

The outcome of the meta-analysis found that PRP injection is more efficacious than control injections (consisting of saline, local anesthetic, or corticosteroid) in reducing tendinopathy pain. Specifically, injection of PRP resulted in statistically significant lower pain severity relative to control in 8 of the 18 groups, whereas 10 of the 18 groups

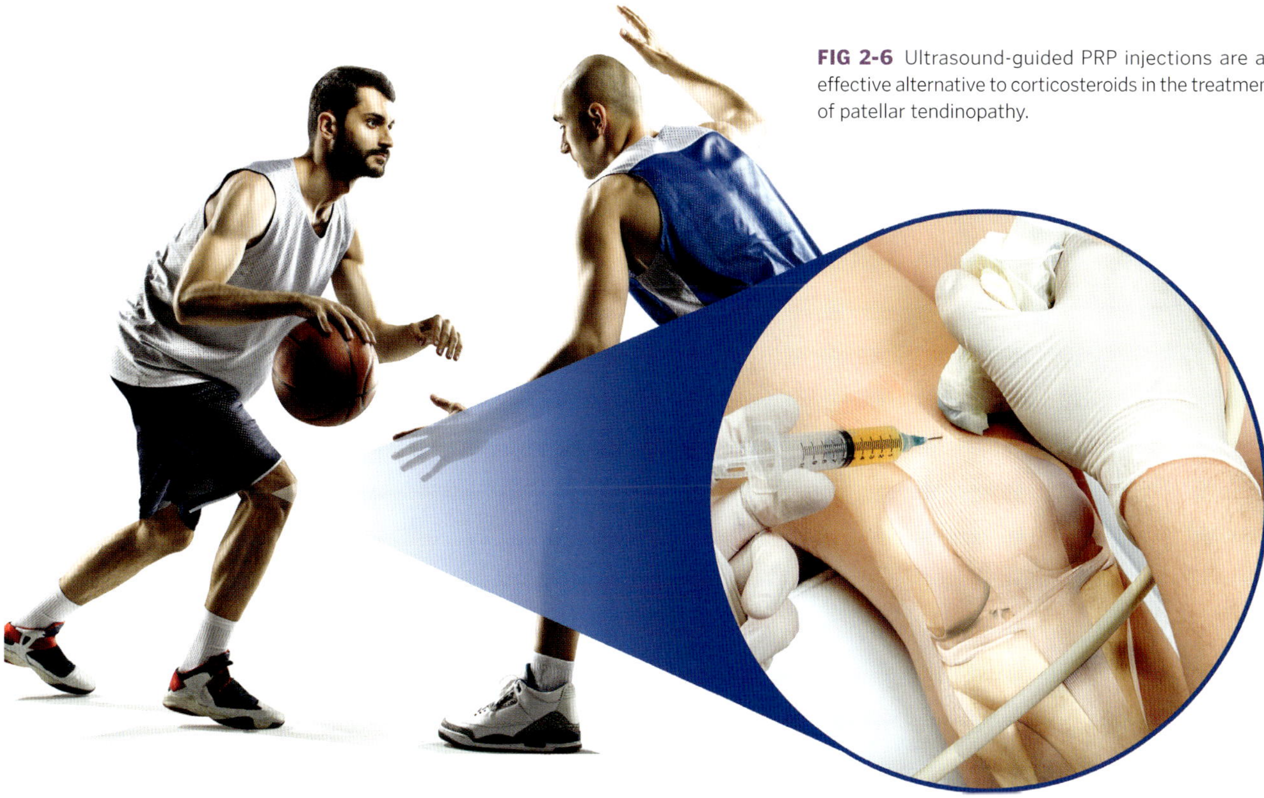

FIG 2-6 Ultrasound-guided PRP injections are an effective alternative to corticosteroids in the treatment of patellar tendinopathy.

reported no differences. No groups reported greater efficacy of the control injections over PRP. Single injection protocols were used in more than 80% of the studies. The only variable that was found to influence PRP efficacy was female sex: A large treatment effect was observed in the groups with more women than men, affirming an earlier study with a similar outcome.[21]

PRP injections for tendinopathy may be a biologic alternative to the more conventional tendon-degrading corticosteroids and risky surgery (Fig 2-6). A 2019 systematic review and meta-analysis of the most effective nonsurgical options for the treatment of patellar tendinopathy concluded that multiple PRP injections may offer better and more stable results than eccentric exercises at long-term follow-up and can be considered a suitable option.[22]

Dentistry

There have been three systematic reviews and meta-analyses of RCTs on the efficacy of autologous blood concentrates in the treatment of intrabony defects in chronic periodontitis patients performed since 2016. The most recent, published in 2019, included 12 clinical trials in which a total of 523 patients (754 sites) were enrolled. The mean duration of follow-up ranged from 3 to 12 months. All 12 of the RCTs evaluated the additive effectiveness of platelet-rich fibrin when used with open flap debridement (OFD) compared with OFD alone. Analysis of the pooled data found that adjunctive autologous blood concentrate yielded significantly superior probing depth (PD) reduction (weighted mean difference [WMD] 1.01) and a higher gain in clinical attachment level (CAL; WMD 1.29). Data regarding both of these outcome measures were available in all 12 studies. Data regarding the other outcome measures—gingival marginal level (GML), defect depth reduction (DDR), and percentage of bone defect fill—were reported in only 8 of the 12 studies. Nevertheless, the meta-analysis demonstrated that autologous blood concentrate was superior to OFD with respect to the change in GML at the 9-month follow-up; that more DDR was achieved in the experimental (autologous blood concentrate) group compared with the OFD group; and that a higher percentage of fill defects in bone were achieved in the autologous blood concentrate

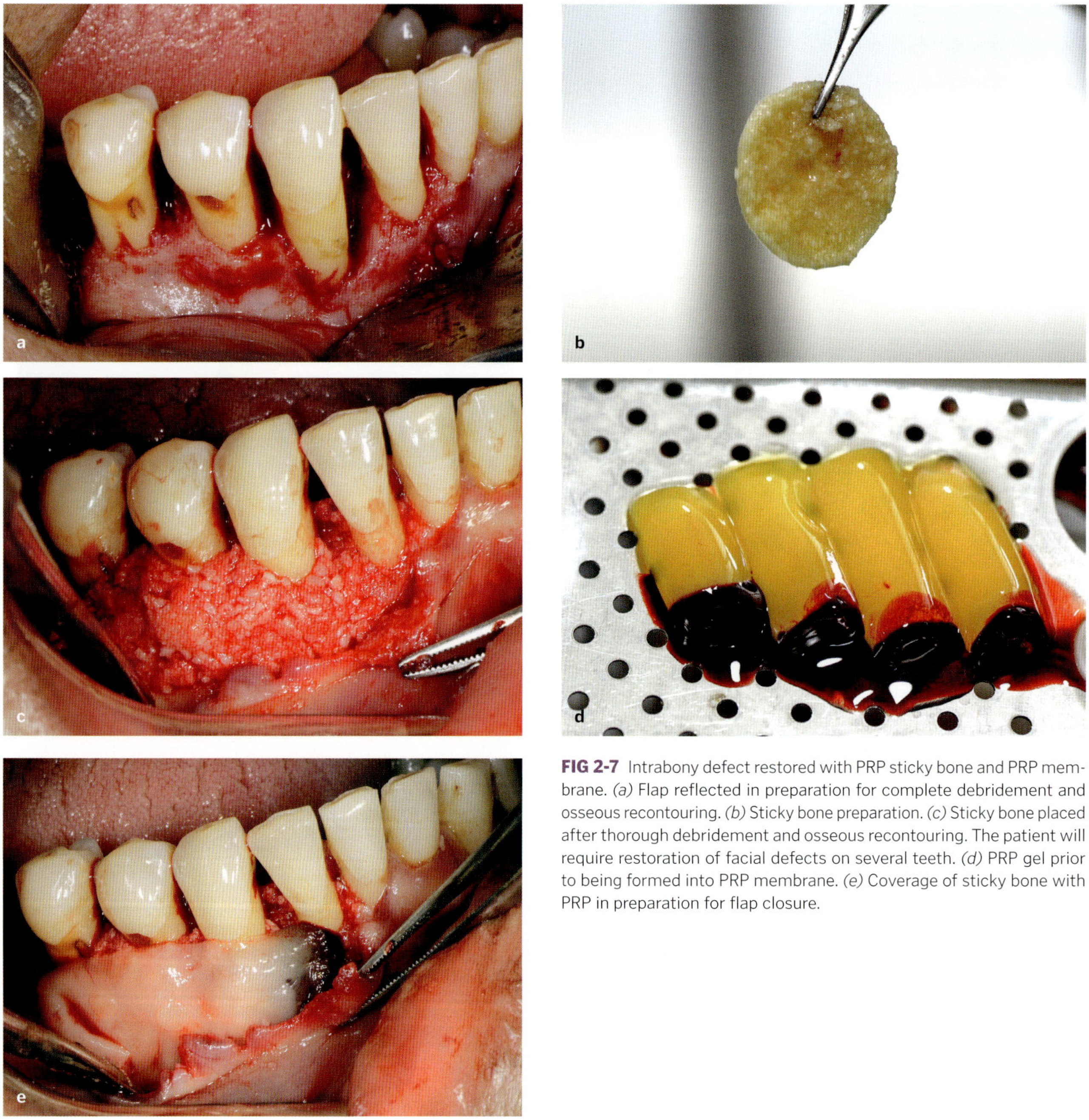

FIG 2-7 Intrabony defect restored with PRP sticky bone and PRP membrane. *(a)* Flap reflected in preparation for complete debridement and osseous recontouring. *(b)* Sticky bone preparation. *(c)* Sticky bone placed after thorough debridement and osseous recontouring. The patient will require restoration of facial defects on several teeth. *(d)* PRP gel prior to being formed into PRP membrane. *(e)* Coverage of sticky bone with PRP in preparation for flap closure.

group compared with the OFD group. All of these results mirrored those of the two earlier systematic reviews, which clearly showed that adjunctive autologous blood concentrate was significantly beneficial to the treatment of intrabony defects in terms of PD reduction, CAL gain, and bone fill[23] (Fig 2-7).

Ophthalmology

One of the most common conditions affecting the general population, dry eye disease (DED) is a multifactorial condition that involves a breakdown in the tear film, the natural layer of tears that coats the front of the eye (Fig 2-8).

FIG 2-8 Breakdown of the tear film, which protects the cornea and conjunctiva, causes irritation and vision changes in patients with Sjögren syndrome and lupus. PRP drops are being used in place of artificial tears with equal or better results.

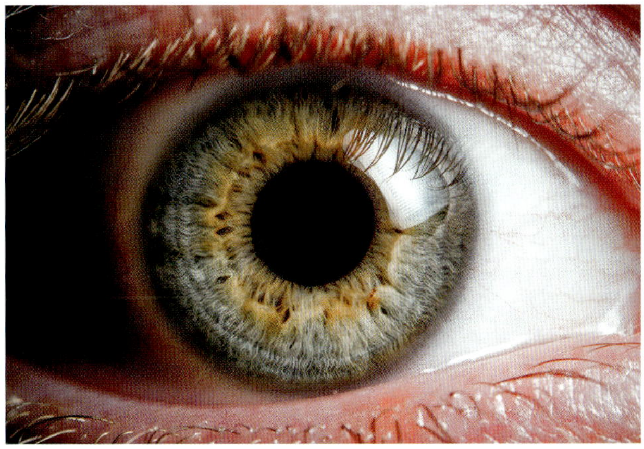

The tear film protects the cornea and conjunctiva from damage that would result from exposure to the air and contributes significantly to the focusing power of the eye. Breakdown of the tear film leads not only to symptoms of irritation but also to unstable and intermittently changing vision. The most common symptoms of DED are burning, stinging, tearing, sandy or gritty feeling, intermittent sharp pain, and photophobia.

In 2017, Alio et al reported the results of a large-scale prospective consecutive study that evaluated the use of PRP eye drops for the treatment of moderate to severe cases of dry eye. The first-line therapy for DED is artificial tears, which provide symptomatic relief but do not resolve underlying causes. For severe cases, topical corticosteroids or cyclosporine A are used to improve symptoms. The theoretical advantages of PRP over artificial tears are its cost-effectiveness, lack of preservatives, nonimmunogenicity, and demonstrated healing qualities.

A total of 368 patients completed therapy consisting of one drop of PRP self-administered in both eyes six times per day for 6 weeks. Improvement was evaluated by the ocular surface disease index (OSDI) self-administered questionnaire, corneal fluorescein staining (CFS), and corrected distance visual acuity. Follow-up visits at the end of each 6-week round of treatment included an interview with the patient assessing the presence of ocular symptoms and ophthalmologic examination to register all clinical findings.

Of the 368 patients treated in the study, 232 (63%) had evaporative DED (EDED), which is associated with Meibomian gland dysfunction and eyelid problems (eg, infrequent blinking). The other 136 patients (37%) had aqueous-deficient DED (ADDED), which is caused by an insufficient volume of tears produced by the lacrimal glands and obstruction of the lacrimal ducts. This form is associated with autoimmune diseases such as Sjögren syndrome, lupus, and rheumatoid arthritis.

The pretreatment OSDI scores averaged 42.9 (on a scale of 0 [no pain] to 100) for the EDED patients and 69.6 for the ADDED patients. Both groups decreased significantly after treatment to 18.1 and 43.2, respectively. These differences were statistically significant. Changes in visual acuity were measured with Snellen charts on a conventional projection screen with far distance correction and without using any eye drops 1 hour before measurement: 106 patients (28.8%) experienced an improvement of one or more lines of vision, but no improvement in vision was observed in 246 patients (66.9%), and 16 patients (4.3%) lost one line of vision. All patients enrolled in the study initially showed superficial punctate keratopathy as measured by CFS. After the treatment, a decrease in the area with staining was found in 280 patients (76.1%), 78 patients (21.2%) did not show any changes, and 10 patients (2.7%) showed worsening in the extension of punctate keratopathy.

The authors concluded that PRP eye drops appear to be an attractive alternative option for the treatment of

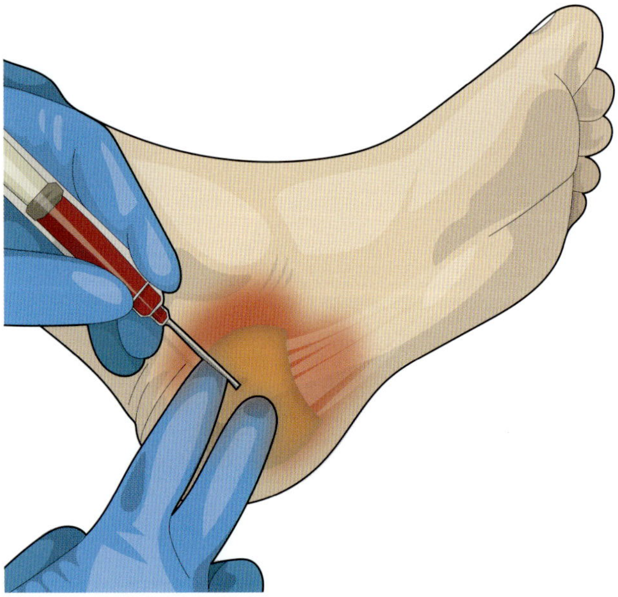

FIG 2-9 Local injections of PRP can offer long-term pain relief to patients with chronic plantar fasciitis that does not respond to standard treatment.

both evaporative and aqueous-deficient DED. However, RCTs are needed to further assess this treatment versus commercial artificial teardrops.[24]

Heat-inactivated PRP as an eye drop therapy has also been used to treat corneal wounds in patients with autoimmune diseases. Heat-inactivated platelet-rich growth factor eye drops significantly increased the proliferation and migration of ocular surface cells while reducing complementary activity and decreasing significantly the presence of immunoglobulin E.[25]

Podiatry

Plantar fasciitis is a chronic disabling disease that affects many people. The cause of pain is tissue degeneration rather than inflammation, and standard treatment options include NSAIDs, corticosteroid injections, orthosis, and physiotherapy.

In 2018, Ling and Wang carried out a systematic review and meta-analysis of RCTs investigating the effects of PRP in the treatment of plantar fasciitis. The main outcomes were changes in VAS score, American Orthopaedic Foot and Ankle Society (AOFAS) score, and Roles-Maudsley score, all of which are standard measures of the efficacy of treatment for ankle and hindfoot injuries. The authors reported that PRP was associated with greater changes in VAS and AOFAS scores than the other treatments, but there was no significant difference found between the treatments in the Roles-Maudsley score. They concluded that PRP was as effective as other treatments in terms of pain and functional results and that it should be considered an alternative approach for patients with plantar fasciitis[26] (Fig 2-9).

Box 2-4 Effects of PRP injection on plantar fasciitis compared with conventional treatment[27]

- Superior to steroid treatment
- Resulted in less pain and better function when combined with shockwave therapy
- Reduced plantar fascia thickness

Another systematic review and meta-analysis, this one carried out in 2017, compared PRP with steroid injections for treatment of plantar fasciitis. Nine RCTs were included in the meta-analysis. Although the evidence was limited owing to the small sample size, the results demonstrated that PRP is superior to steroid treatments for long-term pain relief (Box 2-4).[27] A 2013 randomized trial for treatment of unilateral chronic plantar fasciitis involving 54 subjects ranging in age from 29 to 71 years compared autologous conditioned plasma injection with extracorporeal shockwave and conventional treatments. Treatment with autologous conditioned plasma or with extracorporeal shockwave in addition to conventional treatments resulted in less pain and better function than conventional treatment alone. Plantar fascia thickness was also reduced in the group treated with autologous conditioned plasma.[28]

Veterinary Medicine

A 2014 study focused on the multimodal application of nonpharmaceutical therapies with pharmaceutical ones to reduce human and animal pain, noting that in addition to cold therapy, heat therapy, and acupuncture, extracorporeal shockwave therapy and PRP are nonpharmaceutical therapies growing in popularity due to their attempts to use tissue healing and regeneration as modes of pain management.[29] The potential healing effect of growth factors on tendons, the aorta, and major arteries is the

FIG 2-10 *(a and b)* Injection of PRP and mesenchymal stem cells into the fetlock joint has been shown to be an effective treatment modality for injured tendons in horses.

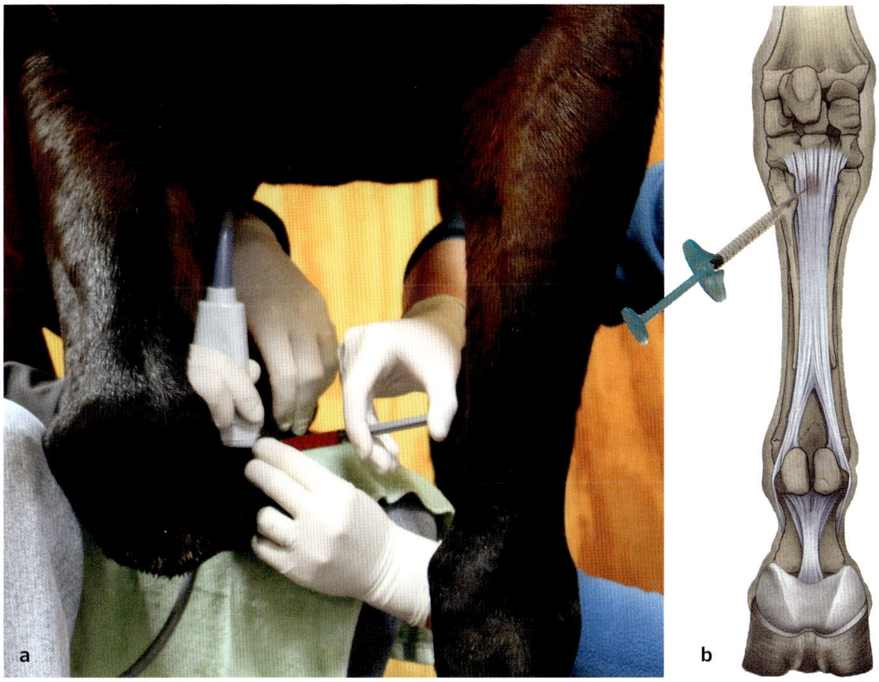

focus of another 2014 study, with a primary emphasis on the effect of TGF-β on heritable conditions of soft tissues. Researchers in the study note that several bone morphogenetic proteins have received US Food and Drug Administration (FDA) approval for the treatment of nonhealing fractures or have also been the focus of clinical studies involving bone fractures and tendon injuries. Controlling the growth factors' angiogenicity as well as their interaction was the focus of the research, which notes the efficacy of PRP and mesenchymal stem cells in successful treatment of horse tendon injuries.[30]

A 2014 study involving tenodesmic injuries in 150 competition horses concluded that PRP growth factor application led to successful tendon formation, allowing 81% of the horses to resume prior activities due to the return of normal morphology and function in the injured tendons.[31] Applying induced mesenchymal stem cells with PRP to horses with degenerative disease in the fetlock joint resulted in marked improvement up to a year after treatment (Fig 2-10). This result is significant because degenerative joint disease accounts for a great number of compromised equine athletes.[32]

Neurosurgery

The role of PRP in promoting spinal fusion after spinal surgery has been studied by many researchers over the past 15 years. Several controlled studies have found a high fusion rate in the PRP groups compared to controls.[33-35] A review and meta-analysis of RCTs was published by *Advances in Orthopedics* in 2020. Manini et al identified a total of 14 studies (5 retrospective and 9 prospective) that together enrolled 713 participants. An independent assessment found that the quality of all included studies was high, scoring more than 8 stars on the Newcastle-Ottawa scale (NOS). The following outcome measures were used: new bone formation in terms of Hounsfield units, pain reduction in terms of VAS score, and time to bone fusion in terms of odds ratio between the two groups at the final follow-up.

Based on the results of their meta-analysis, the authors found that PRP promotes bone fusion when the concentration of platelets is very high. The studies that used concentrations of platelets that were at least five times higher than in peripheral blood obtained superior bone

Box 2-5 Role of PRP in promoting spinal fusion after spinal surgery compared with controls[36]

- Promotes superior bone fusion
- Promotes superior bone formation
- Achieves bone fusion more quickly
- Hastens recovery and shortens the time for rehabilitation

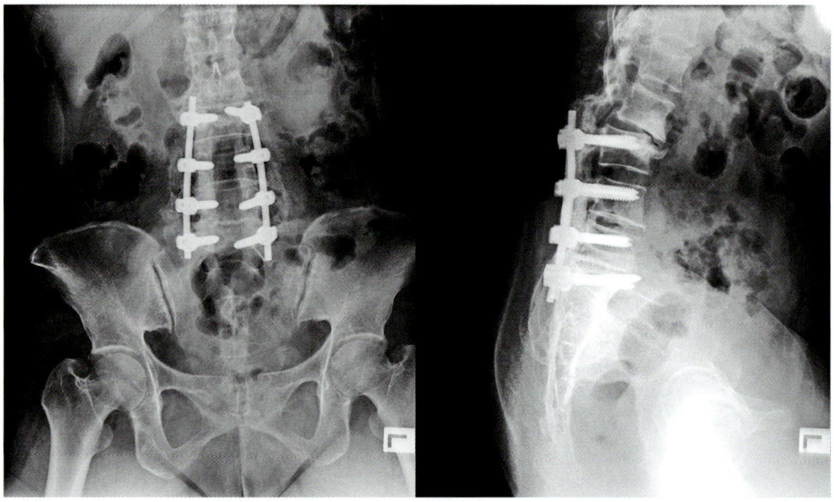

FIG 2-11 The use of PRP during spinal fusion surgery can hasten patient recovery and shorten rehabilitation time.

fusion in participants who received PRP compared to those in the control group. Interestingly, six of the included studies did not measure the concentration of platelets in their PRP, and their bone fusion rates were no better in the PRP groups than in the controls. Manini et al also found that new bone formation was superior in the PRP groups in the four studies that made this data available, and this difference was statistically significant. Time to bone fusion was shorter in the PRP group compared to the control group, as determined by measuring bone density on computed tomography (CT) scans (Box 2-5).[36]

The authors concluded that PRP can be employed to hasten recovery and shorten the time for rehabilitation and the use of orthopedic devices following spinal fusion surgery (Fig 2-11). They also pointed to the need to find a stable carrier of PRP that would allow for continuous release of growth factors for a long period.

Implant-associated spinal infection appears to be better controlled when PRP is used as an adjunct therapy. A 2014 study suggests that the local application of PRP to pressure ulcers in patients with spinal cord injuries may prove more efficacious than saline dressing, noting additionally that more thorough, large-scale trials are warranted due to the several advantages inherent in PRP therapy, including simple preparation, biocompatibility, and low cost.[37]

A more recent study reveals that the combination of PRP and debridement can have a positive healing effect on nonhealing fistulas in patients with spinal cord injuries. A main advantage to the PRP therapy over conventional therapies includes the avoidance of recurrent surgical interventions. Researchers used an MRI-control investigation to demonstrate that fistulas were completely healed after 3 weeks, with no negative side effects or allergic reactions to the PRP therapy.[38]

Other Clinical Applications

A number of other clinical uses of PRP, described below, are being pursued with the same hopes of finding a simple, noninvasive, and effective therapy for wound healing and repair.

Spinal applications

Low back pain (LBP) affects the daily lives of 31 million Americans at any given time and is the most common cause of disability among Americans between the ages of 45 and 65 years (Fig 2-12). Although most cases are self-limiting, a subset of patients experience chronic pain lasting for 6 months or longer. Chronic lumbar discogenic pain, a degenerative disease, constitutes approximately 40% of cases of LBP.

Spine pathology is a confounding problem for physicians because noninvasive imaging methods such as CT scans or MRI have limited ability to determine the exact source of pain. Discography is a somewhat controversial diagnostic procedure that involves injection of a sterile liquid in one or more intervertebral discs to induce pain in affected areas. Typical strategies for managing

FIG 2-12 Intradiscal injection of PRP shows promise in reversing or slowing progressive intervertebral disc degeneration, which is the leading cause of LBP.

this condition include physical therapy, oral medications, epidural injection therapy (steroids or anesthetics), and spinal fusion.

Regenerative therapies such as intradiscal injection of mesenchymal stem cell/bone marrow aspirate or PRP show promise in reversing or slowing progressive intervertebral disc degeneration (IDD). In 2017, Basso et al conducted a review of the clinical trials that had been performed to date (excluding preclinical and in vitro studies). Their aim was to evaluate whether these promising strategies had the potential to be used in clinical practice. Between 2010 and 2016, a total of seven articles were published on the use of regenerative therapies in IDD. Of the seven, three involved the use of PRP.[39]

In 2016, Tuakli-Wosornu et al performed a prospective double-blind RCT to determine whether a single injection of PRP could improve pain and function in patients with IDD. Forty-seven participants (29 in the treatment group, 18 in the control group) were enrolled, and the follow-up rate was 92%. Data for all participants were collected at baseline, 1, 4, and 8 weeks; those in the treatment group were followed for 1 year. No adverse events were reported. Participants in the treatment group showed statistically significant improvements at 8 weeks compared to the control group with regard to pain, function, and patient satisfaction. Patients in the treatment group maintained significant improvements in function through the 1-year follow-up.[40]

Another prospective trial was performed in 2016 by Levi et al to assess changes in pain and function in patients who received intradiscal injection of PRP. Twenty-two patients were enrolled and followed for 6 months after treatment. Positive results—defined as 50% improvement in pain and 30% decrease in disability—were obtained in 14% of patients after 1 month, 32% after 2 months, and 47% after 6 months.[41]

The very first clinical trial performed to assess the efficacy and feasibility of intradiscal PRP injections (Fig 2-13), in 2012, also obtained very encouraging results. At 1 month of follow-up, pain (as measured by VAS) decreased on average from 7.1 to 1.8, and disability (as measured by a standard questionnaire) decreased from 11 to 3.2. These results were sustained for at least 6 months after treatment.[1]

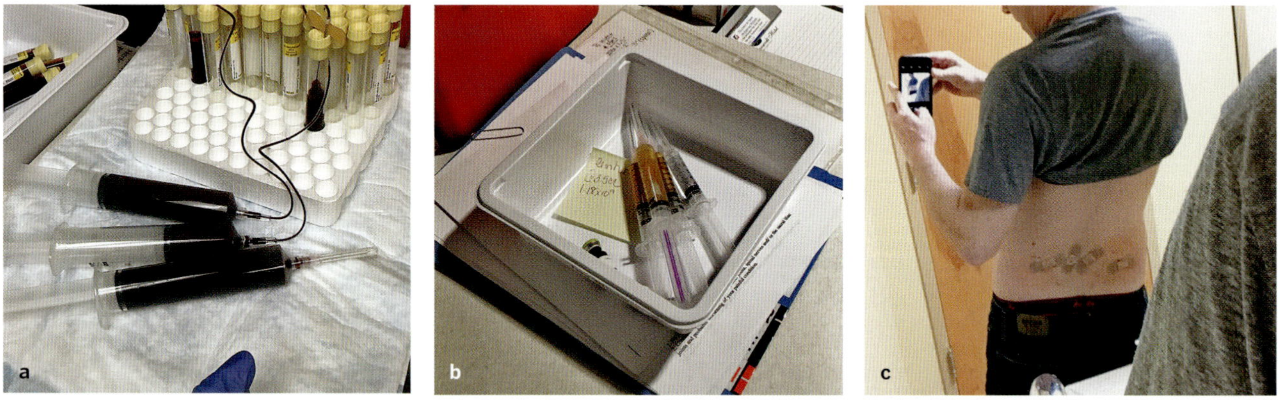

FIG 2-13 *(a to c)* Injections of PRP liquid into this patient's (a dentist) intervertebral disc helped him with improved pain and function.

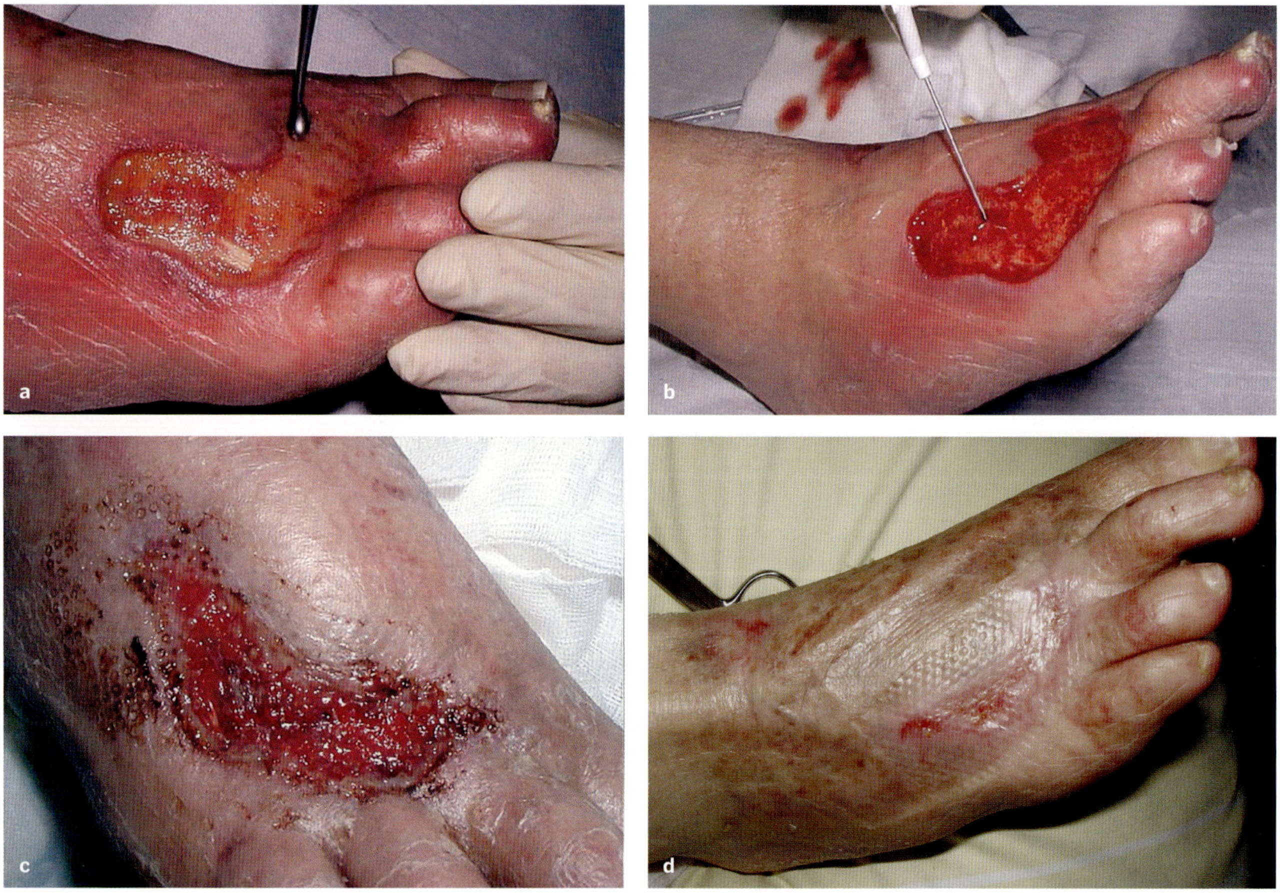

FIG 2-14 Diabetic foot ulcers remain a problem in health care systems universally due to the complications associated with amputation, so there is a profound need for a minimally invasive, relatively inexpensive, efficient, and reliable treatment option, such as one provided by applying PRP. *(a)* Unresolved diabetic foot ulcer of 18 months' duration. Appearance upon presenting to the wound care clinic for debridement. *(b)* Application of PRP gel. *(c)* Postoperative healing at 3 weeks. *(d)* Postoperative healing at 3 months. (Reprinted with permission from Marx and Garg.[42])

Chronic wound healing

The healing properties of PRP in cutaneous wounds is well documented, perhaps no better than in the literature on chronic conditions such as diabetic ulcers (Fig 2-14). Diabetic foot ulcer is a major complication of diabetes mellitus; 88% of all lower leg amputation is related to diabetic foot ulcer.[43] A 2018 randomized controlled multicenter study with a double-blind design was carried out by Goda et al to evaluate the efficacy and safety of PRP

Box 2-6 Effects of PRP dressing on diabetic foot ulcers compared with placebo[44]

- Achieved significantly higher healing rate
- Achieved complete healing faster

Box 2-7 Effects of PRP gel on skin graft reconstruction for tissue loss secondary to ulcerative infections and skin ulcers compared with placebo[45]

- Achieved nearly 50% faster healing
- Achieved greater overall skin graft success
- Resulted in no adverse effects

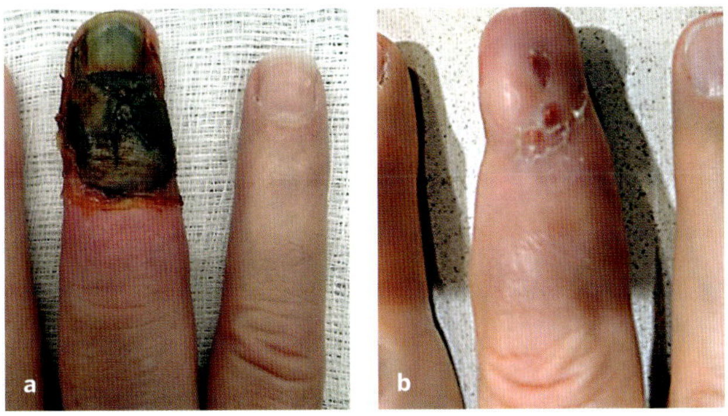

FIG 2-15 *(a and b)* Injection of PRP liquid saved this patient's finger from amputation as a result of gangrene infection.

for diabetic foot ulcer.[44] The study enrolled a total of 50 patients, who were divided into two groups. One group was treated with a dressing made of PRP in gel form, and the other group was treated with a dressing made of PPP in gel form. The healing rate of the PRP group was found to be significantly higher (0.66 ± 0.04) than that of the PPP group (0.49 ± 0.03). In addition, the rate of complete healing was significantly higher (21/25, 84%) in the PRP group than in the PPP group (13/25, 52%). These results were statistically significant (Box 2-6).[44]

Diabetes mellitus—along with advanced age and heart disease—is also a predisposing factor in necrotizing soft tissue infections (Fig 2-15) such as necrotizing fasciitis and Fournier gangrene, which occur primarily in lower limbs. These infections are extremely complex and challenging for both surgeons and patients alike. A prospective RCT was conducted in 2017 to investigate whether PRP gel could accelerate healing after skin graft reconstruction for tissue loss secondary to soft tissue infection. Autologous liquid PRP was combined with thrombin to create the gel. A total of 27 patients were randomized in two groups: the experimental group (n = 14) underwent a split-thickness skin graft (STSG) combined with PRP/thrombin gel, whereas the control group underwent STSG alone. The results showed that patients in the experimental group had a mean complete healing time that was almost 50% less than that of the patients in the control group (37.9 days vs 73.7 days). In addition, the rate of skin graft success was 89.6% vs 79.8% in the experimental vs control group. No adverse reactions were reported by patients in either group. The authors concluded that the combination of PRP/thrombin gel and STSG significantly improved clinical outcomes and shortened the wound healing time for necrotizing soft tissue infections (Box 2-7).[45]

Topical use of platelet concentrates can promote skin wound healing, as shown in the use of fibrin-based products in plastic surgery to reduce scarring and to promote wound closure. Adding platelet growth factors enhances these positive results, including treating chronic skin ulcers. Autologous blood concentrate products that include greater concentrations of leukocytes have been particularly effective in reducing infection and promoting wound healing.[1]

Ulcerative wounds most often considered chronic wounds include pressure wounds, venous leg wounds, arterial wounds, neurotrophic wounds, and diabetes-induced foot ulcers. PRP's fibrin and growth factor components may help in the healing process of such wounds.

Autologous blood products have been used as one of an assortment of possible healing therapies to treat a variety of chronic wounds, such as chronic nonhealing ulcers. Therapies have included hydrocolloid, alginate, hydrogel, and dextranomer dressings; polyurethane foam and membrane dressings; semipermeable polyurethane membrane dressings; and flax dressings. As part of an alternative wound healing accelerator, PRP has been combined with light-emitting diode therapy.[46]

General surgery

PRP is often used as one of several adjuncts to wound healing therapy. For example, it has been used in conjunction with stem cells from adipose tissue and hyaluronic acid for tissue regeneration in loss of substance on the lower limbs. The wounds in this particular case study involved lesions: ulcerative, dystrophic, with substance loss, and with differentiating etiopathogenesis. All 30 patients showed moderate improvement in less than a month, with nearly half healing in about 6 weeks, and nearly 60% with complete healing within 90 days. Lower cost, fewer medications, and greater patient comfort were also the results of the study.[47]

Because of the ability of cytokines present in platelet alpha granules to aid in wound healing, researchers have used PRP gel in a variety of clinical settings (for example, to treat chronic ulcers and, when more traditional therapies have failed, in orthopedic, dental, vascular, and cardiothoracic surgery), including as a topical application for amputations or traumatic loss of finger, with significant positive results, particularly in restoring esthetics. The researchers' decision to use PRP gel over a 5-year period on 115 patients stems from a number of factors, including low cost, easy preparation and availability, and patient comfort and safety.[48]

For many of the same reasons that researchers extended the use of PRP gel treatment to trauma and amputation wounds, others have successfully applied the gel for healing in chronic cutaneous radiation dermatitis, with the adjunct benefits of increasing favorable patient compliance and increasing the cost/benefit ratio from a lower number of medications and hospital visits.[49] Patients with recurrent fistulas and high transsphincteric and suprasphincteric fistulas run the risk of fistula recurrence or increases in fistulas, which have been successfully treated with tissue adhesives and plugs to seal the internal orifice and autologous growth factors in PRP to close the lumen of the fistula.[50-58]

Conclusion

While this chapter focuses on the clinical application of PRP in the significant and wide-reaching medical and surgical areas of cardiothoracic and orthopedic treatment (as well as a number of miscellaneous clinical applications), separate chapters discuss the clinical uses of PRP in plastic surgery and in implant, periodontal, and maxillofacial surgeries. A separate chapter is also devoted to the discussion of PRP preparation and applications of autologous blood concentrate products. As this chapter has shown, the literature frequently calls not only for high-level studies to better measure the efficacy of PRP clinical applications but also for the complementary need to spell out the application of platelet concentrates for the growing variety of clinical applications in wound healing and repair across the medical spectrum.

References

1. Dohan Ehrenfest DM. L-PRP/L-PRF in esthetic plastic surgery, regenerative medicine of the skin and chronic wounds. Curr Pharm Biotechnol 2012;13:1266–1277.
2. Modarressi A. Platelet rich plasma (PRP) improves fat grafting outcomes. World J Plast Surg 2013;2:6–13.
3. Del Corso M, Vervelle A, Simonpieri A, et al. Current knowledge and perspectives for the use of platelet-rich plasma (PRP) and platelet-rich fibrin (PRF) in oral and maxillofacial surgery part 1: Periodontal and dentoalveolar surgery. Curr Pharm Biotechnol 2012;13:1207–1230.
4. Simonpieri A, Del Corso M, Vervelle A, et al. Current knowledge and perspectives for the use of platelet-rich plasma (PRP) and platelet-rich fibrin (PRF) in oral and maxillofacial surgery part 2: Bone graft, implant and reconstructive surgery. Curr Pharm Biotechnol 2012;13:1231–1256.
5. Ferrari M, Zia S, Valbonesi M, et al. A new technique for hemodilution, preparation of autologous platelet-rich plasma and intraoperative blood salvage in cardiac surgery. Int J Artif Organs 1987;10:47–50.
6. Lubkowska A, Dolegowska B, Banfi G. Growth factor content in PRP and their applicability in medicine. J Biol Regul Homeost Agents 2012;26(2 suppl 1):3S–22S.
7. van der Wal M, Boks R, Wijers-Hille M, Hofland J, Takkenberg J, Bogers AJ. The effect of pre-operative blood withdrawal, with or without sequestration, on allogeneic blood product requirements. Perfusion 2015;30:643–649.
8. Zhai Q, Wang Y, Yuan Z, Zhang R, Tian A. Effects of platelet-rich plasmapheresis during cardiovascular surgery: A meta-analysis of randomized controlled clinical trials. J Clin Anesth 2019;56:88–97.
9. Bloemen S, Hemker C, Weerwind P, Maessen J, Mochtar B. Preoperative thrombin generation is predictive for the risk of blood loss after cardiac surgery: a research article. J Cardiothorac Surg 2013;8:154.
10. Vu TD, Pal SN, Ti LK, et al. An autologous platelet-rich plasma hydrogel compound restores left ventricular structure, function and ameliorates adverse remodeling in a minimally invasive large animal myocardial restoration model: A translational approach: Vu and Pal "Myocardial Repair: PRP, Hydrogel and Supplements". Biomaterials 2015;45:27–35.
11. Patel AN, Selzman CH, Kumpati GS, McKellar SH, Bull DA. Evaluation of autologous platelet rich plasma for cardiac surgery: Outcome analysis of 2000 patients. J Cardiothorac Surg 2016;11:62.
12. Han C, Na Y, Zhu Y, et al. Is platelet-rich plasma an ideal biomaterial for arthroscopic rotator cuff repair? A systematic review and meta-analysis of randomized controlled trials. J Orthop Surg Res 2019;14:183.
13. Abrams GD, Frank RM, Fortier LA, Cole BJ. Platelet-rich plasma for articular cartilage repair. Sports Med Arthrosc 2013;21:213–219.
14. Wong S, Ning A, Lee C, Feeley BT. Return to sport after muscle injury. Curr Rev Musculoskelet Med 2015;8:168–175.

15. Charousset C, Zaoui A, Bellaïche L, Piterman M. Does autologous leukocyte-platelet-rich plasma improve tendon healing in arthroscopic repair of large or massive rotator cuff tears? Arthroscopy 2014;30:428–435.
16. Figueroa D, Figueroa F, Calvo R, Vaisman A, Ahumada X, Arellano S. Platelet-rich plasma use in anterior cruciate ligament surgery: Systematic review of the literature. Arthroscopy 2015;31:981–988.
17. Mishra A, Harmon K, Woodall J, Vieira A. Sports medicine applications of platelet rich plasma. Curr Pharm Biotechnol 2012;13:1185–1195.
18. Shen L, Yuan T, Chen S, Xie X, Zhang C. The temporal effect of platelet-rich plasma on pain and physical function in the treatment of knee osteoarthritis: Systematic review and meta-analysis of randomized controlled trials. J Orthop Surg Res 2017;12:16.
19. Laudy AB, Bakker EW, Rekers M, Moen MH. Efficacy of platelet-rich plasma injections in osteoarthritis of the knee: A systematic review and meta-analysis. Br J Sports Med 2015;49:657–672.
20. Ye Y, Zhou X, Mao S, Zhang J, Lin B. Platelet rich plasma versus hyaluronic acid in patients with hip osteoarthritis: A meta-analysis of randomized controlled trials. Int J Surg 2018;53:279–287.
21. Miller LE, Parrish WR, Roides B, Bhattacharyya S. Efficacy of platelet-rich plasma injections for symptomatic tendinopathy: Systematic review and meta-analysis of randomised injection-controlled trials. BMJ Open Sport Exerc Med. 2017;3(1):e000237.
22. Andriolo L, Altamura SA, Reale D, Candrian C, Zaffagnini S, Filardo G. Nonsurgical treatments of patellar tendinopathy: Multiple injections of platelet-rich plasma are a suitable option: A systematic review and meta-analysis. Am J Sports Med 2019;47:1001–1018.
23. Li A, Yang H, Zhang J, Chen S, Wang H, Gao Y. Additive effectiveness of autologous platelet-rich fibrin in the treatment of intrabony defects. Medicine 2019;98:e14759.
24. Alio J, Rodriguez AE, Ferreira-Oliveira R, Wrobel-Dudzinska D, Abdelghany AA. Treatment of dry eye disease with autologous platelet-rich plasma: A prospective, interventional, non-randomized study. Ophthalmol Ther 2017;6:285–293.
25. Anitua E, Muruzabal F, De la Fuente M, Merayo-Lloves J, Orive G. Effects of heat-treatment on plasma rich in growth factors-derived autologous eye drop. Exp Eye Res 2014;119:27–34.
26. Ling Y, Wang S. Effects of platelet-rich plasma in the treatment of plantar fasciitis. A meta-analysis of randomized controlled trials. Medicine 2018;97:e12110.
27. Yang W-Y, Han Y-H, Cao X-W, et al. Platelet-rich plasma as a treatment for plantar fasciitis. A meta-analysis of randomized controlled trials. Medicine 2017;96:e8475.
28. Chew KT, Leong D, Lin CY, Lim KK, Tan B. Comparison of autologous conditioned plasma injection, extracorporeal shockwave therapy, and conventional treatment for plantar fasciitis: A randomized trial. PM R 2013;5:1035–1043.
29. Corti L. Nonpharmaceutical approaches to pain management. Top Companion Anim Med 2014;29:24–28.
30. Halper J. Advances in the use of growth factors for treatment of disorders of soft tissues. Adv Exp Med Biol 2014;802:59–76.
31. Scala M, Lenarduzzi S, Spagnolo F, et al. Regenerative medicine for the treatment of teno-desmic injuries of the equine. A series of 150 horses treated with platelet-derived growth factors. In Vivo 2014;28:1119–1123.
32. Broeckx S, Zimmerman M, Crocetti S, et al. Regenerative therapies for equine degenerative joint disease: A preliminary study. PLoS One 2014;9:e85917.
33. Kubota G, Kamoda H, Orita S, et al. Platelet-rich plasma enhances bone union in posterolateral lumbar fusion: A prospective randomized controlled trial. Spine J 2019;19:e34–e40.
34. Tarantino R, Donnarumma P, Mancarella C, et al. Posterolateral arthrodesis in lumbar spine surgery using autologous platelet-rich plasma and cancellous bone substitute: An osteoinductive and osteoconductive effect. Global Spine J 2014;4:137–142.
35. Hartmann EK, Heintel T, Morrison RH, Weckbach A. Influence of platelet-rich plasma on the anterior fusion in spinal injuries: A qualitative and quantitative analysis using computer tomography. Arch Orthop Trauma Surg 2010;130:909–914.
36. Manini DR, Shega FD, Guo C-F, Wang Y-X. Role of platelet-rich plasma in spinal fusion surgery: Systematic review and meta-analysis. Adv Orthop 2020;2020:8361798.
37. Singh R, Rohilla RK, Dhayal RK, Sen R, Sehgal PK. Role of local application of autologous platelet-rich plasma in the management of pressure ulcers in spinal cord injury patients. Spinal Cord 2014;52:809–816.
38. Biglari B, Reitzel T, Swing T, et al. A pilot study on the effectiveness of platelet-rich plasma and debridement for the treatment of nonhealing fistulas in spinal cord-injured patients. Adv Skin Wound Care 2015;28:123–128.
39. Basso M, Cavagnaro L, Zanirato A, et al. What is the clinical evidence on regenerative medicine in intervertebral disc degeneration? Muscoloskeletal Surg 2017;101:93–104.
40. Tuakli-Wosornu YA, Terry A, Boachie-Adjei K, et al. Lumbar intradiskal platelet-rich plasma (PRP) injections: A prospective, double-blind, randomized controlled study. PM R 2016;8:1–10.
41. Levi D, Horn S, Tyszko S, Levin J, Hecht-Leavitt C, Walko E. Intradiscal platelet-rich plasma injection for chronic discogenic low back pain: Preliminary results from a prospective trial. Pain Med 2016;17:1010–1022.
42. Marx RE, Garg AK. Dental and Craniofacial Applications of Platelet-Rich Plasma. Chicago: Quintessence, 2005.
43. Alvarsson A, Sandgren B, Wendel C, Alvarsson M, Brismar K. A retrospective analysis of amputation rates in diabetic patients: Can lower extremity amputations be further prevented? Cardiovasc Diabetol 2012;11:18.
44. Goda AA, Metwally M, Ewada A, Ewees H. Platelet-rich plasma for the treatment of diabetic foot ulcer: A randomized, double-blind study. Egypt J Surg 2018;37:178–184.
45. Hersant B, SidAhmed-Mezi M, Bosc R, Meningaud JP. Autologous platelet-rich plasma/thrombin gel combined with split-thickness skin graft to manage postinfectious skin defects: A randomized controlled study. Adv Skin Wound Care 2017;30:502–508.
46. Skórkowska-Telichowska K, Czemplik M, Kulma A, Szopa J. The local treatment and available dressings designed for chronic wounds. J Am Acad Dermatol 2013;68:e117–e126.
47. Cervelli V, De Angelis B, Lucarini L, et al. Tissue regeneration in loss of substance on the lower limbs through use of platelet-rich plasma, stem cells from adipose tissue, and hyaluronic acid. Adv Skin Wound Care 2010;23:262–272.
48. Balbo R, Avonto I, Marenchino D, Maddalena L, Menardi G, Peano G. Platelet gel for the treatment of traumatic loss of finger substance. Blood Transfus 2010;8:255–259.
49. Iervolino V, Di Costanzo G, Azzaro R, et al. Platelet gel in cutaneous radiation dermatitis. Support Care Cancer 2013;21:287–293.
50. Kucharczyk A, Kołodziejczak M, Sudoł-Szopinska I, Bielecki K. Autologous growth factors used for the treatment of recurrent fistula-in-ano preliminary results. Tech Coloproctol 2014;18:317–318.
51. Cintron JR, Park JJ, Orsay CP, et al. Repair of fistulas-in-ano using fibrin adhesive: Long-term follow-up. Dis Colon Rectum 2000;43:944–950.
52. Ventakesh KS, Ramanujam P. Fibrin glue application in the treatment of recurrent anorectal fistulas. Dis Colon Rectum 1999; 42:1136–1139.
53. Knighton DR, Ciresi KF, Fiegel VD, Austin LL, Butler EL. Classification and treatment of chronic nonhealing wounds. Successful treatment with autologous platelet-derived wound healing factors (PDWHF) Ann Surg 1986;204:322–330.
54. Sentovich SM. Fibrin glue for all anal fistulas. J Gastrointest Surg 2001;5:158–161.

55. van der Hagen SJ, Baeten CG, Soeters PB, van Gemert WG. Autologous platelet derived grow factors (platelet-rich plasma) as an adjunct to mucosal advancement flap in high crypto glandular peri-anal fistulae: A pilot study. Colorectal Dis 2011;13:215–218.
56. Tsai CH, Hsu HC, Chen YJ, Lin MJ, Chen HT. Using the growth factors-enriched platelet glue in spinal fusion and its efficiency. J Spinal Disord Tech 2009;22:246–250.
57. Hartmann EK, Heintel T, Morrison RH, Weckbach A. Influence of platelet-rich plasma on the anterior fusion in spinal injuries: A qualitative and quantitative analysis using computer tomography. Arch Orthop Trauma Surg 2010;130:909–914.
58. Zurita M, Otero L, Aguayo C, et al. Cell therapy for spinal cord repair: optimization of biologic scaffolds for survival and neural differentiation of human bone marrow stromal cells. Cytotherapy 2010;12:522–537.

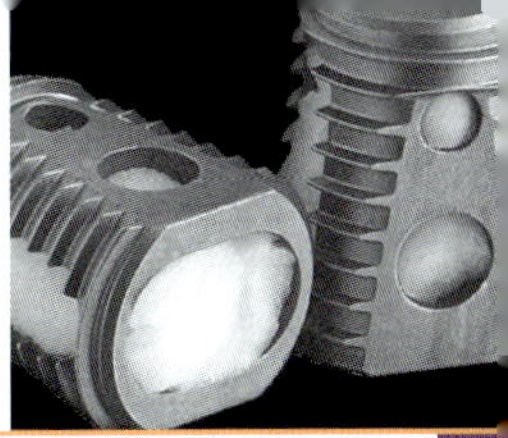

3

Biologic Growth Factors, PRP, and Bone Morphogens in Bone Regeneration Procedures

Some of the most promising research currently underway on bone regeneration involves applications of biologic growth factors and bone morphogens. In general, these biologic proteins are intimately involved in regulating the cellular events that occur during repair of bone and other tissues in the body (Fig 3-1). Such events include cell proliferation, chemotaxis, mitosis, differentiation, and matrix synthesis. Growth factors bind to specific receptors on target cell surfaces, leading to a complex cascade of intracellular events that culminates in bone formation. The basic theory is that by clinically applying an added number of these bone-building "orchestrators" to a wound site, they may in effect jump-start and perhaps even improve the body's normal osteoregenerative potential. Some researchers even speculate that in the not-too-distant future, growth factors and bone morphogens may eventually make the use of other bone graft materials—even autogenous bone—obsolete.

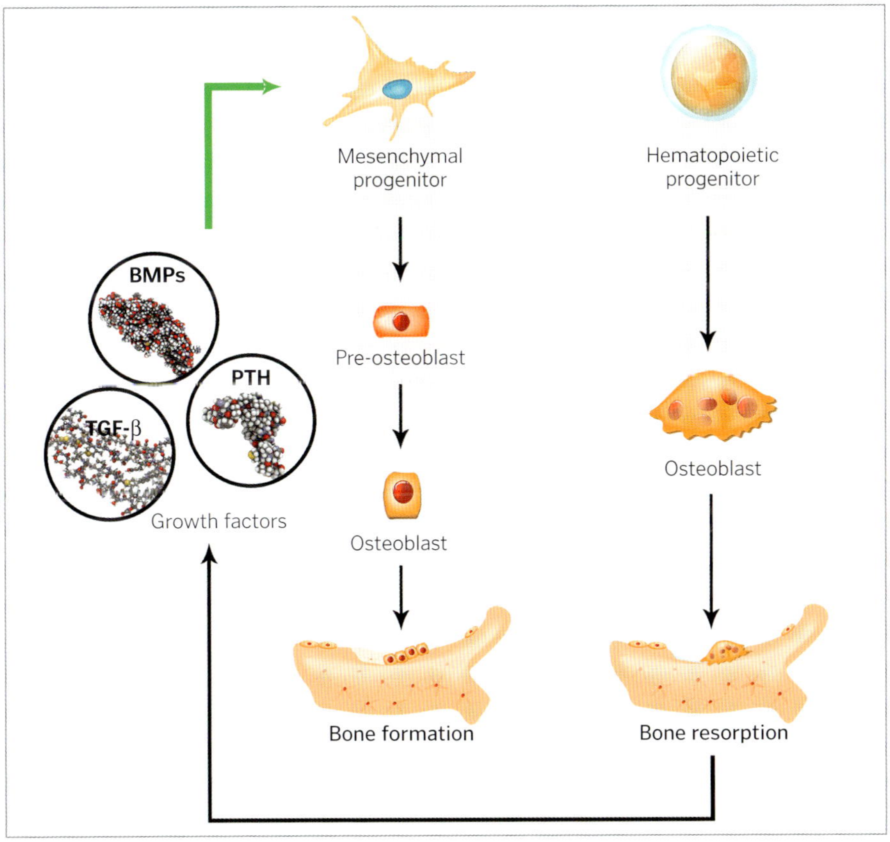

FIG 3-1 Bone formation during development, remodeling, and repair. BMPs, bone morphogenetic proteins; TGF-β, transforming growth factor β; PTH, parathyroid hormone.

The potential also exists for achieving earlier and better osseointegration of dental implants. Such biologic treatments have gained a foothold in the market for both orthopedic and oral and maxillofacial applications. Infuse Bone Graft (Medtronic) received US Food and Drug Administration (FDA) approval in early 2002 for use in spinal fusion surgery. This product consists of two collagen sponges that can be soaked in recombinant human bone morphogenetic protein-2 (rhBMP-2) and placed into the defect or inserted into a titanium cage for spinal surgery.

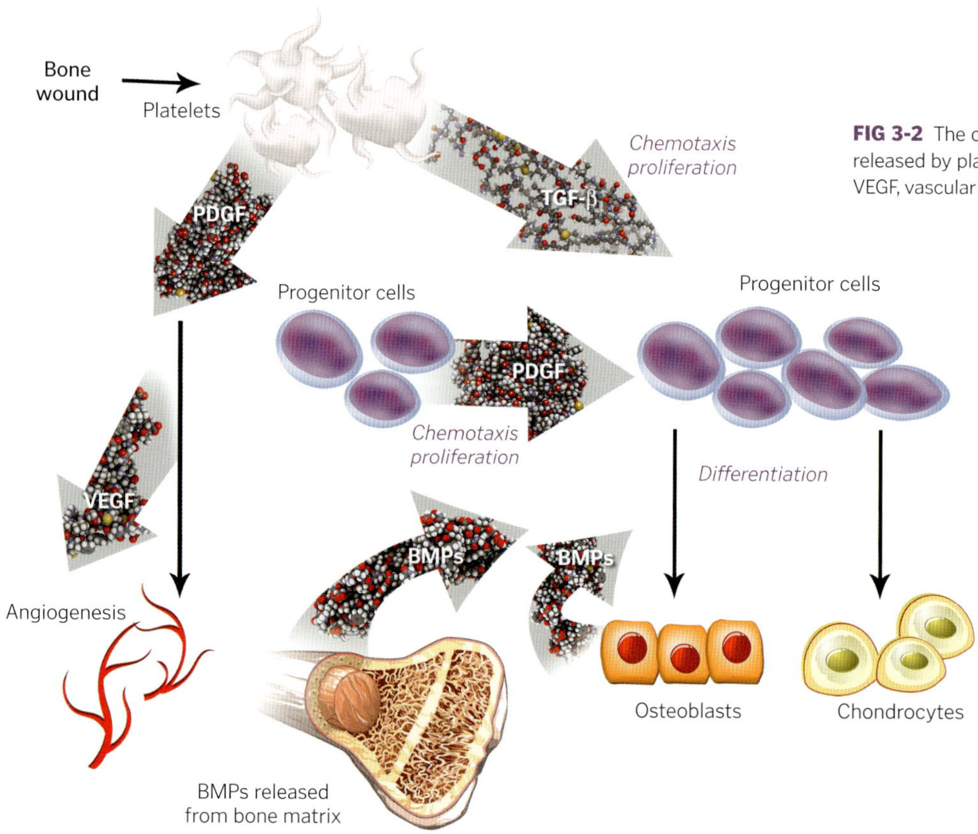

FIG 3-2 The cascade of growth factors released by platelets during bone repair. VEGF, vascular endothelial growth factor.

In 2004, Infuse was FDA-approved for treating tibial shaft fractures, and in 2007 it was cleared for oral defects associated with extraction sockets. Several other individual growth factors and their specific functions in bone healing have also been identified and tested for their ability to affect bone growth. These include platelet-derived growth factor (PDGF), insulinlike growth factor (IGF), fibroblastic growth factor (FGF), and transforming growth factor β (TGF-β) (Fig 3-2). This chapter describes their function in bone formation and provides information on the current state of research for each factor.

Platelet-Derived Growth Factor

A molecule that plays a key role in the regulation of tissue repair, PDGF is secreted from aggregated platelets after tissue injury. It functions by binding to surface receptors of cells of mesenchymal origin (including osteoblasts and osteoclasts) and participates in the development and remodeling of bone and other types of tissue. PDGF has five isoforms, but only PDGF-BB binds with all of the potential PDGF receptors.

Several clinical and experimental studies have investigated the therapeutic value of PDGF in the treatment of periodontal bony defects. In an animal model, Simion et al investigated the outcome of vertical ridge augmentation by combining rhPDGF with deproteinized cancellous bovine bone blocks. Their results showed that rhPDGF significantly enhances new bone formation when combined with xenograft bone blocks compared to when xenograft bone blocks are used alone.[1] Similarly, Urban et al demonstrated that rhPDGF in combination with bone mineral and barrier membrane enhances new bone formation around periodontal bone defects. PDGF has also been shown to regenerate bone in peri-implant defects.[2] McAllister et al assessed whether PDGF-BB could be used to regenerate bone around dental implants following tooth extraction. Histologic assessment showed that the use of rhPDGF-BB with either a xenograft or beta tricalcium

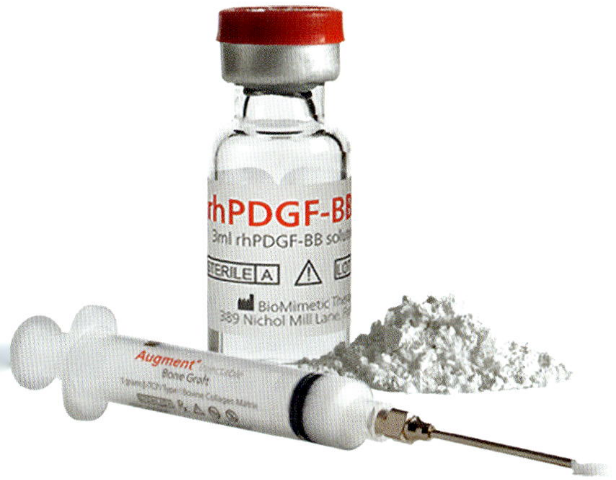

FIG 3-3 Augment Injectable is used to fill bone defects in ankle and hindfoot fusion procedures.

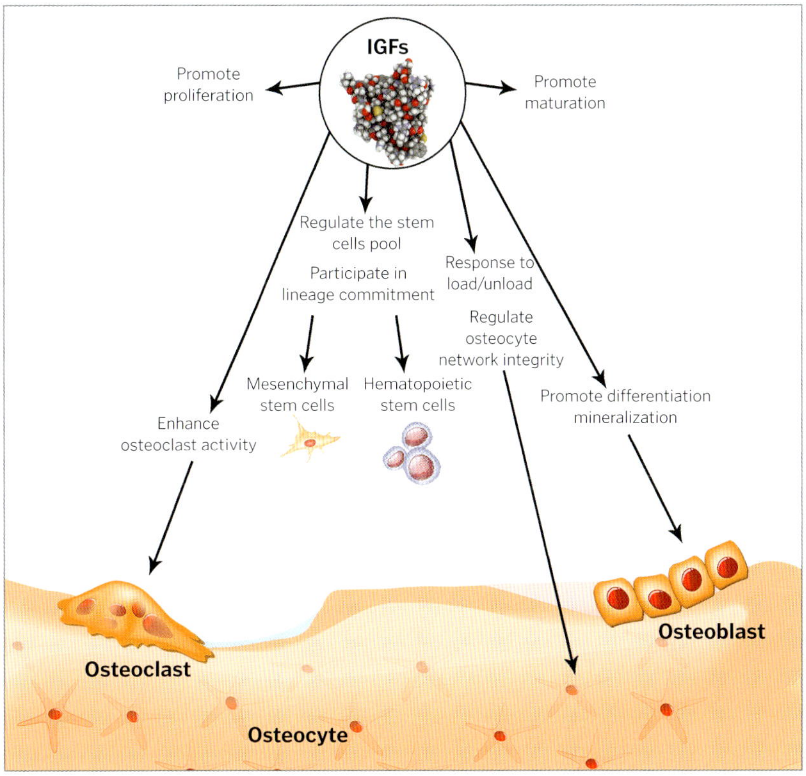

FIG 3-4 Role of IGF-1 in the regulation of bone growth and development. Changes in IGF-1 levels correlate with changes in cortical and cancellous bone properties. IGF-1 regulates matrix deposition and mineralization via its effects on osteoblasts, osteocytes, and osteoclasts.

phosphate (β-TCP) resulted in uneventful socket healing. At the restorative phase of treatment, all implants demonstrated 100% osseointegration.[3]

Augment Injectable (Wright), another biologic bone grafting product, is used to fill bone defects in ankle or hindfoot fusion procedures (Fig 3-3). It incorporates rhPDGF-BB with a blend of collagen and β-TCP, both of which are biologics that encourage the regrowth of the patient's own bone.

Augment is intended for use as an alternative to autograft in arthrodesis (ie, fusion procedures) of the ankle and/or hindfoot in patients with osteoarthritis, posttraumatic arthritis, rheumatoid arthritis, psoriatic arthritis, avascular necrosis, joint instability, joint deformity, congenital defect, or joint arthroplasty. The original form of the graft, which was FDA approved in 2018, came in two parts—a cup containing the β-TCP granules and a vial containing the growth factor solution—that had to be mixed by the surgeon at least 10 minutes before its introduction to the graft site. Augment Injectable is instead premixed and therefore much easier for surgeons to use.

Insulinlike Growth Factor

As the name indicates, IGFs are biochemically and functionally similar to insulin.[4] They are primarily produced by the liver and circulate in the vascular system.[5]

There are two types of IGF (IGF-1 and IGF-2), and they have similar functions but are independently regulated. Although its direct or indirect effects on bone remodeling are not fully understood, IGF-1 appears to stimulate bone formation by increasing cellular proliferation and differentiation as well as bone matrix production (Fig 3-4).

Recent studies of IGF-1 and its receptors in mice and in vitro cell cultures reveal the complex actions of these molecules in skeletal cells. Bone is constantly remodeled by bone-resorbing osteoclasts and bone-forming osteoblasts. The osteoclasts are derived from hematopoietic stem cells (HSCs), while osteoblasts differentiate from bone marrow mesenchymal stem cells (MSCs). Bone formation and resorption are tightly balanced to maintain the structural integrity of the skeleton and to produce calcium to meet systemic demands.

Table 3-1 Tissue applications of FGF

Target tissue	Subfamily of FGF	Materials/ carriers	In vivo/ in vitro	Animal/ cell	Functions/ effects
Bone	FGF1	Hydroxyapatite-fibrin scaffold	In vivo	Rat	Osteogenic markers, bone regeneration
	FGF2	Hyaluronate scaffold	In vitro	BMSCs	Osteogenic markers, mineralization
		Gelatin hydrogel	In vivo	Rabbit	Mineralization, bone regeneration
		Hydroxyapatite porous granules	In vitro	MC3T3-E1	Cell proliferation, osteoblast differentiation
		Collagen-bioactive glass	In vivo	Rat	Bone regeneration
		Titanium-based metals/Matrigel	In vivo/ In vitro	Rat	Bone regeneration
		Hydroxyapatite/collagen scaffold	In vivo	Rabbit	Bone regeneration, cartilage regeneration
		Titanium implant-melatonin	In vivo	Rat	Osseointegration
Dental	FGF2	Gelatin microsphere	In vivo	Dog	Periodontal regeneration
		Tricalcium phosphate	In vivo	Dog	Alveolar tissue regeneration

BMSC, bone marrow–derived mesenchymal stem cell. (Adapted from Gao et al.[9])

Studies of global IGF-1 knockout mice that survived to adulthood show decreased bone formation rates but higher trabecular bone mass, indicating reduction in both osteoblastic and osteoclastic activities and demonstrating the role of IGF-1 signaling in bone remodeling.

A number of in vitro and in vivo studies have demonstrated the importance of IGF-1 signaling for normal osteoblast development and function. According to Bikle et al, animals that overexpress IGF-1 have shown higher osteocyte lacunae occupancy and an increased bone formation rate, bone volume, and bone mineral density, yet without any change in total osteoblasts or osteoclasts.[6] These findings suggest that osteoblast-derived IGF-1 primarily exerts its anabolic effects by enhancing osteoblast function and osteocyte survival.[6–8]

Fibroblast Growth Factors

FGFs are so-named because of their general growth-promoting effects on most fibroblastic cell types. This growth factor—which occurs in both acidic and basic forms and is stored in bone —also stimulates angiogenesis for vascular invasion of bone, wound healing, and cell migration. FGFs bind FGF receptors and regulate developmental pathways, controlling events in the early embryo through development of multiple organ systems. FGF receptors are expressed on many different cell types and regulate key cell behaviors, such as proliferation, differentiation, and survival, which make FGF signaling susceptible to subversion by cancer cells.

Recent studies of FGF have focused on mutations related to disease. Some germline FGF mutations that have been identified in human disease include hereditary deafness; Kallmann syndrome, a developmental disorder characterized by anosmia and hypogonadism; lacrimo-auriculo-dento-digital (LADD) syndrome, which is characterized by hearing loss, dental anomalies, and lacrimal and salivary gland hypoplasia; and hypophosphatemic rickets.

FGFs have a variety of biologic functions, including roles in mitogenesis, cellular migration, differentiation, angiogenesis, and wound healing. Because of their capacity to perform multiple functions, FGFs have been used in the regeneration of skin, blood vessels, muscle, adipose tissue, tendons and ligaments, cartilage, bone, teeth, and nerves (Table 3-1).[9] When administered directly to the

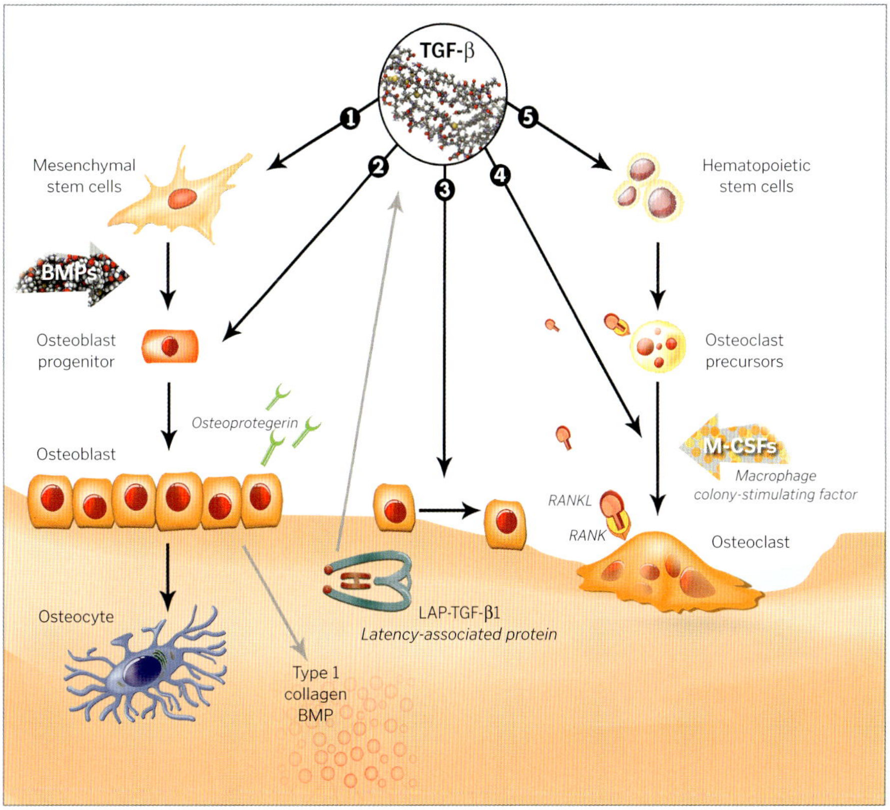

FIG 3-5 The five major roles of TGF-β1 in osteoimmunity.

site of wounds, FGFs rapidly lose their functional biologic capacity. Therefore, a wide range of synthetic and natural biomaterials have been studied to try to identify one that can carry and deliver FGFs to targeted regions and allow their release at a controllable and sustainable rate.[10]

In dentistry, many applications of FGFs are found in periodontal regeneration. For example, periodontal ligament tissue has been shown to be regenerated by the action of FGF2 contained in a gelatin carrier. At 8 weeks after implantation in nonhuman primates, the FGF2-gelatin group showed significant regeneration of periodontal tissues in a dose-dependent manner.[11] Shirakata et al demonstrated that treatment with FGF2 induced the promotion of periodontal healing in two-wall intrabony defects in dogs, suggesting that they were a possible candidate for replacement of enamel matrix derivative or PDGF combined with TCP.[12]

In dental implant therapy, the use of FGF2 has been shown to influence the formation of bone around titanium-based metal. FGF2 suspended in Matrigel (Corning) was administered to the titanium surface and then implanted in ovariectomized rats. The Matrigel was shown to prolong the lifespan of FGF2 upon sustained release for up to 21 days. The implant samples after 3 months were found to induce significant enhancement in new bone formation and mechanical stability on the Matrigel-FGF2 group when compared to groups treated with FGF2 or Matrigel alone.[9]

Transforming Growth Factor β

TGF-β is a multifunctional growth factor synthesized by many cell types, and almost every cell type can be stimulated by at least one of the various TGF-β molecules. It is one of the major growth factors present in bone (as well as platelets) and is structurally related to, but functionally different from, the BMPs. Generally, TGF-β is a weak mitogen for osteoblasts. It has been shown to be chemotactic for bone cells and may increase or decrease their proliferation depending on various conditions. It has also been shown to stimulate type 1 collagen synthesis.[5,13]

TGF-β signaling plays several fundamental roles in bone regeneration and bone fracture healing (Fig 3-5), as many studies have demonstrated. Dubon et al have shown that TGF-β induces the migration and mobilization of bone marrow–derived mesenchymal stem cells (BMSCs) to maintain bone homeostasis.[14] Zhao et al have found that

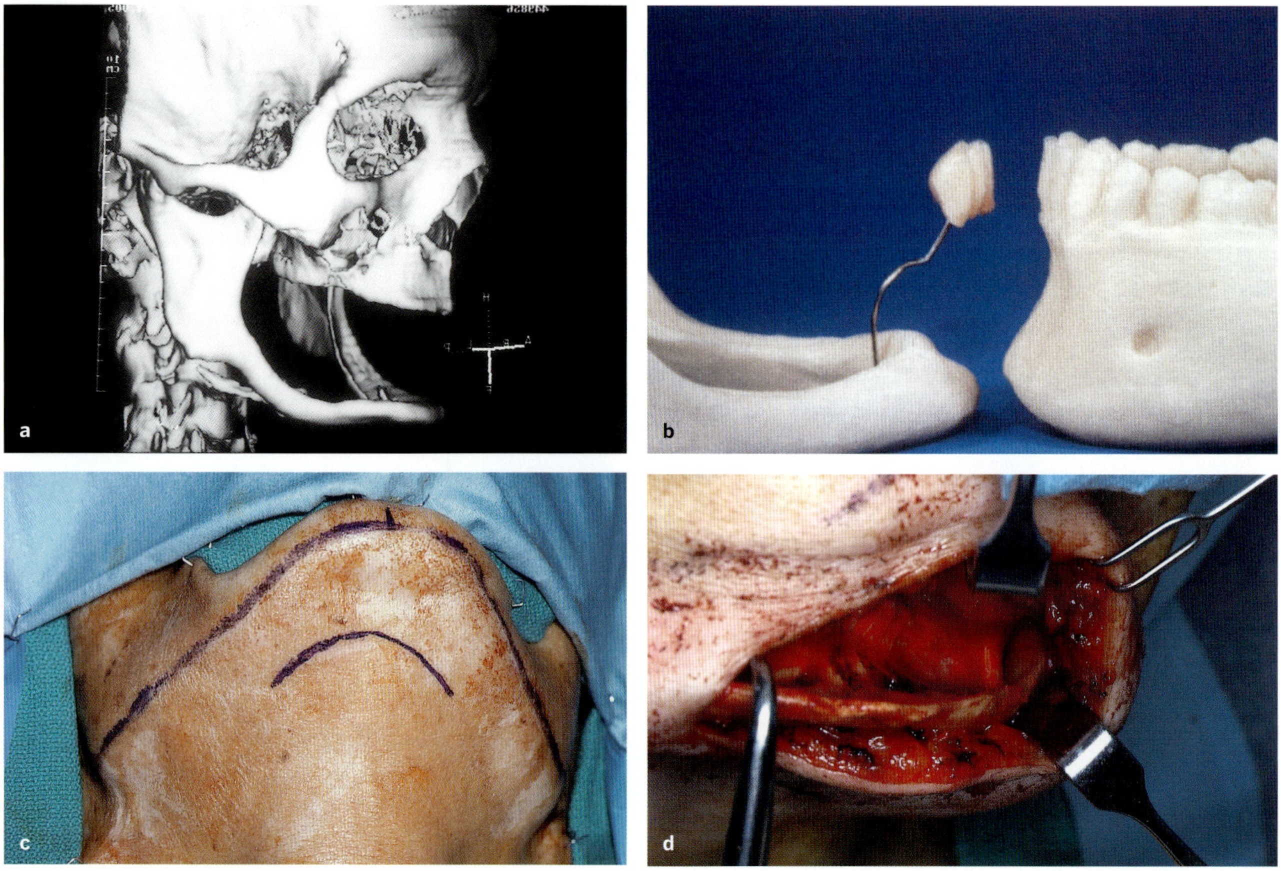

FIG 3-6 Rehabilitation of patient with severely resorbed mandible using autogenous bone graft combined with PRP to obtain 15 mm of new bone. *(a)* The resorbed mandible represents the remnant of the inferior border, which is dense cortical bone. Note that the mandibular canal is unroofed by the resorption process so that the mental nerve now emerges in the second molar region. *(b)* The severely resorbed mandible represents a significant and yet often underappreciated loss of vertical bone height and loss of vertical dimension. *(c)* Reconstruction of the severely resorbed mandible is approached through a transcutaneous curvilinear incision in the submental triangle. *(d)* The periosteum is reflected from the labial and crestal areas bilaterally to the retromolar areas. In this 4-mm mandible, the mental nerves emerge from the resorbed crestal bone, and the hypertrophied genial tubercles are noted in the midline. *(e)* The first dental implant is placed 5 mm anterior to the emergence of the mental nerve, and each succeeding implant is placed parallel to and 1 cm apart from center to center. *(f)* Autogenous cancellous bone and marrow is then placed around all of the dental implants and compacted with bone packers to the height of the implant cover screws. *(g)* Once the liquid PRP covering gel is in situ, the consistency that it provides allows contouring and sculpting of the graft into a ridge form around the implants. *(h)* The implants seen at 4 months postsurgery through an intraoral crestal incision and now ready for the prosthetic phase. Note the excess bone growth even up to, above, and over the cover screw. *(i)* Severely resorbed mandible that has been grafted according to the tent-pole procedure, 1 week after surgery. Note the placement of the implants into the inferior border and the absence of mineralization, as one would expect in an autogenous cancellous marrow graft at this time. *(j)* Six months after graft placement and 3 months after loading the dental implants with a prosthesis, the mineralization of the graft is more apparent, and 15 mm of bone height has been gained. Note the development of a new foramen and mandibular canal in the bone graft. *(k)* Pretreatment profile of patient with inadequate mandibular bone height. Note the loss of facial vertical dimensions and the pointed chin appearance often referred to as a "witch's chin." *(l)* Correction of facial vertical dimension and a normalized nose-to-chin relationship and contour as a result of the tent-pole grafting procedure. (Reprinted with permission from Marx and Garg.[19])

patients with bone fractures have an increased expression of TGF-β1 in their blood.[15] Moreover, the important roles of TGF-β1 and TGF-β3 in osteoblast differentiation and bone formation are widely recognized.[16,17] During osteoclast bone resorption, TGF-β is released and activated to recruit MSCs through SMAD signaling.[18]

Platelet-Rich Plasma

As an alternative source of growth factors, PRP has become an increasingly popular clinical tool for several types of surgery, including oral bone-regenerative procedures (Fig 3-6).[19] PRP is a concentrated autologous source

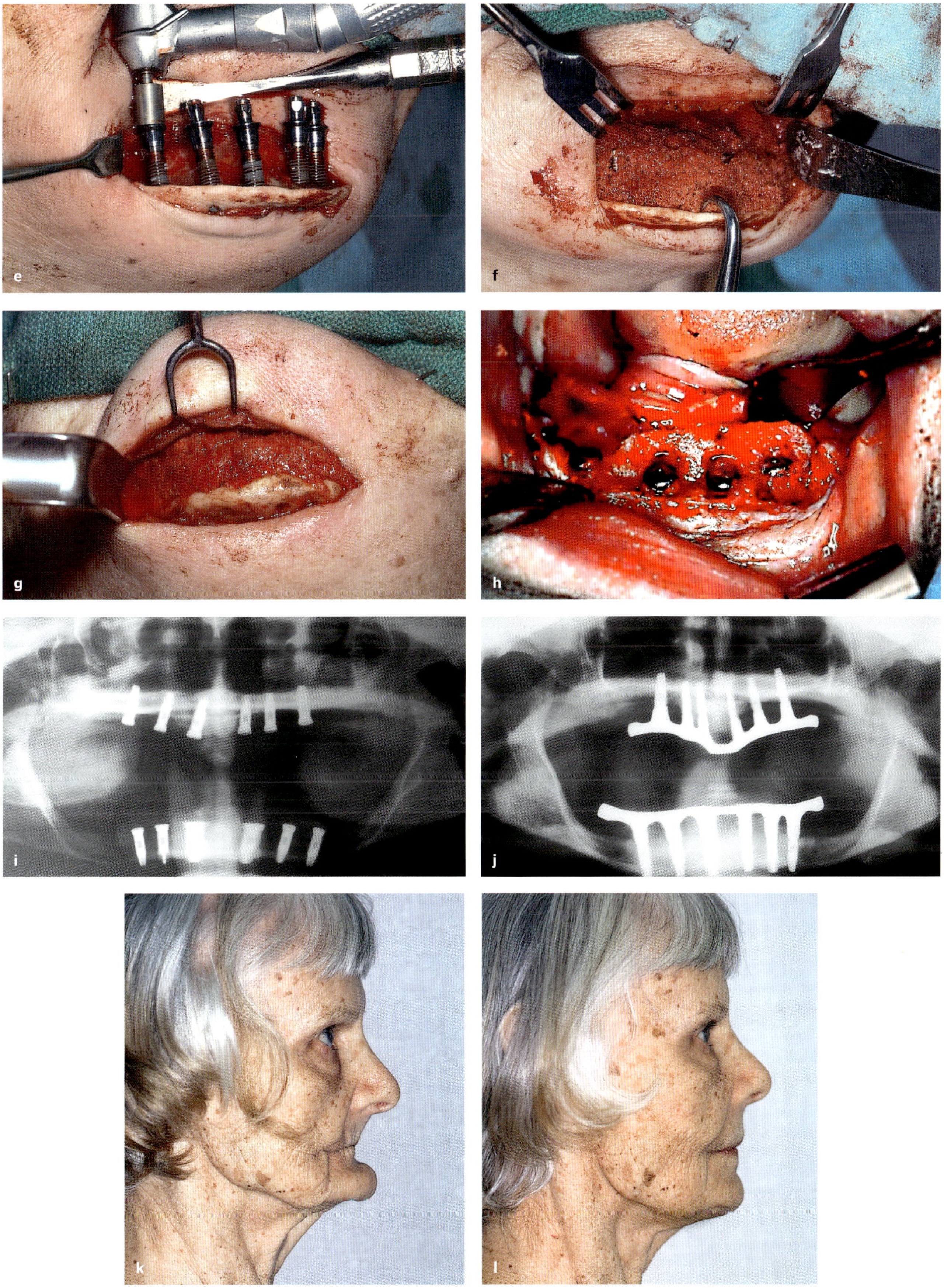

Table 3-2 Summary of growth factors contained in PRP

Growth factor	Function
TGF-β	• Stimulates undifferentiated mesenchymal cell proliferation • Regulates endothelial, fibroblastic, and osteoblastic mitogenesis • Regulates collagen synthesis and collagenase secretion • Regulates mitogenic effects of other growth factors • Stimulates endothelial chemotaxis and angiogenesis • Inhibits macrophage and lymphocyte proliferation
FGF	• Promotes growth and differentiation of chondrocytes and osteoblasts • Mitogenetic for mesenchymal cells, chondrocytes, and osteoblasts
PDGF	• Mitogenetic for mesenchymal cells and osteoblasts • Stimulates chemotaxis and mitogenesis in fibroblast, glial, and smooth muscle cells • Regulates collagenase secretion and collagen synthesis • Stimulates macrophage and neutrophil chemotaxis
Epithelial growth factor (EGF)	• Stimulates endothelial chemotaxis or angiogenesis • Regulates collagenase secretion • Stimulates epithelial or mesenchymal mitogenesis
VEGF	• Increases angiogenesis and vessel permeability • Stimulates mitogenesis for endothelial cells
Connective tissue growth factor (CTGF)	• Promotes angiogenesis, cartilage regeneration, and fibrosis and platelet adhesion
IGF-1 and IGF-2	• Chemotactic for fibroblasts and stimulates protein synthesis • Enhances bone formation
Platelet factor 4 (PF4)	• Stimulates the initial influx of neutrophils into wounds • Chemoattractant for fibroblasts
Interleukin 8 (IL-8)	• Proinflammatory mediator • Aids in recruitment of inflammatory cells
Keratinocyte growth factor (KGF)	• Promotes endothelial cell growth, migration, adhesion, and survival • Stimulates angiogenesis

of several growth factors, particularly PDGF, TGF-β1, and TGF-β2, as well as vascular endothelial growth factor (VEGF), IGF, and other growth factors possibly contained within platelets (Table 3-2).

As elucidated in the following chapters of this book, PRP can be used to enhance graft material, to form a growth factor–rich membrane,[20] or to enhance a traditional barrier membrane. PRP's fibrinogen component also makes it an excellent hemostatic tool, tissue sealant, wound stabilizer, and graft condenser through the creation of a gel-like substance that allows for sculpting and excellent adherence in defects.

It has been demonstrated radiographically that adding PRP to graft material significantly accelerates the rate of bone formation and improves trabecular bone density as compared with sites treated only with autogenous graft material[4,19,20] (Fig 3-7). Widespread clinical reports and recent published research also suggest that PRP significantly enhances soft tissue healing; reduces bleeding, edema, and scarring; and decreases patients' self-reported pain levels postoperatively.[21–26] PRP growth factors are particularly attractive for use in patients with conditions that typically reduce the success of bone

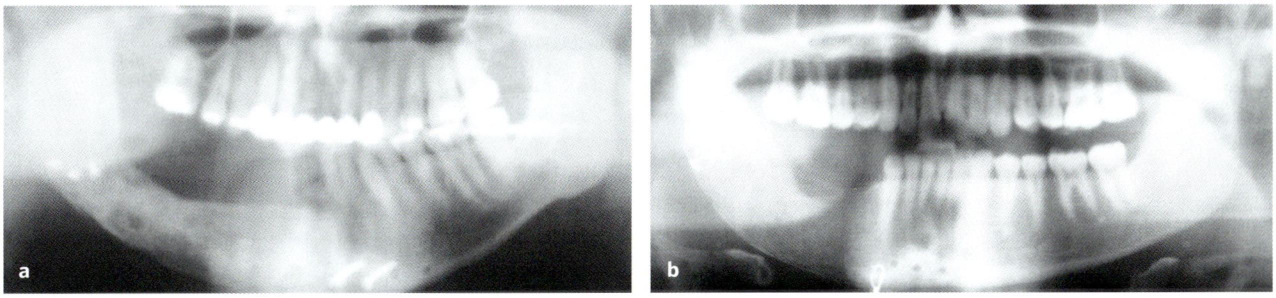

FIG 3-7 *(a)* Panoramic radiograph of a hemimandibular autogenous bone graft at 4 months showing excellent bone formation and continuity interrupted by areas of immaturity. *(b)* Panoramic radiograph of a size-, age-, and time-matched hemimandibular autogenous bone graft enhanced with PRP showing greater bone density indicative of advanced maturity. (Reprinted with permission from Marx and Garg.[19])

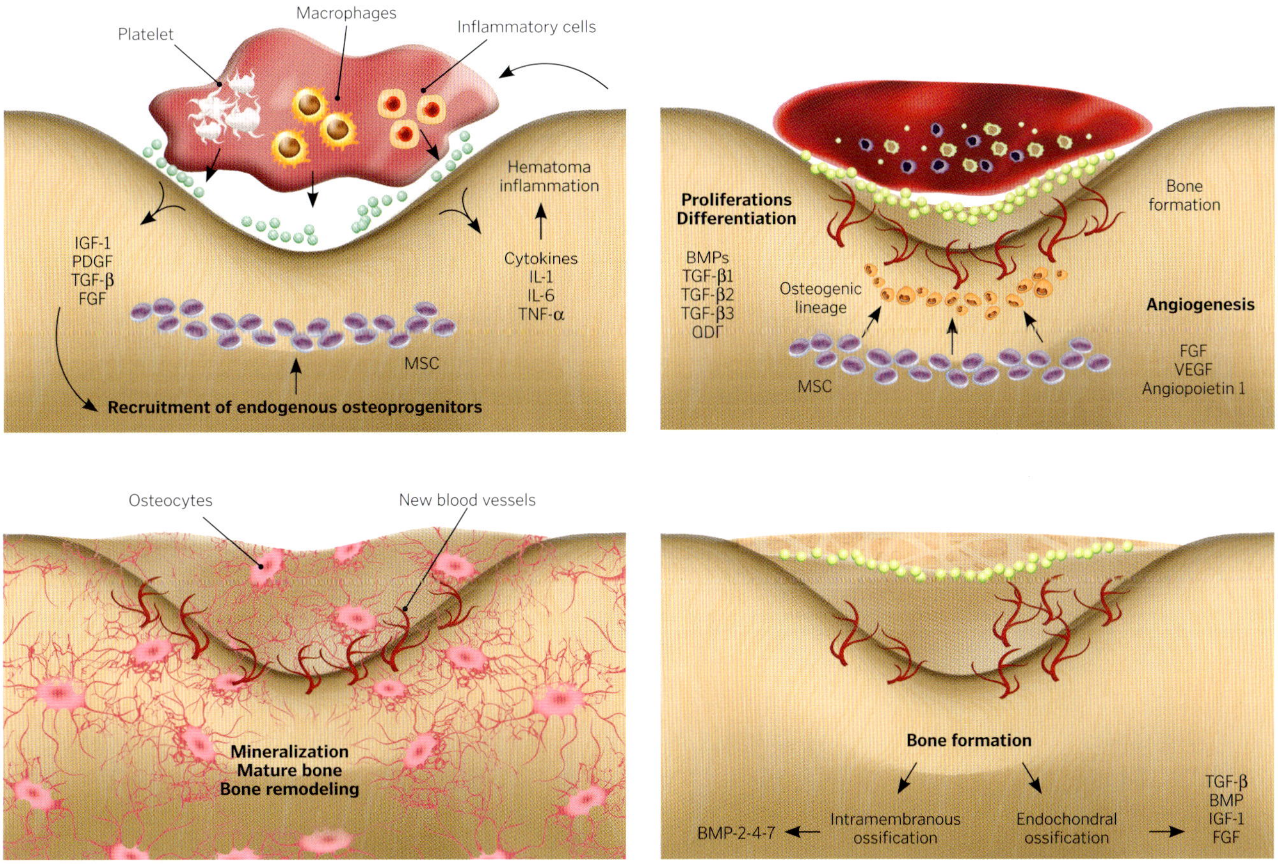

FIG 3-8 Temporal evolution of bone regeneration following PRP application. GDF, growth differentiation factor.

grafts and osseointegration, including those with an edentulous and severely atrophic maxilla, osteoporosis, and tissues that have been scarred and altered by prior dental disease.[27]

The concentrated levels of PDGF and TGF-β in PRP (in addition to the presence of other growth factors and proteins) appear to provide benefits that lead to more rapid and effective bone regeneration. The beneficial effects of these growth factors can perhaps best be understood through a description of the components and biology of PRP (Fig 3-8).

Specific studies of PRP have identified at least three important factors in the alpha granules of the sequestered platelets: PDGF, TGF-β1, and TGF-β2.[28-30] In addition, other

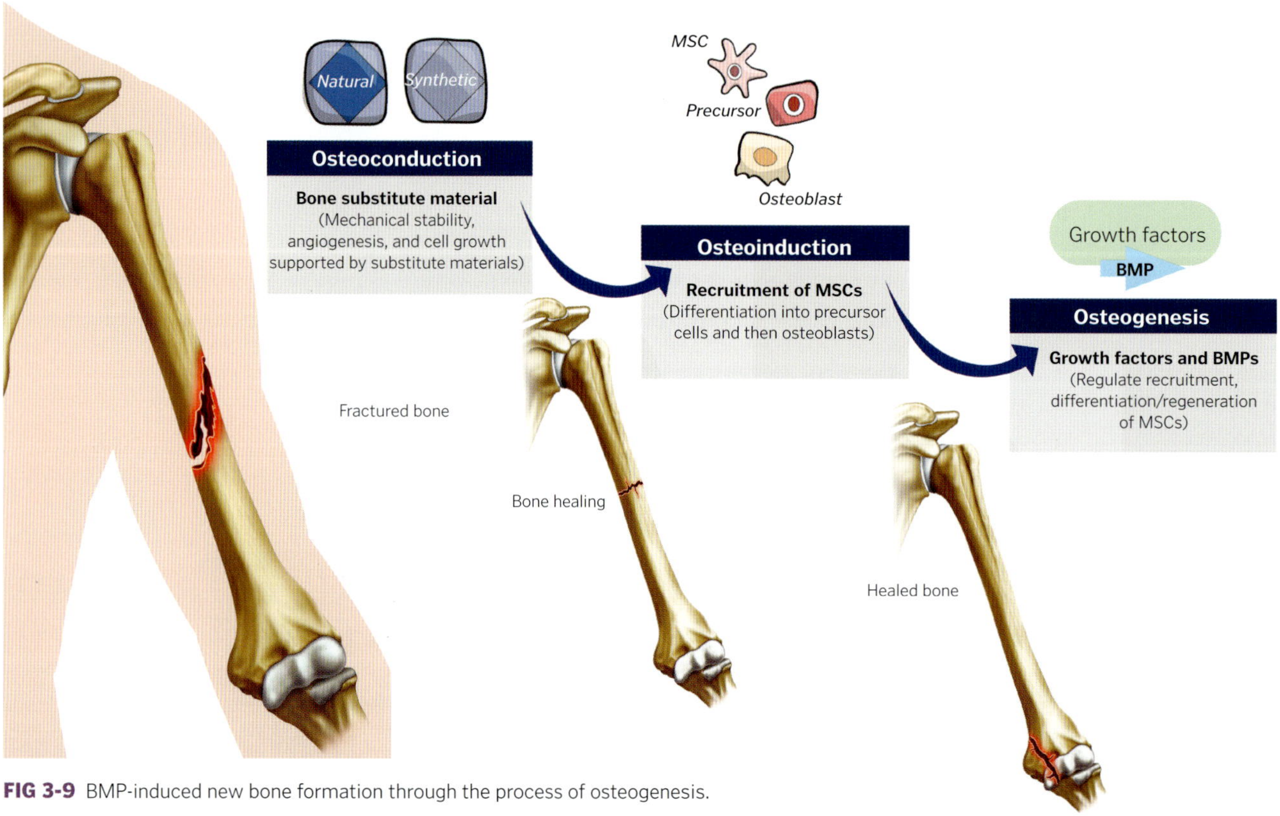

FIG 3-9 BMP-induced new bone formation through the process of osteogenesis.

studies have documented the presence of VEGF and IGF-1 in platelets from peripheral human blood tests.[31–35]

PRP accelerates and expands cells' wound healing response and acts biochemically to set the rate and amount of regeneration in bone. Its activity lasts only 10 days, but it has long-lasting effects.

A graft-surrounding matrix is formed by fibrin, fibronectin, and vitronectin. PDGFs have a mitogenic effect on osteoblasts, endothelial cells, and MSCs. The latter are also acted upon mitogenetically and angiogenetically by TGF-β isomers, which induce osteoblastic differentiation as well. While capillary ingrowth is promoted by VEGF, the lack of epithelial cells renders epithelial growth factor (EGF) inert. Osteoprogenitor cell mitosis begins within 72 hours, and capillary buds appear. In the entire first phase of bone graft healing (about 2½ to 3 weeks), the graft is penetrated by capillaries, and osteoprogenitor cells proliferate rapidly. During this phase, cell instability and infection are common, with the potential for lysing and arresting the development of wound healing. Obviously, prevention of infection and contamination are essential, as is graft stability.[19]

Bone Morphogenetic Proteins

Although BMPs are often referred to as a category of growth factors, they are actually a distinct group of proteins. Bone-related growth factors exist primarily in bone matrix and are released during remodeling or in response to trauma. When this occurs, the growth factors modulate or stimulate the neighboring osteoprogenitor (differentiated) cells that are already present in the area, inducing and aiding bone formation. Because they rely on the presence of local osteoprogenitor cells, however, growth factors cannot induce bone growth in ectopic sites. Their effect is also limited in large bony defects.[36]

The main advantage of BMPs, which are also found in extracellular bone matrix, is that they do not require existing osteoprogenitor cells to form bone. BMPs are osteoinductive factors that can stimulate mesenchymal cells to differentiate into cartilage- and bone-forming cells[37] (Fig 3-9). Unlike true growth factors, however, they are not mitogenic for many cells and cell types.[38] In embryonic development, BMPs affect the direction of cell differentiation and also act as positional signals,

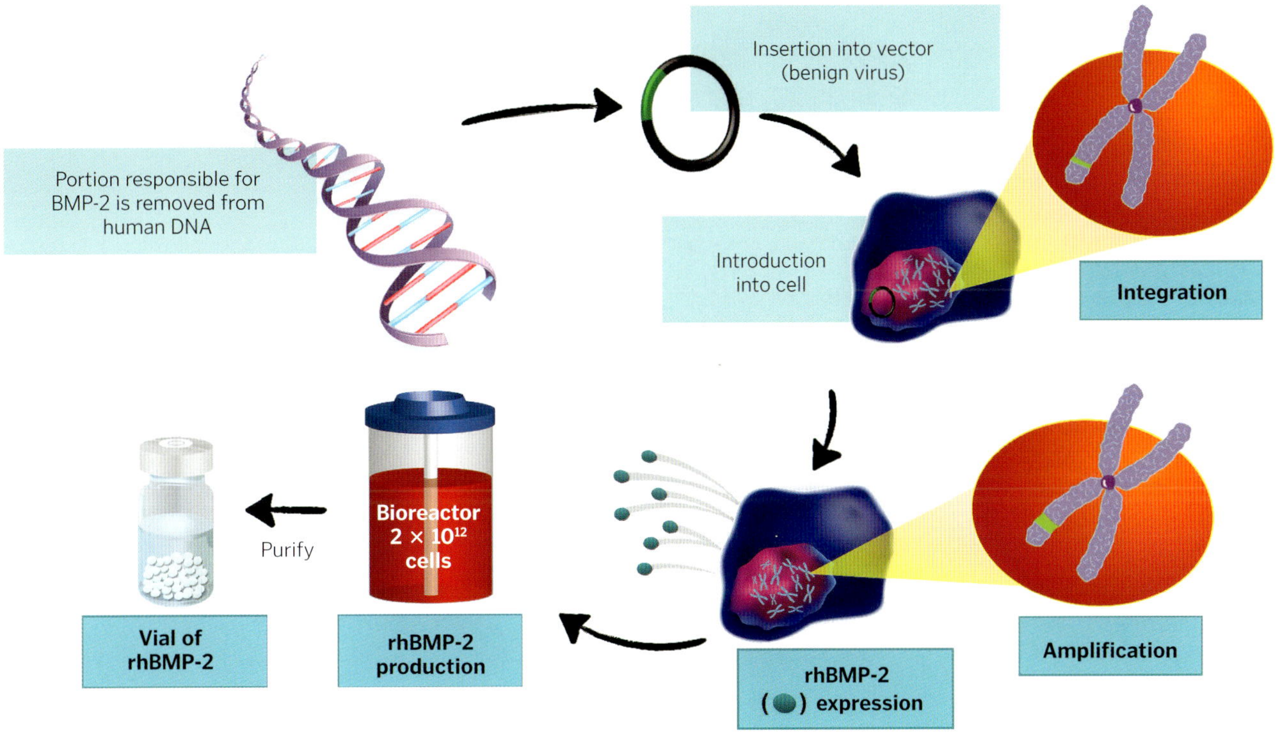

FIG 3-10 rhBMP-2 is genetically engineered by inserting the human BMP-2 protein into a benign virus vector, which is then amplified to produce BMP-2 for human use.

providing information needed to form a growth pattern that culminates in the adult form. Whether the ability to form such a pattern also exists in adult bone regeneration is still unclear, but the properties of BMPs essentially allow them to create an unlimited amount of new bone.[37]

The development of bone can occur in two ways. With intramembranous ossification, mesenchymal cells differentiate directly to bone. This occurs in the flat bones of the craniofacial skeleton. With endochondral ossification, cartilage provides a blueprint for bone morphogenesis, and this ephemeral blueprint is subsequently replaced by bone. This is how most bones in the human skeleton are formed, so BMPs appear to be capable of influencing both types of bone formation.[39–41]

Marshall Urist is credited with the earliest work that isolated BMPs during the 1960s. He discovered that this extractable protein in demineralized bone matrix was responsible for inducing bone formation when implanted ectopically in an animal.[42]

Early on, isolating and purifying BMPs was difficult because only trace amounts exist in bone. In 1988, Wang et al made a major advancement by identifying three different BMPs in bovine bone, characterizing their amino acid sequences, and then isolating recombinant clones for each one that were used to screen the corresponding human DNA libraries.[43] This led to the discovery of rhBMP-1, -2a, and -3 (Fig 3-10). Since then, the number of known BMPs has increased to at least 20, each of which has a slightly different amino acid structure.[44] Most of these BMPs are related to one another and are classified as part of the TGF-β superfamily because of their amino acid sequences.

rhBMP-2

BMPs can stimulate the development of new blood vessels as well as the proliferation, differentiation, and migration of stem cells from the surrounding tissues into cartilage- and bone-forming cells in a wound bed. As stated earlier, the primary activity of rhBMP-2 appears to be differentiating mesenchymal precursor cells into mature osteoblasts and/or chondroblasts. In addition, rhBMP-2 is chemotactic for some osteoblastic-type cells. rhBMP-2 has been shown to induce the complete sequence of endochondral ossification. Its application results in local induction of cartilage,

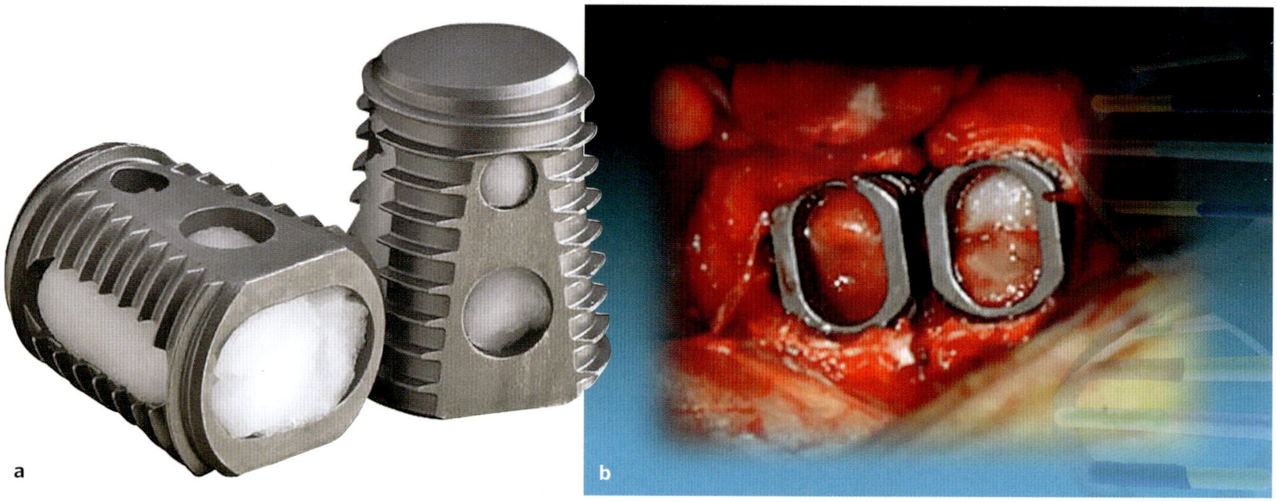

FIG 3-11 *(a and b)* Infuse Bone Graft consists of two parts: a solution containing rhBMP-2 and an absorbable collagen sponge (ACS). The protein is a (genetically engineered) version of a natural protein found in small quantities in the body.

which is subsequently replaced by bone and marrow. It also plays a particularly important role in the early stage of osteoinduction.[45] As increasing amounts of rhBMP-2 are implanted into a site, the time required for bone formation to occur decreases. It has also been shown that large amounts of rhBMP-2 result in concurrent formation of cartilage and bone. These findings indicate that rhBMP-2 may influence both the direct bone formation pathway and the endochondral sequence.[40]

The role of BMPs in bone formation has been studied extensively for more than two decades, and both in vitro and in vivo studies have demonstrated the osteogenic capacity of rhBMP-2. In 2002, the FDA approved the use of an rhBMP-2 product (Infuse) for anterior lumbar interbody fusions (Fig 3-11). Infuse consists of 1.5 mg/mL of rhBMP-2 delivered on a bovine collagen carrier.

Several preclinical studies published around that time demonstrated the viability of applying rhBMP-2 to repair oral and maxillofacial osseous defects as an alternative to autogenous bone[46-48] (Fig 3-12). Since then, a number of clinical studies have demonstrated the effects of rhBMP-2 on periodontal regeneration, bone augmentation procedures, and peri-implantitis defects in addition to socket sinus grafting (Fig 3-13).

In 2003, a randomized controlled clinical and histomorphometric study was performed by Jung et al to test whether the addition of rhBMP-2 to a xenogeneic bone substitute would improve bone volume, density, or maturation in guided bone regeneration procedures. Thirty-four implants were placed in 11 partially edentulous patients. Half of the bone defects were augmented with Bio-Oss (Geistlich), a bovine-derived bone mineral, and the other half with Bio-Oss coated with rhBMP-2. After 6 months, the test sites showed a statistically significant increase in maturation and graft-bone contact compared to the control sites.[49]

The first clinical study performed in the United States focused on the use of rhBMP-2 rather than graft material for cases requiring maxillary sinus augmentation. In 2005, Boyne et al evaluated 11 of 12 patients who underwent sinus elevation with rhBMP-2 in an absorbable collagen sponge (ACS) carrier. All patients experienced induced bone formation, averaging 8.5 mm in height. The regenerated maxillary bone in 8 of the 11 patients was judged suitable for accommodating implants following the BMP treatment. Histologic analysis of bone cores removed during implant placement showed a density similar to that of native bone.[50]

In 2007, the FDA approved the use of Infuse for maxillary sinus grafting and bone grafting procedures associated with extraction sockets using the same ACS delivery system (Fig 3-14). Much of the current research on BMP-2 focuses on finding alternate methods of delivery. BMP is a late-acting growth factor that has a very short half-life.

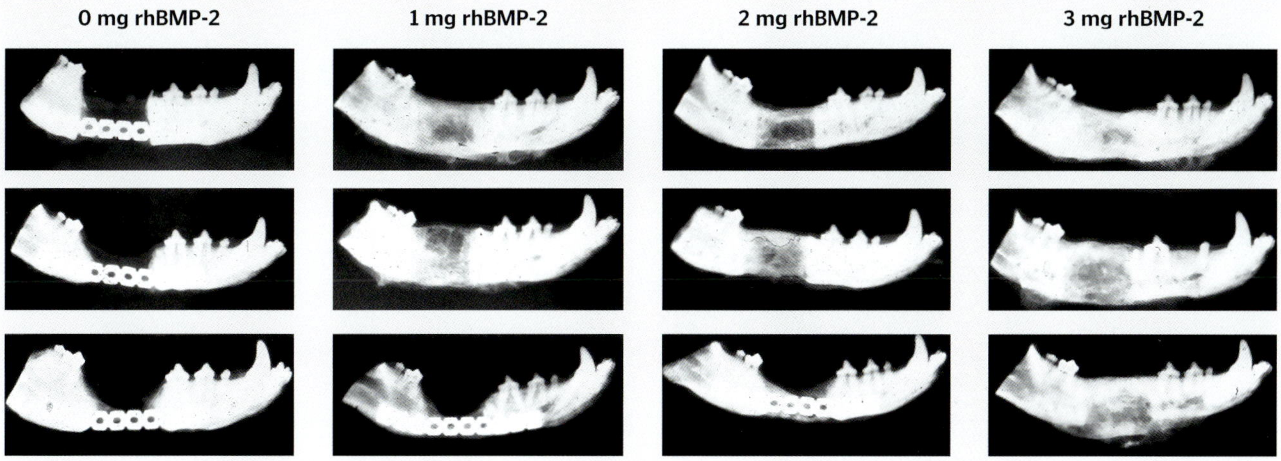

FIG 3-12 Dog mandibles from rhBMP-2 dose-response study. In this study, critical-sized defects were created in the mandibles of 12 dogs that were then treated with 0 mg (control group), 1 mg, 2 mg, or 3 mg of rhBMP-2. Radiographic results demonstrate that closure of the defects was achieved only in the dogs that were treated with 3 mg of rhBMP-2.

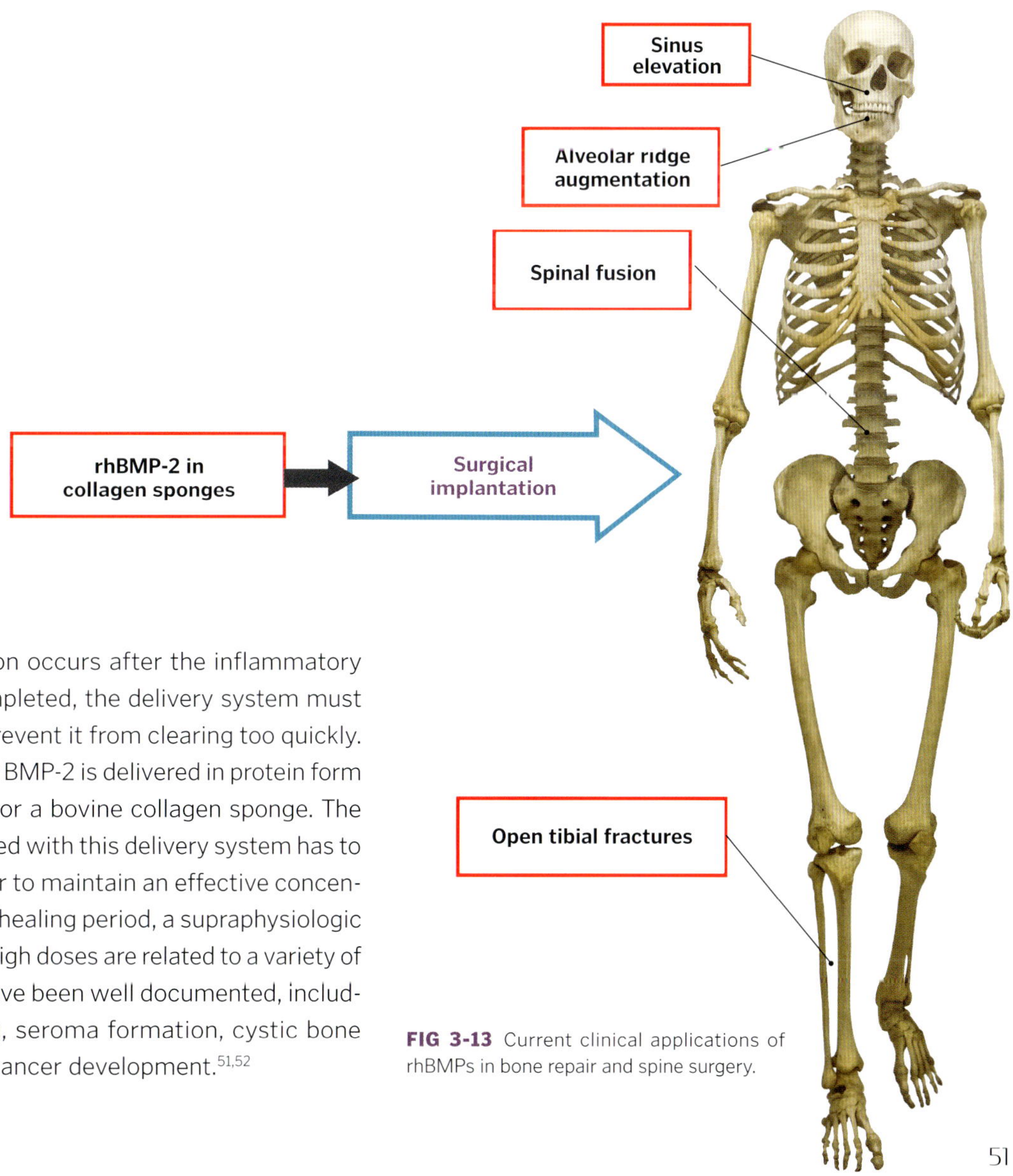

Because all of its action occurs after the inflammatory healing process is completed, the delivery system must be designed so as to prevent it from clearing too quickly. Currently, recombinant BMP-2 is delivered in protein form with bone substitutes or a bovine collagen sponge. The main problem associated with this delivery system has to do with dosing. In order to maintain an effective concentration throughout the healing period, a supraphysiologic dose is required. Such high doses are related to a variety of adverse effects that have been well documented, including extensive swelling, seroma formation, cystic bone lesion formation, and cancer development.[51,52]

FIG 3-13 Current clinical applications of rhBMPs in bone repair and spine surgery.

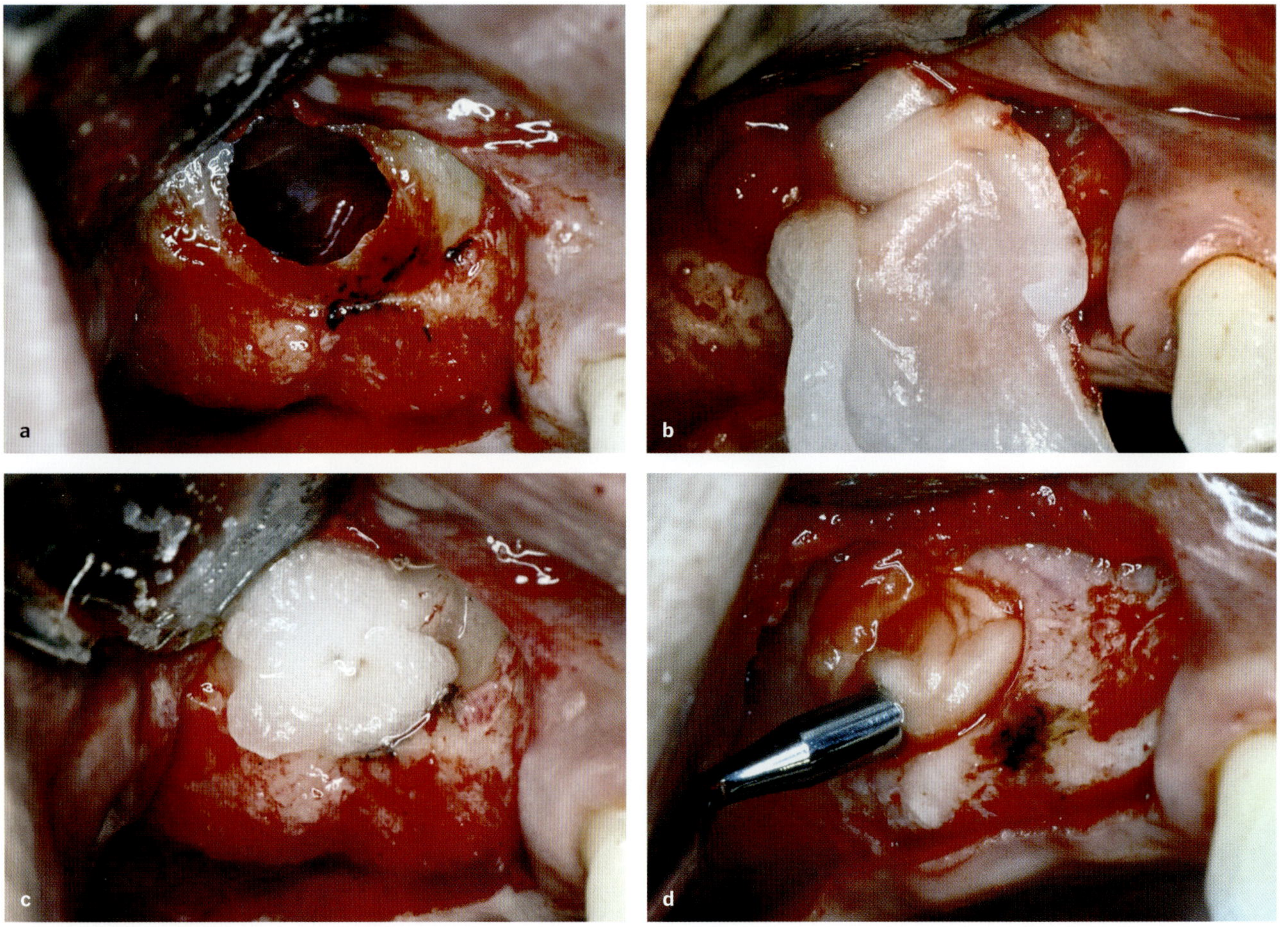

FIG 3-14 Application of rhBMP-2/ACS in a sinus elevation procedure. *(a)* A window in the lateral wall of the maxilla provides access to the maxillary sinus cavity and reveals an intact sinus membrane, which has been carefully elevated to avoid tears and to contain the rhBMP-2. *(b)* After careful elevation of the sinus membrane, the collagen sponge saturated with rhBMP-2 is introduced into the sinus cavity. *(c)* The surgeon introduces the collagen membrane strips one by one, making sure that no air pockets (empty spaces) are created. *(d)* A modified amalgam plugger (3- to 5-mm-diameter tip) is ideal for packing the rhBMP-2–laden collagen sponge strips into the cavity.

Since their discovery in 1965, BMPs have been subject to significant clinical interest for their medical applications. Understanding their biochemical structure and physiologic role as members of the TGF-β family has shed light on their role as initiators of bone growth.

Gene Therapy

Gene delivery is a relatively new system being explored for bone regeneration in dentistry and other fields. Gene delivery involves the use of vectors to transfer a BMP-2–encoded gene into the defect site. With this technology, BMP-2 is synthesized in vivo, and therefore secretion can be sustained at physiologic concentrations for weeks or even months. This delivery system mimics the action of BMP-2 in the wound healing process, avoiding the adverse effects related to high doses of rhBMP-2. The production of BMP-2 in vivo induces osteogenic differentiation and mineralization of the site.

In vivo gene delivery directly transfers target genes into the host, either locally or systemically. Ex vivo delivery is cell based: Cells are harvested from the host, transduced with the vector, and then administered into the defect (Fig

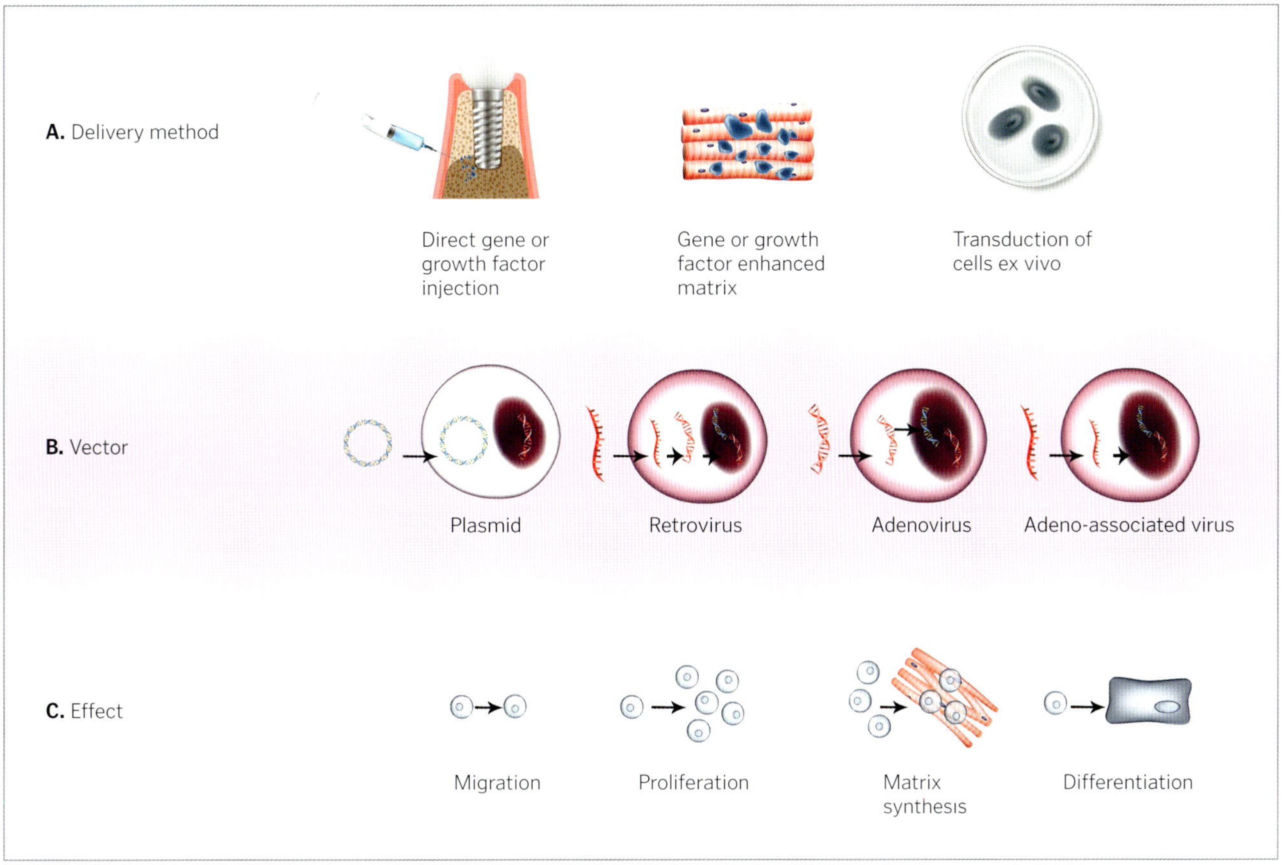

FIG 3-15 In the FDA-approved protocol, rhBMP-2 is delivered locally via a collagen matrix. Ex vivo and in vivo methods of gene delivery of BMP-2 are now being explored using retro- and adenovirus vectors to achieve cell migration, proliferation, and differentiation.

3-15). Both in vivo and ex vivo methods of gene delivery have been used successfully, although each has limitations. Studies have demonstrated a higher efficacy of gene transfer for ex vivo gene therapy than for in vivo delivery. Also, there have been some concerns raised about safety related to the high titers of viral or nonviral vectors used for in vivo delivery. However, ex vivo gene delivery methods are expensive and time-consuming.

For ex vivo BMP-2 gene delivery, MSCs are frequently used as gene carriers. These include bone marrow stromal cells, muscle-derived cells, adipose-derived stem cells, periodontal ligament stem cells, and fibroblasts.[53-58] In two animal studies, bone regeneration via ex vivo BMP-2 gene delivery produced complete defect closure in calvarial defects within 4 weeks.[54,55]

In dentistry, one of the advantages of BMP-2 gene therapy is the ability to regenerate bone that is load bearing during treatment due to the prolonged secretion of BMP-2 at low concentrations. Recently, preclinical translation of BMP-2 gene therapy has begun, and maxillary sinus–related and dental implant–related models have been utilized. Periodontal regeneration via BMP-2 gene delivery has also been attempted, although a small animal model was used. To achieve BMP-2 gene-activated bone regeneration in the oral and maxillofacial region, efficient vector systems with reduced immunogenicity and a stable cell supply, such as the umbilical cord, gingival tissue, or adipose tissue, are critical factors for gene delivery.

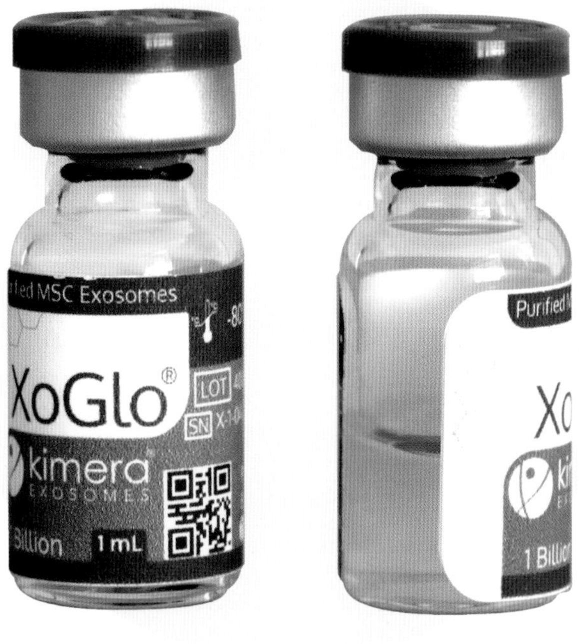

FIG 3-16 Many investigators have shifted their focus from MSCs to exosomes as potential targets for altering specific regulatory cell functions. XoGlo (BioRich Medical) is a purified MSC-derived exosome product containing many growth factors.

Mesenchymal Stem Cells

Another relatively new cell-based therapeutic paradigm involves a novel application of MSCs. Because they are found in many areas of the human body (eg, the liver, bone marrow, placenta, amniotic fluid, and dental pulp), MSCs have been the focus of research in cell therapy for more than 30 years. Transplanted MSCs, which are pluripotent, have been shown to improve conditions such as myocardial infarction, kidney injury, liver fibrosis, and bone joint injury. Human umbilical cord MSCs have been favored by researchers in cell transplantation therapy over other sources (such as bone marrow) because they can be obtained noninvasively and easily cultured in vitro. However, studies have found that after MSC transplantation, the number of cells remains very small, and local cell differentiation is insufficient to support complex pathologic changes.

Exosomes

Researchers have recently found that MSC-derived exosomes contain a variety of bioactive substances, including cytokines and growth factors, that influence tissue responses to injury, infection, and disease. Exosomes are carriers and transfer agents of intracellular information, and many investigators have shifted their focus from MSCs to exosomes as potential targets for altering specific regulatory cell functions.

In the field of bone-regenerative medicine, experimental applications of umbilical cord–derived MSC exosomes have been drawing significant attention (Fig 3-16). In rat models, exosomes have been shown to promote angiogenesis and fracture healing through upregulation of VEGF, to reduce steroid-induced femoral head necrosis through increased levels of VEGF and BMP-2, to repair cartilage defects, and, in a mouse model, to have therapeutic and preventive effects on osteoporosis.[59]

References

1. Simion M, Rocchietta I, Kim D, Nevins M, Fiorellini J. Vertical ridge augmentation by means of deproteinized bovine bone block and recombinant human platelet-derived growth factor-BB: A histological study in dog model. Int J Periodontics Restorative Dent 2006;26:415–423.
2. Urban I, Caplanis N, Lozada JL. Simultaneous vertical guided bone regeneration and tissue regeneration in the posterior maxilla using recombinant human platelet-derived growth factor: A case report. J Oral Implantol 2009;35:251–256.
3. McAllister BS, Haghighat K, Prasad HS, Rohrer MD. Histologic evaluation of recombinant human platelet derived growth factor-BB after use in extraction socket defect: A case series. Int J Periodontics Restorative Dent 2010;30:365–373.
4. Marx RE, Carlson ER, Eichstaedt RM, Schimmele SR, Strauss JE, Georgeff KR. Platelet-rich plasma: Growth factor enhancement for bone grafts. Oral Surg Oral Med Oral Pathol Oral Radiol Endod 1998;85:638–646.

5. Rose LF, Rosenberg E. Bone grafts and growth and differentiation factors for regenerative therapy: A review. Pract Proced Aesthet Dent 2001;13:725–734.
6. Bikle DB, Tahimic C, Chang W, Wang Y, Anastassios P, Barton ER. Role of IGF-I signaling in muscle bone interactions. Bone 2015;80:79–88.
7. Hock JM, Centrella M, Canalis E. Insulin-like growth factor I has independent effects on bone matrix formation and cell replication. Endocrinology 1988;122:254–260.
8. Hill PA, Tumber A, Meikle MC. Multiple extracellular signals promote osteoblast survival and apoptosis. Endocrinology 1997;138:3849–3858.
9. Gao Y, Zhu S, Luo E, Li J, Feng G, Hu J. Basic fibroblast growth factor suspended in Matrigel improves titanium implant fixation in ovariectomized rats. J Controlled Release 2009;139:15–21.
10. Yun Y-R, Won JE, Jeon E, et al. Fibroblast growth factors: Biology, function, and application for tissue regeneration. J Tissue Eng 2010;1(1):218142.
11. Takayama S, Murakami S, Shimabukuro Y, Kitamura M, Okada H. Periodontal regeneration by FGF-2 (bFGF) in primate models. J Dent Res 2001;80:2075–2079.
12. Shirakata Y, Taniyama K, Yoshimoto T, et al. Regenerative effect of basic fibroblast growth factor on periodontal healing in two-wall intrabony defects in dogs. J Clin Periodontol 2010;37:374–381.
13. Cochran DL, Wozney JM. Biological mediators for periodontal regeneration. Periodontol 2000 1999;19:40–58.
14. Dubon MJ, Yu J, Choi S, Park K-S. Transorming growth factor β induces bone marrow mesenchymal stem cell migration via noncanonical signals and N-cadherin. J Cell Physiol 2018;233:201–213.
15. Zhao L, Jiang S, Hantash BM. Transforming growth factor β1 induces osteogenic differentiation of murine bone marrow stromal cells. Tissue Eng Part A Feb 2010;16:725–733.
16. Haghighizadeh E, Shahrezaee M, Sharifzadeh SR, Momeni M. Transforming growth factor-β3 relation with osteoporosis and osteoporotic fractures. J Res Med Sci 2019;22:24.
17. Song N, Kwon S, Kim S, et al. Sulfuretin induces osteoblast differentiation through activation of TGF-β signaling. Mol Cell Biochem 2015;410:55–63.
18. Crane JL, Xian L, Cao X. Role of TGF-β signaling in coupling bone remodelling. In: Feng XH, Xu P, Lin X (eds). TGF-β Signaling. Methods in Molecular Biology, vol 1344. New York: Humana Press, 2015.
19. Marx RE, Garg AK. Dental and Craniofacial Applications of Platelet-Rich Plasma. Chicago: Quintessence, 2005.
20. Garg AK, Gargenease D, Peace I. Using a platelet-rich plasma to develop an autologous membrane for growth factor delivery in dental implant therapy. Dent Implantol Update 2000;11:41–44.
21. Marx RE. Platelet-rich plasma: A source of multiple autologous growth factors for bone grafts. In: Lynch SE, Genco RJ, Marx RE (eds). Tissue Engineering: Application in Maxillofacial Surgery and Periodontics. Chicago: Quintessence, 1999:71–82.
22. Tischler M. Platelet-rich plasma. The use of autologous growth factors to enhance bone and soft tissue grafts. N Y State Dent J 2002;68:22–24.
23. Petrungaro PS. Using platelet-rich plasma to accelerate soft tissue maturation in esthetic periodontal surgery. Compend Contin Educ Dent 2001;22:729–732,734,736.
24. Marx RE. Clinical Effects of Platelet-Rich Plasma on Soft Tissue Healing. Presented at the First Symposium on Platelet-Rich Plasma and Its Growth Factors, Lake Buena Vista, FL, 28 February–2 March 2002.
25. Krauser JT. PRP and PepGen P-15: Case Report on a Bilateral Sinus Graft. Presented at the First Symposium on Platelet-Rich Plasma and Its Growth Factors, Lake Buena Vista, FL, 28 February–2 March 2002.
26. Misch DM. The Use of Platelet-Rich Plasma in Oral Reconstruction with Dental Implants. Presented at the First Symposium on Platelet-Rich Plasma and Its Growth Factors, Lake Buena Vista, FL, 28 February–2 March 2002.
27. Marx RE, Garg AK. Bone graft physiology with use of platelet-rich plasma and hyperbaric oxygen. In: Jensen OT (ed). The Sinus Bone Graft. Chicago: Quintessence, 1999:183–189.
28. Bowen-Pope DF, Malpass TW, Foster DM, Ross R. Platelet-derived growth factor in vivo: Levels, activity, and rate of clearance. Blood 1984;64:458–469.
29. Wickenhauser C, Hillienhof A, Jungheim K, et al. Detection and quantification of transforming growth factor beta (TGF-beta) and platelet-derived growth factor (PDGF) release by normal human megakaryocytes. Leukemia 1995;9:310–315.
30. Ledent E, Wasteson A, Berlin G. Growth factor release during preparation and storage of platelet concentrates. Vox Sang 1995:68:205–209.
31. Hartmann K, Baier TG, Loibl R, Schmitt A, Schonberg D. Demonstration of type I insulinlike growth factor receptors on human platelets. J Recept Res 1989;9:181–198.
32. Stuart CA, Meehan RT, Neale LS, Cintron NM, Furlanetto RW. Insulin-like growth factor-I binds selectively to human peripheral blood monocytes and B-lymphocytes. J Clin Endocrinol Metab 1991;72:1117–1122.
33. Kooijman R, Willems M, De Haas CJ, et al. Expression of type I insulin-like growth factor receptors on human peripheral blood mononuclear cells. Endocrinology 1992:131:2244–2250.
34. Taylor VL, Spencer EM. Characterisation of insulin-like growth factor-binding protein-3 binding to a novel receptor on human platelet membranes. J Endocrinol 2001;168:307–315.
35. Auernhammer CJ, Fottner C, Engelhardt D, Bidlingmaier M, Strasburger CJ, Weber MM. Differential regulation of insulin-like growth factor-(IGF) I and IGF-binding protein (IGFBP) secretion by human peripheral blood mononuclear cells. Horm Res 2002;57:15–21.
36. Lee MB. Bone morphogenetic proteins: Background and implications for oral reconstruction. A review: J Clin Periodontol 1997;24:255–265.
37. Reddi A, Cunningham NS. Initiation and promotion of bone differentiation by bone morphogenetic proteins. J Bone Miner Res 1993;8(suppl 2):S499–S502.
38. Salata LA, Franke-Stenport V, Rasmusson L. Recent outcomes and perspectives of the application of bone morphogenetic proteins in implant dentistry. Clin Implant Dent Relat Res 2002;4:27–32.
39. Reddi AH. Bone morphogenesis and modeling: Soluble signals sculpt osteosomes in the solid state. Cell 1997;89:159–161.
40. Vozney JM. The potential role of bone morphogenetic proteins in periodontal reconstruction. J Periodontol 1995;66:506–510.
41. Wang EA, Rosen V, DAiessandro JS, et al. Recombinant human bone morphogenetic protein induces bone formation. Proc Natl Acad Sci U S A 1990:87:2220–2224.
42. Urist MR. Bone: Formation by autoinduction. Science 1965;150: 893–899.
43. Wang EA, Rosen V, Cordes P, et al. Purification and characterization of other distinct bone-inducing factors. Proc Natl Acad Sci U S A 1988:85:9484–9488.
44. Vehof JW, Haus MT, de Ruijter AE, Spauwen PH, Jansen JA. Bone formation in transforming growth factor beta-I-loaded titanium fiber mesh implants. Clin Oral Implants Res 2002;13:94–102.
45. Okubo Y, Bessho K, Fujimura K, Kusumoto K, Ogawa Y, Iizuka T. Expression of bone morphogenetic protein in the course of osteoinduction by recombinant human bone morphogenetic protein-2. Clin Oral Implants Res 2002;13:80–85.
46. Boyne PJ. Application of bone morphogenetic proteins in the treatment of clinical oral and maxillofacial osseous defects. J Bone Joint Surg Am 2001;83-A suppl 1(pt 2):S146–S150.

47. Wikesjö UM, Sorensen RG, Wozney JM. Augmentation of alveolar bone and dental implant osseointegration: Clinical implications of studies with rhBMP-2. J Bone Joint Surg Am 2001;83-A suppl 1(pt 2):S136–S145.
48. Cochran DL, Schenk R, Buser D, Wozney JM, Jones AA. Recombinant human bone morphogenetic protein-2 stimulation of bone formation around endosseous dental implants. J Periodontol 1999;70:139–150.
49. Jung RE, Glauser R, Schärer P, Hämmerle CH, Sailer HF, Weber FE. Effect of rhBMP-2 on guided bone regeneration in humans: A randomized, controlled clinical and histomorphometric study. Clin Oral Implant Res 2003;14:556–568.
50. Boyne PJ, Lilly LC, Marx RE, et al. De novo bone induction by recombinant human bone morphogenetic protein-2 (rhBMP-2) in maxillary sinus floor augmentation. J Oral Maxillofac Surg 2005;63:1693–1707.
51. Carragee EJ, Hurwitz EL, Weiner BK. A critical review of recombinant human bone morphogenetic protein-2 trials in spinal surgery: Emerging safety concerns and lessons learned. Spine J 2011;11:471–491.
52. Wikesjö UM, Huang YH, Xiropaidis AV, et al. Bone formation at recombinant human bone morphogenetic protein-2-coated titanium implants in the posterior maxilla (type IV bone) in nonhuman primates. J Clin Periodontol 2008;35:992–1000.
53. Park J, Ries J, Gelse K, et al. Bone regeneration in critical size defects by cell-mediated BMP-2 gene transfer: A comparison of adenoviral vectors and liposomes. Gene Ther 2003;10:1089–1098.
54. Lee JY, Musgrave D, Pelinkovic D, et al. Effect of bone morphogenetic protein-2-expressing muscle-derived cells on healing of critical-sized bone defects in mice. J Bone Joint Surg Am 2001;83:1032–1039.
55. Lee JY, Peng H, Usas A, et al. Enhancement of bone healing based on ex vivo gene therapy using human muscle-derived cells expressing bone morphogenetic protein 2. Hum Gene Ther 2002;13:1201–1211.
56. Xu L, Sun X, Bai J, et al. Reosseointegration following regenerative therapy of tissue-engineered bone in a canine model of experimental peri-implantitis. Clin Implant Dent Relat Res 2016;18:379–391.
57. Park SY, Kim KH, Gwak EH, et al. Ex vivo bone morphogenetic protein 2 gene delivery using periodontal ligament stem cells for enhanced re-osseointegration in the regenerative treatment of peri-implantitis. J Biomed Mater Res Part A 2015;103:38–47.
58. Lu CH, Chang YH, Lin SY, Li KC, Hu YC. Recent progresses in gene delivery-based bone tissue engineering. Biotechnol Adv 2013;31:1695–1706.
59. Phinney DG, Pittenger MF. Concise review: MSC-derived exosomes for cell-free therapy. Stem Cells 2017;35:851–858.

4

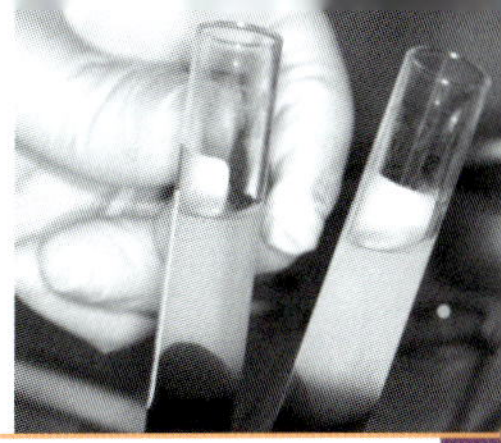

Eight Forms of Autologous Blood Concentrate: Preparation and Clinical Applications

This chapter details the processes and procedures for preparing PRP in each of its eight configurations. The author is the codeveloper (with Dr Robert Marx) of the first PRP formula ever used in clinical practice. Since then, he has developed seven additional formulas that deliver the same potent clinical effects yet are configured uniquely for specific surgical and therapeutic indications.

Biology of Blood

If you have never used PRP, the following information will give you some insight into its biologic composition and properties. This will help you understand how the platelets in blood behave and how to activate them to release their growth factors for optimal clinical advantage.

In general terms, blood has four primary components: red blood cells (also known as *erythrocytes*), white blood cells (also known as *leukocytes*), platelets, and fibrin (Fig 4-1). Platelets become activated when they clot—this is when they release their growth factors (platelet-derived growth factor [PDGF], epithelial growth factor [EGF], etc), which accelerate healing and tissue regeneration. (See chapter 1 for a detailed description of each type of growth factor found in PRP and the role it plays in healing and tissue regeneration.)

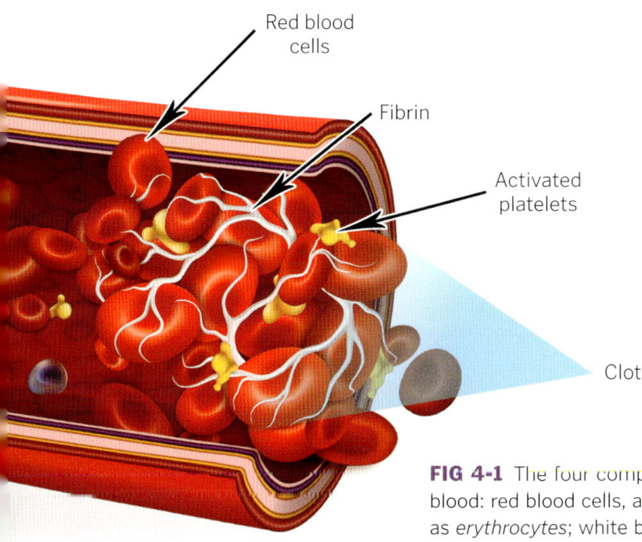

FIG 4-1 The four components of blood: red blood cells, also known as *erythrocytes*; white blood cells, also known as *leukocytes*; activated platelets; and fibrin.

The first step in making any form of PRP is to separate each of these components. To begin, the patient's blood is put into a 10-mL test tube that is in turn placed into a PRP centrifuge. The centrifuge is programmed to control for speed (rpm) and time (minutes) before it begins its cycle. At the end of the cycle, the whole blood will have been separated into its individual components, with the most dense (weightiest) components at the bottom and the thinnest ones at the top. Each layer is identifiable visually, as shown in Fig 4-2.

For PRP development, our primary focus is on the middle layer containing the platelets, white blood cells, and fibrin. In a 10-mL test tube, the middle layer makes up just 1 mL or 10% of the entire contents. In other words, the middle 1 mL contains 10 times the concentration of fibrin, white blood cells, and platelets that would be found in a sample of whole blood. If we used the middle layer plus the top layer (or 5 mL in total), we would only get two times the concentration, diluting (and reducing the benefits of) the growth factors. So for maximum concentration, we need to use 1 to 2 mL of the middle layer.

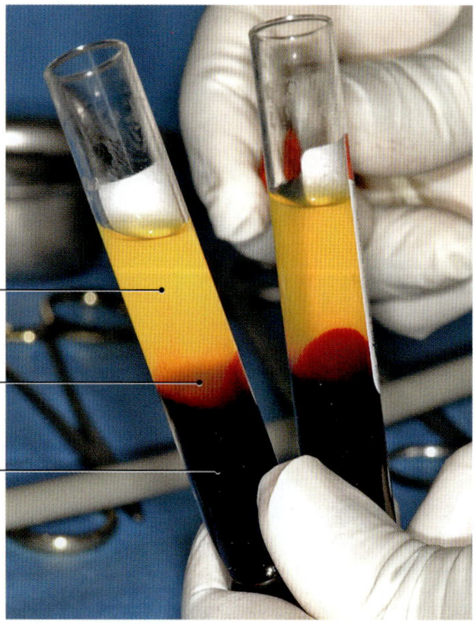

FIG 4-2 Appearance of whole blood following centrifugation.

The layers are defined as follows:

- **Top:** Plasma, or serum, which consists mainly of proteins, collects at the top of the tube. It is pale yellow in appearance and represents approximately 55% of the total volume.
- **Middle:** White blood cells, platelets, and fibrinogen form a narrow cream-colored layer in the middle of the test tube known as the *buffy coat*. It represents approximately 6% of the total volume.
- **Bottom:** Red blood cells, which carry the blood's oxygen, get packed into the bottom of the test tube. Representing approximately 41% of the total volume, this distinctively red layer is known as the *hematocrit*.

Each of these components—platelets, white blood cells, and fibrin—performs a distinct role in the action of PRP. Platelets—which are simply fragments of mesenchymal stem cells that develop from megakaryoblasts—have no nuclei and hence cannot replicate, limiting their maximum life span to just 10 days.[1] They migrate to the site of an injury in response to the exposure to air; this is the mechanism that induces them to clot. (Glass will also induce platelets to clot. Clinically, a glass test tube will enhance clotting, while a plastic test tube will minimize clotting.)

When platelets clot, they change their configuration from a discoid shape to a pseudopodal formation, which creates a sort of mesh that entraps strands of fibrin. Hence, the clot roughly consists of half platelets and half fibrin. The fibrin in PRP serves as a cell scaffold and is a kind of intercellular highway that the platelets creep along to build bone.

When we separate (spin) platelets too fast, they become damaged and incapable of releasing fully activated growth factors. Conversely, if we spin them too slowly, they will not separate into layers. The optimal separation rate ranges from 1,000 to 3,000 rpm for most office centrifuges. As the platelets clot, their growth factors are released very quickly—mostly within the first 8 hours, although the process continues for the entire life span of the platelets. The younger, more robust platelets—those that still have a full 10-day life span and thus contain more growth factors—are concentrated just below the middle layer, at the top of the red blood cell layer. To capture these robust platelets, we use the middle layer plus the top (about 0.5 mL) of the red blood cell layer.

The white blood cells in the middle layer also contain some growth factors, but research has shown that their unique contribution to the action of PRP is to reduce the opportunities for development of postoperative infection.[2]

When developing PRP, timing is important. If a clot was allowed to form in a test tube prior to an 8-hour surgery and then added to the surgical site at the end, all of the growth factors would be left in the test tube. Like blood banks, we need to use test tubes containing an anticoagulant (ACD-A) to avoid this scenario. (It is an industry-wide standard in the United States that yellow-top test tubes contain ACD-A, and red-top test tubes contain nothing.) ACD-A binds the calcium contained in the blood and prevents clotting. Once we add the platelet concentrate to the surgical site, it will be exposed to the calcium naturally contained in the patient's blood and quickly begin to clot. This is how we get activated growth factor release in situ.

In the pages that follow, the formula, equipment, and supplies needed for each form of PRP are presented, along with clear, step-by-step instructions for making them. Each formula is followed by a discussion of the rationale and benefits of that particular form of PRP, as well as the specific clinical applications for which it was developed.

PRP Liquid

This original formulation consists of the middle layer, containing white blood cells, fibrin, and platelets, and some of the red blood cells found at the very top of the third layer containing the youngest platelets with the longest lifespan

(Fig 4-3). Because it is made using a yellow-top test tube, we know that it contains an anticoagulant (ACD-A) and therefore will not clot until the platelets are activated by one of the methods described earlier.

PRP liquid is a concentrate of seven autologous growth factors and three cell adhesion molecules; it has a wound healing and regenerative potential that is 10 times greater than whole blood. This is the basis for the other seven forms of PRP described in this chapter, and it is used in medical/dental surgical specialties to reduce postoperative pain, promote tissue repair, and reduce healing times. In dentistry, PRP liquid is used as an adjunct in implant therapy, as well as in some forms of minimally invasive facial rejuvenation therapy.

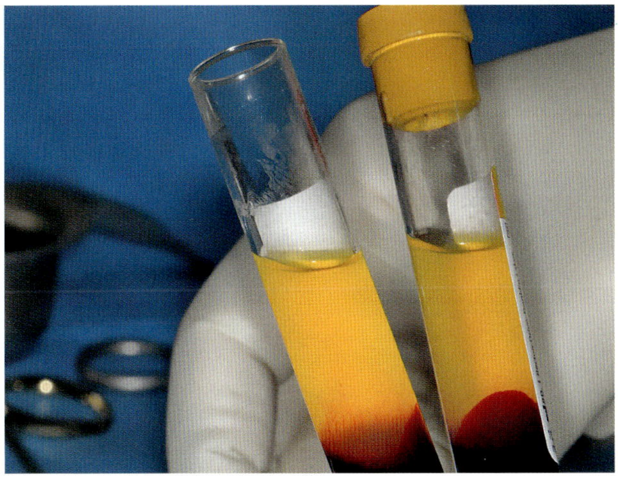

FIG 4-3 Liquid PRP.

Formula for PRP liquid

Yield: 10 mL blood = 2 mL PRP liquid per test tube

Prep: 5–10 min | **Spin:** 12 min* | **Speed:** 2,000 rpm* **Varies based on machine used.*

Uses:

- Hydrating particulate graft material
- Hydrating block grafts
- Hydrating collagen membrane barriers
- Placing into implant osteotomies
- Microneedling/facial rejuvenation procedures

Needed:

- PRP centrifuge
- Phlebotomy kit
- Yellow-top test tubes
- Syringe

Protocol:

1. Draw blood to fill two, four, six, or eight yellow-top vacutainer test tubes, depending on the volume of PRP liquid needed. Each test tube holds 10 mL of blood and will yield 2 mL of liquid PRP.
2. Place the test tubes in the centrifuge immediately.
3. Spin for approximately 12 minutes at about 2,000 rpm. (The exact time and speed will vary depending on your centrifuge device.)
4. Immediately after the spinning cycle is completed, remove the test tubes and place them in the test tube rack.
5. Move the rack to the clinical area if the centrifuge is located in another area of the office.
6. Draw up the top 3–4 mL of liquid from the test tubes and discard it. This is the plasma or serum containing minimal or no growth factors.
7. Draw up the next approximate 2 mL; this is the liquid PRP that you will be using. It should include the yellow liquid portion, what appears to be a white line in the test tube, and the top 0.5 mL of the red layer. (The latter is included only for bone or soft tissue grafting; see notes below.)

Notes:

The last step will differ slightly depending on whether the PRP liquid will be used for facial rejuvenation or bone grafting:

- *Facial rejuvenation:* Draw up only the yellow liquid portion and avoid getting what appears to be the white line in the test tube or any of the red portion below it.
- *Bone or soft tissue grafting:* Draw up the yellow liquid, the white line, and the top 0.5 mL of the red layer. This will give you the concentrated platelets, concentrated fibrin, and concentrated leukocytes (represented by the white line), as well as the youngest and most robust platelets contained in the top of the red portion.

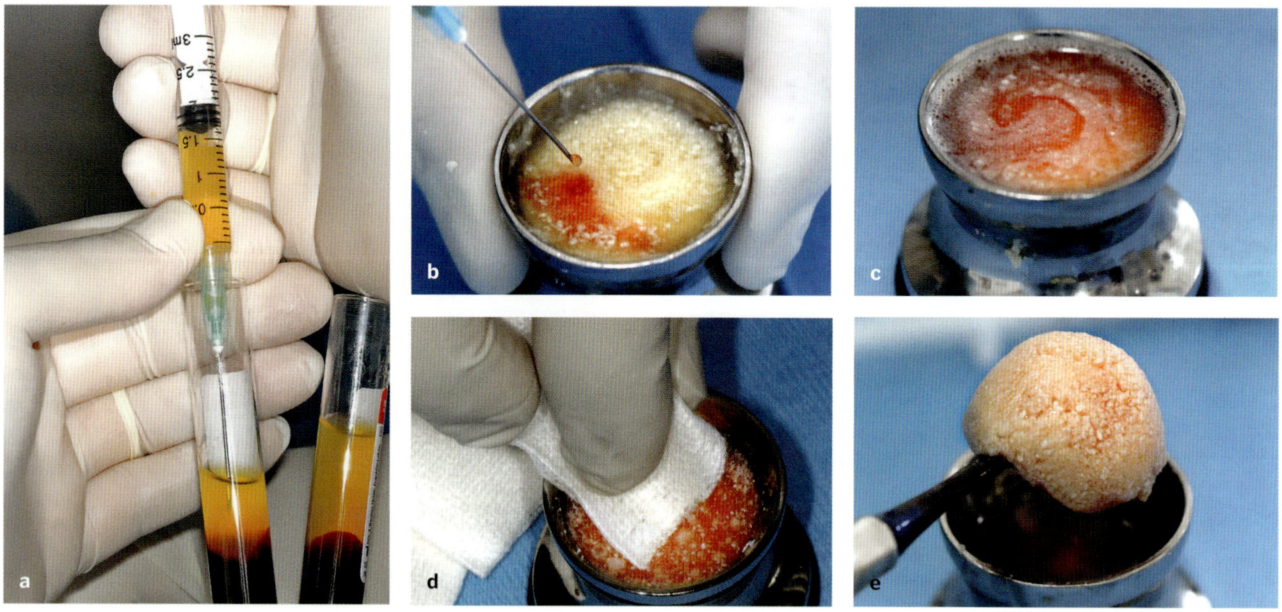

FIG 4-4 Liquid PRP hydrating particulate graft material. *(a)* Drawing up the middle buffy coat layer and the top 0.5 mL of the red blood cell layer into the syringe. *(b)* Adding the contents of the syringe to bone particles. *(c)* Combining the PRP liquid and the bone particles to prepare the composite graft. *(d)* Just prior to use, using gauze pads on the graft to absorb any excess PRP liquid. *(e)* Ready-to-use graft. Because of the presence of ACD-A in the test tube, the graft hydrated with PRP will not begin to clot and hold together until it has been placed in the surgical site.

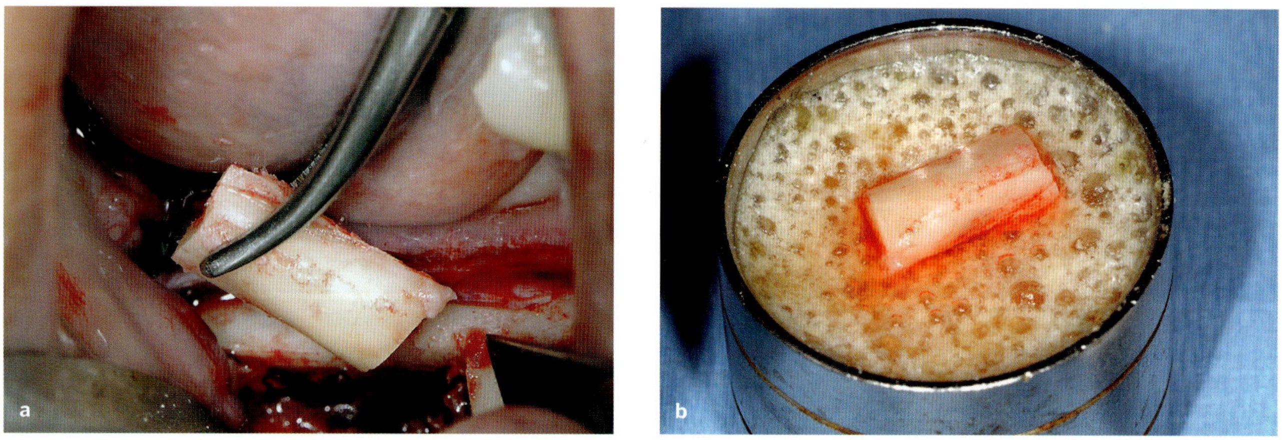

FIG 4-5 PRP liquid hydrating a block graft. *(a)* Harvesting of a block bone graft from the ramus. *(b)* Ramus bone block stored in PRP liquid.

Implantology

PRP liquid can be used in place of saline in many implant-related surgical procedures. For example, PRP liquid can be used to hydrate both particulate bone particles (Fig 4-4) and bone blocks (Fig 4-5) when bone grafting is indicated in patients who lack sufficient bone to support an implant. When substituted for sterile saline, PRP can accelerate healing and increase the amount of bone regenerated at the site. During guided bone regeneration procedures requiring the use of a collagen membrane, the membrane can be soaked in PRP before placement to increase the number of growth factors at the site (Fig 4-6).

Many implant surgeons favor collagen membranes over polytetrafluoroethylene (PTFE) membranes—which cannot absorb PRP—for this very reason. PRP liquid can also be placed into an implant osteotomy to accelerate osseointegration and enhance healing. This can be especially beneficial in older patients, who often heal slowly; in those with diabetes or other medical compromise; and in patients who smoke, which is a known risk factor for osseointegration failure. (See chapter 5 for more on the implant-related applications of PRP.)

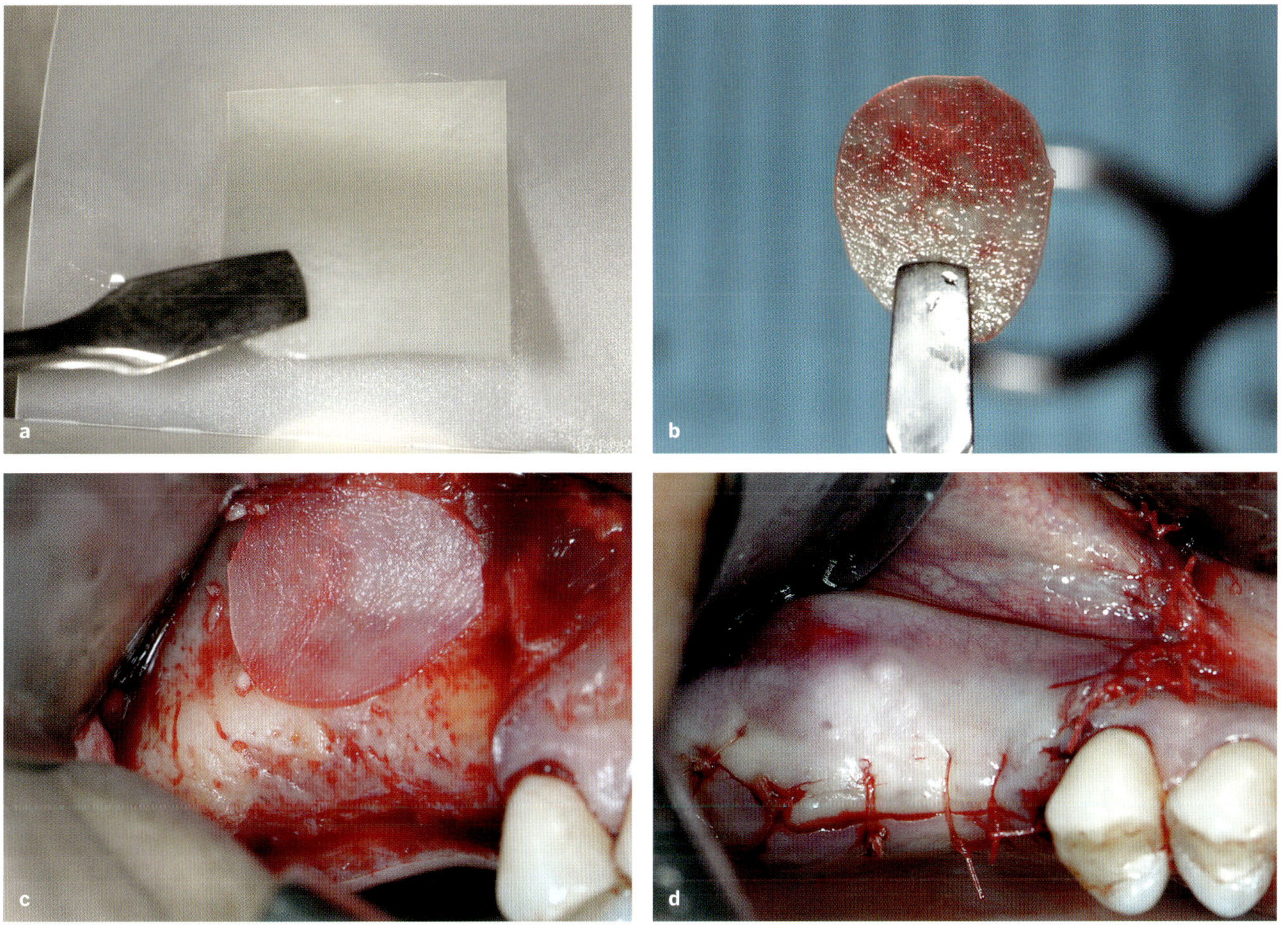

FIG 4-6 PRP liquid hydrating a collagen membrane. *(a)* Commercially available porcine collagen membrane. *(b)* Appearance of the collagen membrane once cut to the appropriate size and soaked in PRP liquid and ready to be placed in the surgical site. *(c)* Placement of the PRP liquid–soaked collagen membrane over a lateral window during sinus elevation. *(d)* Primary closure of the site with resorbable 4-0 sutures.

Dermal microneedling

PRP liquid is now commonly used in many facial rejuvenation procedures as well. As we age, our skin produces less collagen, resulting in the development of age-related lines, wrinkles, and folds. The growth factors in PRP can stimulate new collagen production and help restore the firmness and elasticity of youthful skin.

Skin is not naturally absorbent by design; it serves as a first-line defense against airborne pollutants in the environment or harmful chemicals that we may encounter. This explains why the expensive creams and ointments sold at high-end department stores offer limited benefit. The skin is structured in such a way as to keep things out.

Dermal microneedling is a technique that was developed to penetrate the skin barrier in order to allow the absorption of products specially designed to stimulate cell renewal. Hyaluronic acid, which is naturally produced by the body to regulate skin moisture and elasticity, is one such product. As we age, our body produces less of it and the result is dryness, scaliness, and laxity. Juvéderm (Allergan), Restylane (Galderma), and many other commercial liquid hyaluronic acid products are designed to replenish what is lost as we age.

Liquid PRP, when applied after dermal microneedling, works by a different mechanism. The growth factors in PRP do not simply replenish what is lost; they also stimulate the proliferation of mesenchymal stem cells and fibroblasts, which secrete both hyaluronic acid and type 1 collagen. When applied topically following microneedling treatment, PRP has been shown to improve skin texture, color homogeneity, elasticity, and firmness (Fig 4-7).

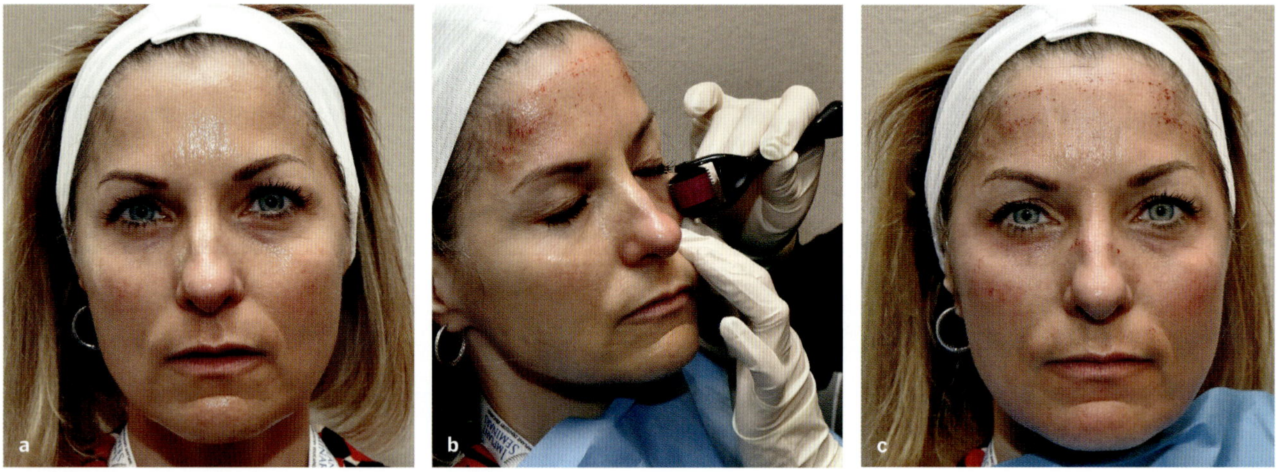

FIG 4-7 PRP liquid in dermal microneedling. *(a)* Pretreatment appearance. Topical anesthesia has been applied to the face in preparation for dermal microneedling. *(b)* Dermal rolling of the face. *(c)* Posttreatment appearance. Liquid PRP was applied to the face after micropunctures from the dermal roller.

In addition, PRP has the significant advantage of being made from the patient's own blood.

Dermal microneedling can be performed by patients at home using consumer-grade microneedling rollers (with needle lengths of 0.2 and 0.3 mm) and over-the-counter skin products. Medical professionals use professional-grade microneedling rollers that have needle lengths of 0.5, 1.0, and 1.5 mm and require a topical anesthetic for pain control, combined with either medical-grade hyaluronic acid or PRP. The formula for liquid PRP is appropriate to use for this procedure, with one modification: Do not include any of the red blood cells from the top of the red layer of the test tube. While this sacrifices a small percentage of the most potent platelets contained in the red portion, the clinical benefit is that the patient does not leave the procedure with a bloody face, which many would find alarming.

Today, many dentists are expanding their practices to include facial rejuvenation procedures such as dermal microneedling and administering dermal filler injections (more on that below).[3] They know their patients' faces better than other doctors, and they have already gained their patients' trust in their clinical and esthetic decision-making abilities. Moreover, some have been using PRP in their dental practices for many years. (See chapter 8 for more on the esthetic applications of PRP.)

Hair regrowth

Liquid PRP is also ideal for stimulating hair regrowth. The most common form of hair loss in adults is called androgenetic alopecia (AGA), also known as *male-pattern baldness*. This condition occurs because of a biochemical process that causes a gradual reduction in the size of hair follicles. While it is generally accepted that genetics influences the development of AGA, the exact reasons why some people are affected and others are not remains unclear. The conventional therapy for AGA is topical minoxidil, a potassium channel blocker that causes hair regrowth via vasodilation. The efficacy of this treatment, which both prevents hair loss and promotes new hair growth, ranges from 30% to 60%.

Microneedling has been found to induce hair regrowth in adults with AGA through stimulation of stem cells and activation of growth factors, which in turn stimulate the dermal papillae. In addition, various studies have demonstrated that the application of PRP or topical minoxidil following microneedling is superior to either of these therapies alone. The procedure (Fig 4-8) is identical to dermal microneedling except that a needle or a dermal microneedling pen device is used in place of a roller in patients with only minor hair loss. (A roller would trap the hair and be useless.) Alternatively, PRP liquid is injected into the individual hair follicles to restimulate growth.

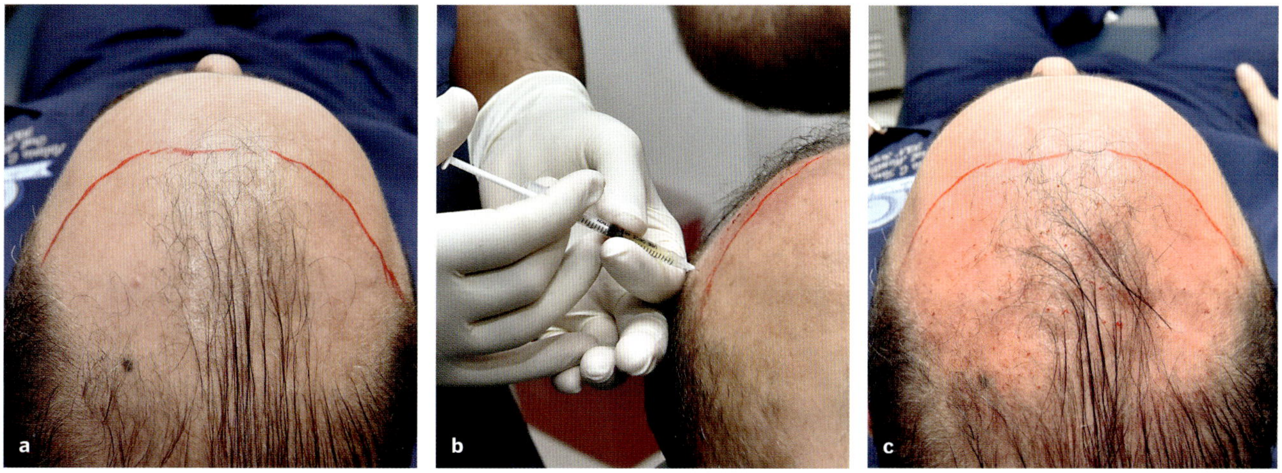

FIG 4-8 Treatment of alopecia with PRP liquid. *(a)* Pretreatment appearance and following application of topical anesthetic in preparation for PRP injections. *(b)* Intra- and subdermal injection of PRP liquid into areas of thinning hair. *(c)* Approximately 100 to 200 injections are made in total at each appointment, with three appointments 7 weeks apart as the optimal treatment protocol.

The procedure can be time-consuming because it requires approximately 200 injections or the stamping of areas slowly with a microneedling pen device. Also, as the needle penetrates the skin, it becomes blunt and must be replaced after about five injections, so about 40 needles are required for a single treatment. A topical anesthetic should be applied 20 minutes prior to treatment. Also, a small amount of sodium bicarbonate can be added to the liquid PRP to neutralize it. Although whole blood has neutral pH, the buffy layer on its own can be somewhat acidic. Adding sodium bicarbonate to the PRP (1:4) eliminates the sting from the injection and increases patient comfort.

Hair grows in phases of approximately 7 weeks. At any given time, one-third of our hair is growing and the other two-thirds is dormant. Therefore, in order to ensure stimulation of all hair follicles while they are active, the procedure should be repeated every 7 weeks, for a total of three treatments. After this initial three-part treatment, the patient should return for one follow-up treatment each year to maintain hair growth. (See chapter 8 for more on the esthetic applications of PRP.)

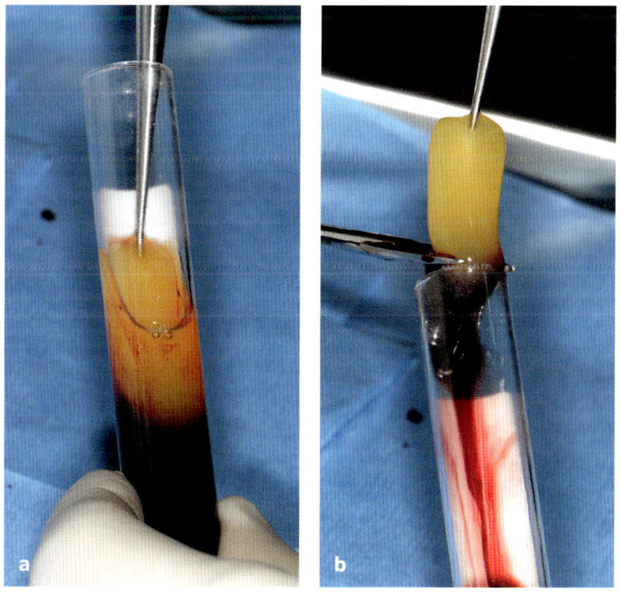

FIG 4-9 *(a and b)* Appearance of PRP gel as it is removed from the test tube.

PRP Gel

PRP gel is simply PRP liquid that is "activated"—that is, allowed to clot (over the approximately 10-minute spin cycle)—before it is added to the surgical site. Clotting can be initiated by the use of a red-top glass test tube (Fig 4-9) rather than a yellow-top plastic test tube. Because it is activated, growth factor release begins immediately in the test tube and the storage container. During the surgery, less than full growth factor release occurs in the surgical site.

The advantage of the gel over the liquid depends on the application. In a crestal sinus elevation procedure, the gel occupies space and thus takes the place of and obviates the need for a bone allograft. In a third molar socket, the addition of the clot provided by PRP gel can reduce the patient's chances of developing an alveolar osteitis.

Formula for PRP gel

Yield: 10 mL blood = 1–3 mL PRP gel per test tube

Prep: 5–10 min | **Spin:** 10 min* | **Speed:** 2,000 rpm* **Varies based on machine used.*

Uses:

- Crestal sinus elevation
- Socket grafting of third molar sites

Needed:

- PRP centrifuge
- Phlebotomy kit
- Red-top test tubes
- Fine tweezers
- Scissors
- Glass or ceramic dappen dish

Protocol:

1. Draw blood to fill two, four, six, or eight red-top vacutainer test tubes, depending on the amount of PRP gel needed. Each test tube holds 10 mL of blood and will yield 1–3 mL of PRP gel.
2. Place the test tubes in the centrifuge immediately.
3. Spin for approximately 10 minutes at about 2,000 rpm. (The exact time and speed will vary depending on your centrifuge device.)
4. Immediately after the spinning cycle is completed, remove the test tubes and place them in the test tube rack.
5. Move the rack to the clinical area if the centrifuge is located in another area of the office.
6. Using the fine long tweezers, after opening and removing the rubber stopper, reach into each of the red-top test tubes and grasp the clot in the middle portion.
7. If there is more than 1 mm of red portion attached, use the scissors to cut this portion off and discard it.
8. Place it in the dappen dish.
9. Repeat step 6 until you have retrieved the PRP gel from each of the test tubes.

Note:

This will give you the concentrated platelets, concentrated fibrin, and concentrated leukocytes (represented by the white line), as well as the youngest and most robust platelets contained in the top of the red portion, in a gel configuration.

Crestal sinus elevation

A crestal sinus elevation is performed to increase the height of the posterior maxilla in an edentulous area where an implant is desired. In this procedure (Fig 4-10), a drill with a safety bur and/or an osteotome is first used to fracture the sinus floor. Care must be taken to avoid perforating the sinus membrane. The membrane is then carefully elevated to make space for placement of the grafting material. In place of autogenous or allograft bone particles, PRP gel is used.

The PRP gel is cut into 3- to 4-mm-wide pieces, which are packed into the sinus as the membrane is being elevated. A condenser is used to condense small pieces,

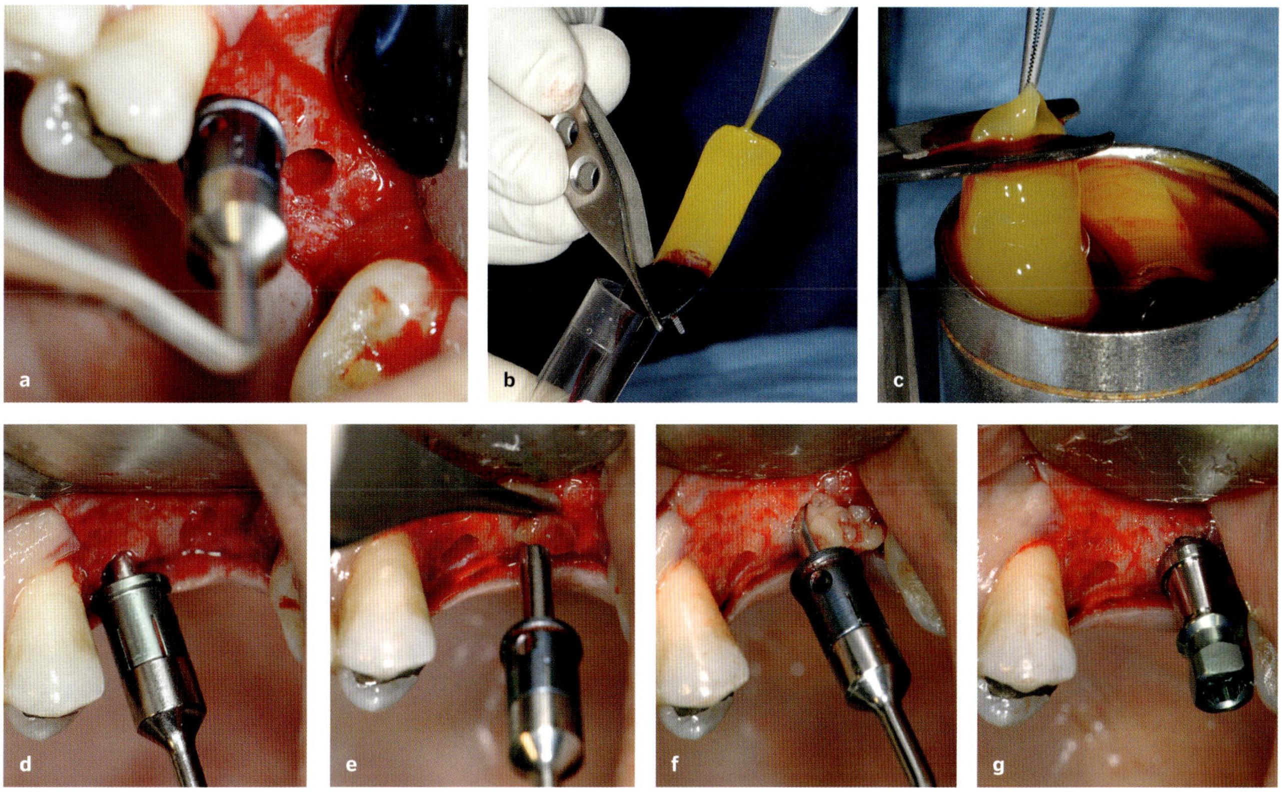

FIG 4-10 PRP gel for crestal sinus elevation. *(a)* Performing and checking the crestal sinus elevation osteotomies. *(b)* Removing the PRP gel from the red-top test tube with tweezers. *(c)* Transferring the PRP gel to a metal container and cutting it into 3- to 4-mm pieces. *(d)* Using a condenser to place the small pieces of PRP gel into the maxillary sinus below the sinus membrane, one at a time to gently elevate the maxillary sinus membrane. *(e and f)* Continued use of a condenser to compact additional small pieces of PRP gel into the space below the sinus membrane, one at a time. *(g)* Implant placement simultaneous to the sinus membrane elevation and graft with PRP gel.

one at a time. A stopper can be placed on the condenser to help avoid inadvertent perforation of the sinus membrane during this process.

The author has found that a single PRP gel clot will generally provide 4 mm of elevation for the implant. However, there are many factors that can affect an individual's clotting ability. For example, a patient's diet has a significant influence on how their blood clots; a high-fat diet often results in relatively large clots as compared to a vegetarian diet, which tends to result in small blood clots. In addition, the length of time the clot sits in the dappen dish waiting to be used also affects its size. The longer it sits, the smaller the clot gets as the thrombin continues to drain from it. (See chapter 7 for more on hard tissue–related applications of PRP.)

Once a sufficient amount of PRP gel has been packed into the sinus and condensed, an implant is placed simultaneously. The site is then sutured closed and allowed to heal for a minimum of 4 to 6 months. A 2005 randomized controlled trial concluded that the concentrated growth factors in PRP can produce more rapid and effective bone regeneration than some bone allografts. In addition, connective and gingival tissues also benefit from the growth factors in PRP, leading to enhanced tissue regeneration.[4]

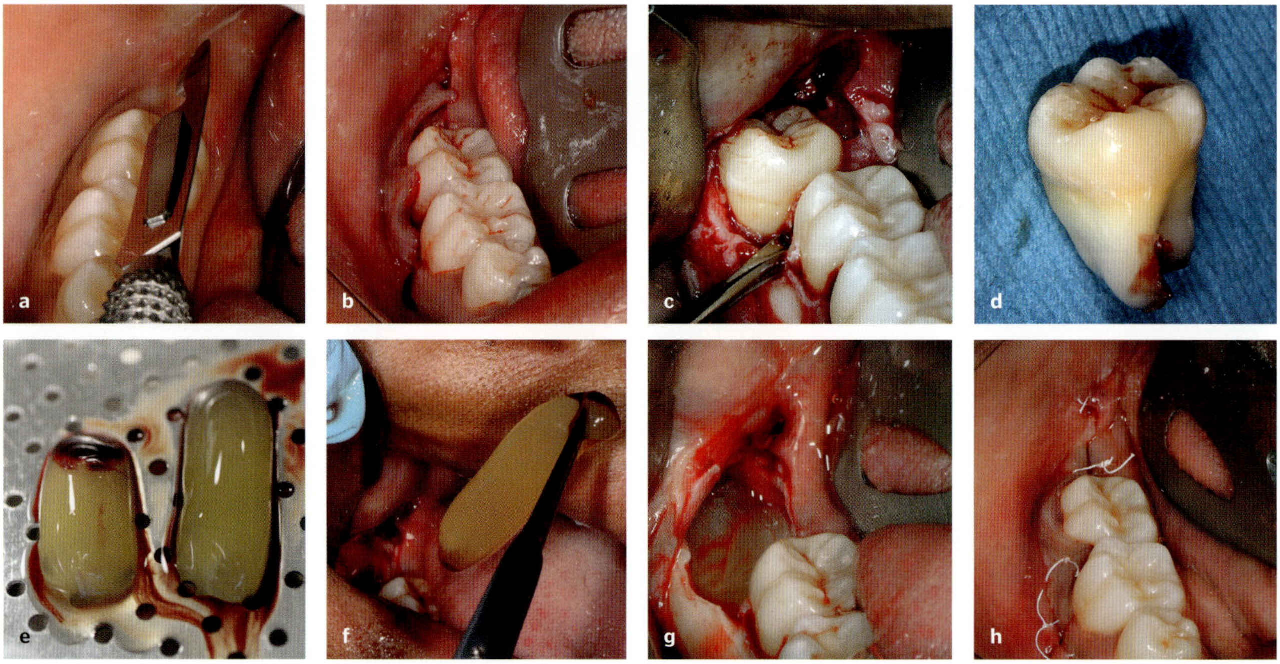

FIG 4-11 PRP gel for socket preservation. *(a)* Exposing the third molar tooth and site. *(b)* Elevating the flap. *(c)* Luxating and elevating the third molar tooth, after removing overlying bone if necessary. *(d)* Extracted third molar. *(e)* PRP gel ready for placement. *(f)* Placing the PRP gel into the third molar extraction socket. *(g)* PRP gel adapts into the extraction socket easily and quickly. *(h)* Suturing of the flap over the PRP gel and socket to obtain good primary closure.

Third molar socket grafting

Surgical removal of mandibular third molars is associated with a number of postoperative complications, particularly in older adults. Alveolar osteitis, for example, is characterized by postoperative pain inside and around the extraction site that gradually increases in severity and is accompanied by partial or total loss of the blood clot in the alveolar socket. Third molar extraction can also result in multiple periodontal defects at the distal root of the second molar.

To avoid these complications, PRP gel can be added to third molar extraction sockets at the time of surgery (Fig 4-11). The increased concentration of growth factors promotes both soft and hard tissue healing as well as angiogenesis, and the white blood cells in the PRP gel inhibit bacterial growth. A 2017 randomized controlled study concluded that these beneficial effects mitigate the likely causes of alveolar osteitis and significantly reduce the incidence of this condition.[5] (See chapter 7 for more on hard tissue–related applications of PRP.)

In another study, a split-mouth design (with and without PRP gel) was used to evaluate soft tissue healing and gingival attachment by assessing for wound dehiscence, probing depth, bone regeneration, and bone density up to 6 months after surgery. The authors reported that on the second postoperative day, 8% of patients in the PRP group versus 92% of those in the non-PRP group showed wound dehiscence in the extraction socket. The PRP group also had a significantly greater decrease in probing depth and a significant increase in bone density compared to the non-PRP group. The authors concluded that PRP contributed to better healing of soft tissue and faster regeneration of bone following third molar extraction surgery.[6,7]

PRP Membrane

PRP membranes are essentially compressed and dehydrated forms of PRP gel (Fig 4-12). To make them, we place PRP gel "clots" into a specially designed box that efficiently extracts liquid from them while preserving the integrity of the fibrin strands holding them together (Fig 4-13). Alternatively, an older method uses gauze to compress and extract the liquid from the gel, leaving just the membrane (Fig 4-14).

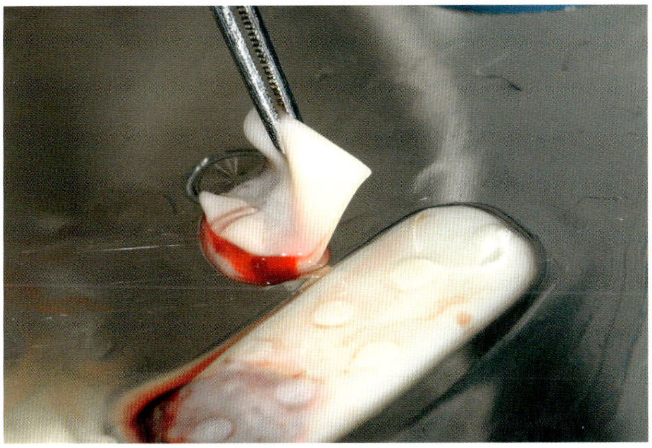

FIG 4-12 PRP membrane.

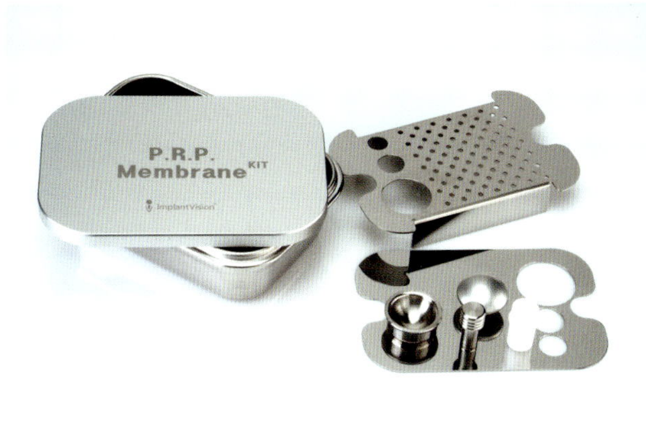

FIG 4-13 Metal box used to uniformly flatten PRP gel into PRP membrane(s).

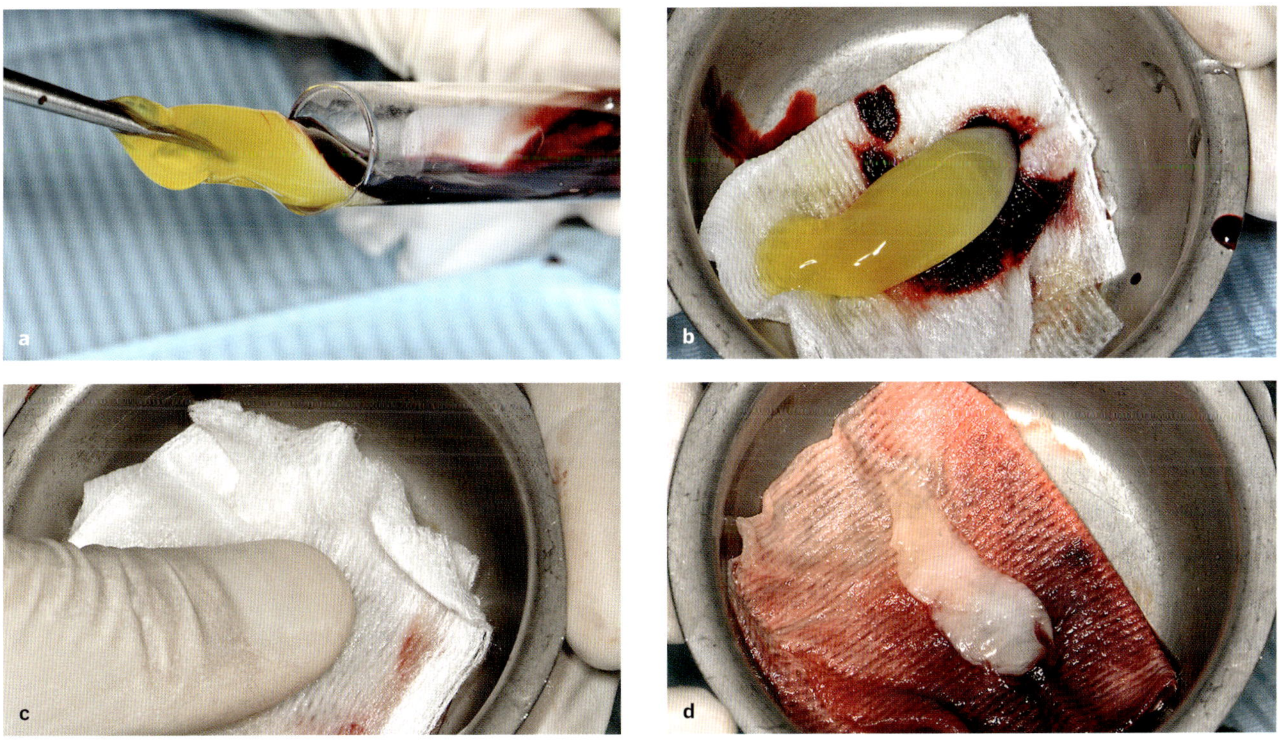

FIG 4-14 PRP membrane preparation with gauze (an older method). *(a)* Removal of the PRP gel from the test tube. *(b)* Placement of the PRP gel in a metal dish lined with gauze. *(c)* Using gauze to extract liquid from the gel. *(d)* Appearance of the compressed gel formed into a membrane. *(e)* PRP membrane is ready for use.

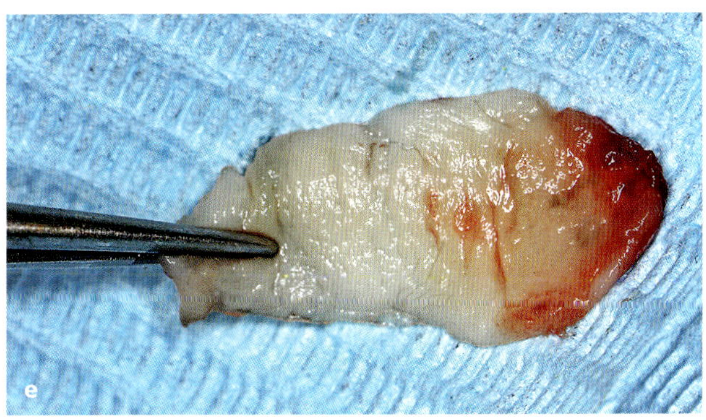

Formula for PRP membrane

Yield: 10 mL blood = 1–3 PRP membranes per test tube

Prep: 5–10 min | **Spin:** 10 min* | **Speed:** 2,000 rpm* *Varies based on machine used.*

Uses:

- Coverage of a socket graft
- Placement on lateral wall of a sinus graft
- Patching of a torn sinus membrane
- Coverage of a block graft
- Use in soft tissue grafting
- Placement into a connective tissue harvest site
- Placement between implant and healing cuff at stage 2 surgery

Needed:

- PRP centrifuge
- Phlebotomy kit
- Red-top test tubes
- Fine tweezers
- Glass dappen dish
- Scissors
- PRP membrane box

Protocol:

1. Draw blood to fill two, four, six, or eight red-top vacutainer test tubes, depending on the number of PRP membranes needed. Each test tube holds 10 mL of blood and will yield one to three PRP membranes.
2. Place the test tubes in the centrifuge immediately.
3. Spin for approximately 10 minutes at about 2,000 rpm. (The exact time and speed will vary depending on your centrifuge device.)
4. Immediately after the spinning cycle is completed, remove the test tubes and place them in the test tube rack.
5. Move the rack to the clinical area if the centrifuge is located in another area of the office.
6. Using the fine long tweezers, after opening and removing the rubber stopper, reach into each of the red-top test tubes and grasp the clots in the middle portion.
7. If there is more than 1 mm of red portion attached, use the scissors to cut this portion off and discard it.
8. Place all of the clots in a PRP membrane box and cover them with the metal insert. Place the lid and leave the box undisturbed for 3–10 minutes or until ready to use.
9. Open the lid of the membrane box and carefully remove the metal insert, which may require gentle force due to the suctioning effect. The PRP membranes will be found on the reverse side of the metal insert or on the mesh grid at the bottom of the box.

Note:

This will give you the concentrated platelets, concentrated fibrin, and concentrated leukocytes (represented by the white line), as well as the youngest and most robust platelets contained in the top of the red portion, in a membrane configuration.

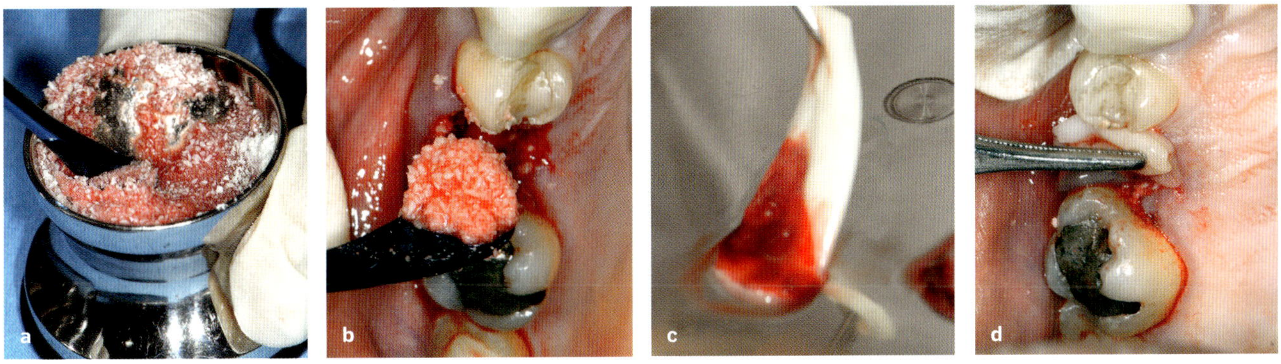

FIG 4-15 PRP membrane for socket grafting. *(a)* Hydration of the particulate bone allograft with PRP liquid. *(b)* Placement of the bone graft material in the socket. *(c)* Removal of the PRP membrane from the PRP membrane box. *(d)* Placement of the PRP membrane over the grafted socket prior to suturing.

PRP membranes are extremely useful in surgical procedures that require the use of a short-term barrier membrane and/or containment of particulate or block bone grafts.

Socket grafting

During immediate implant placement, gaps of more than 2 mm should be grafted to induce osteoinduction. In such cases, PRP liquid is combined with allogeneic bone particles or a bone substitute such as freeze-dried bone allograft (FDBA) to form a composite graft, which is condensed into the socket gap. A PRP membrane is placed over the site to contain the bone graft and provide additional growth factors (Fig 4-15).

Similarly, PRP membranes (and liquid) can be useful in cases where a thin buccal plate has been lost and immediate grafting is necessary. Activated PRP is combined with allogeneic bone particles, or a bone substitute. Within approximately 1 minute, the clot will mature, and the composite PRP–bone graft will become sufficiently malleable to place in the socket and positioned so that it conforms to the buccal spaces. The loosely packed graft can be covered with a PRP membrane to enclose the graft particles, leaving space for an osteoconductive matrix to form. (See chapter 7 for more on hard tissue–related applications of PRP.)

Alveolar ridge splitting/expansion

The ridge splitting/expansion technique (Fig 4-16) is a two-stage surgical procedure for ridge expansion in preparation for implant placement. It begins with a full-thickness mucoperiosteal flap followed by a crestal incision and greenstick fracture of the atrophic ridge. Inserted into the space created by the fracture is a composite graft consisting of PRP liquid combined with particulate bone graft material. One or more PRP membranes are adapted over the grafted site. The mucoperiosteal flap is then repositioned and fixed with nonresorbable sutures. After approximately 6 to 9 months of healing, the site is reopened and prepared for implant placement.

In this procedure, the PRP membranes function as a containment device that will also enhance healing and bone regeneration at a significantly lower cost than collagen membranes.

Perforated sinus membrane repair

Following inadvertent perforation of a sinus membrane, a PRP membrane can also serve as an effective covering to give the tear time to heal (Fig 4-17). Perforation of the sinus membrane is the most common intraoperative complication associated with sinus elevation procedures.[8-10]

Choosing an appropriate membrane to use when repairing a sinus membrane perforation is based on the size of the defect. Tears that are less than 10 mm in diameter heal within about 10 days. Because a PRP membrane will resorb over a period of 10 days as well, it is an ideal choice for small perforations. Tears greater than 10 mm take longer to heal and therefore require a longer-lasting collagen membrane. The PRP membrane can be placed directly over the sinus membrane and should not be in contact with the floor or walls of the sinus cavity. The growth

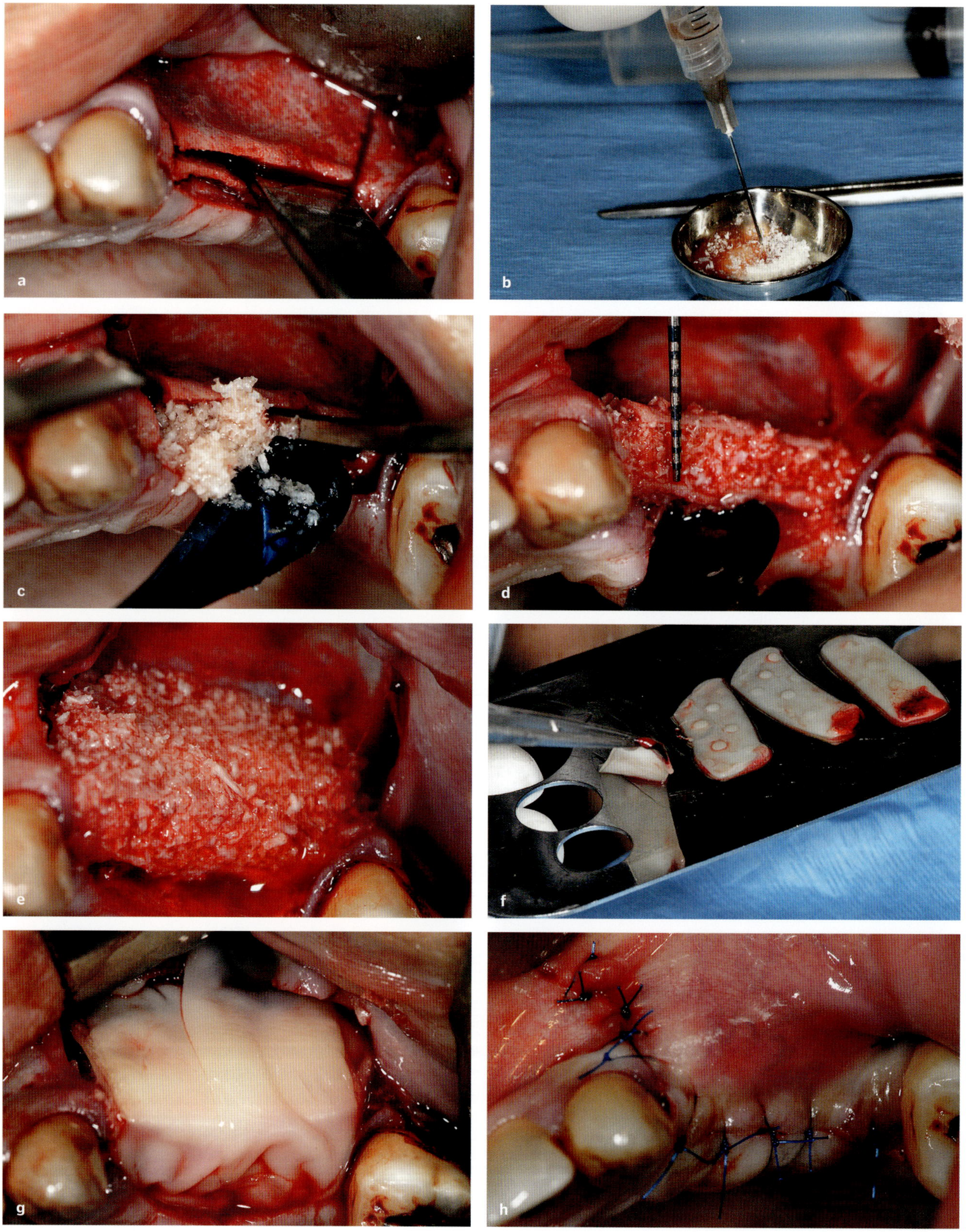

FIG 4-16 PRP membrane for ridge expansion. *(a)* Alveolar ridge expansion. *(b)* Hydration of particulate bone graft material with PRP liquid. *(c)* Placement of the hydrated bone graft material into the ridge split. *(d)* Ridge split and graft prior to membrane placement and suturing. *(e)* Placement of additional bone graft material to smooth out any rough bony edges. *(f)* Removal of the PRP membrane from the PRP membrane box (shown in Fig 4-13). *(g)* Placement of the PRP membrane over the entire grafted site. *(h)* Repositioning of the mucoperiosteal flap, after releasing it, over the site and passive closure.

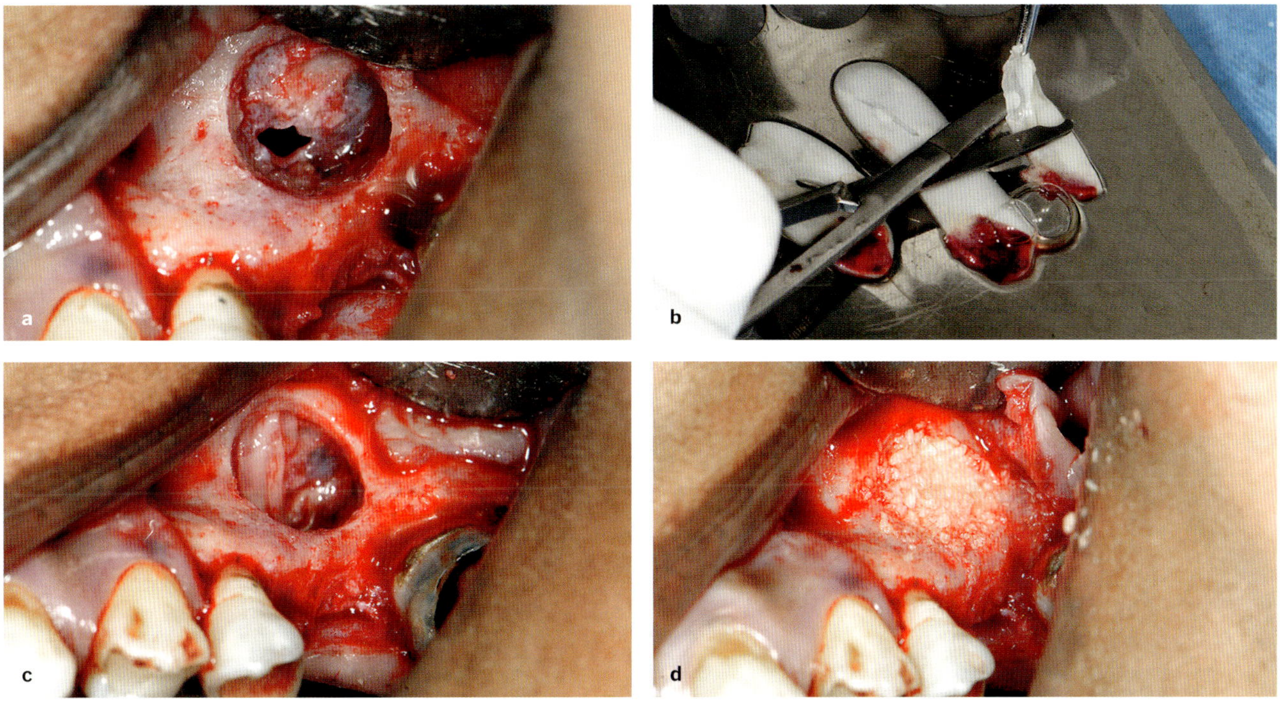

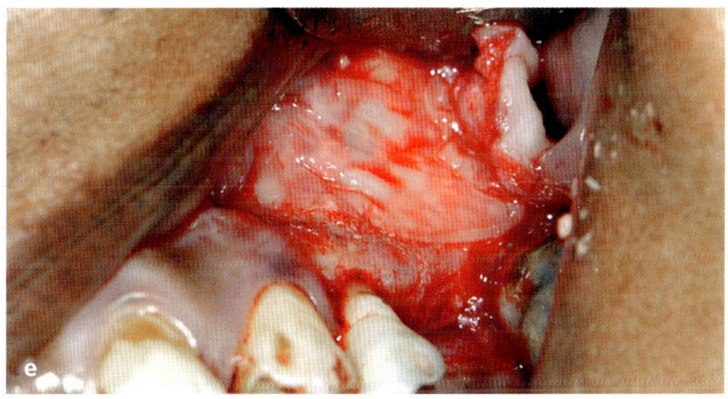

FIG 4-17 PRP membrane for lateral window sinus elevation with membrane perforation. *(a)* Lateral window access to sinus showing perforation of the sinus membrane. *(b)* PRP membrane formed from PRP gel in PRP membrane box. *(c)* Placement of the PRP membrane over the perforated sinus membrane as a patch. *(d)* Bone graft material placed into the maxillary sinus for augmentation of the maxillary sinus. *(e)* PRP membrane placed over the grafted sinus prior to soft tissue flap closure.

factors in the PRP accelerate healing of the membrane and promote angiogenesis. (See chapter 6 for more on soft tissue–related applications of PRP.)

Gingival recession defect grafting

PRP membranes can be an effective substitute for connective tissue grafts, which are considered the gold standard in the treatment of gingival recession defects. In addition to avoiding the painful surgery required to harvest connective tissue from the palate, several published reports have demonstrated earlier vessel formation and tissue formation with PRP membranes compared with connective tissue grafts, with less than 1% difference in the amount of root coverage obtained.[11]

Recently, PRP membranes have been used in place of connective tissue or collagen grafts in the so-called *pinhole* and the *vestibular incision subperiosteal tunnel access (VISTA) techniques* (Fig 4-18). In these minimally invasive root coverage approaches, a small access hole is made in the unattached gingiva, and specialized instruments are used to create a subperiosteal tunnel, which also extends underneath each papilla. PRP membranes are then introduced to cover the root dehiscence and bulk up the soft tissue. Finally, the mucogingival complex is advanced coronally and stabilized in the new position with a horizontal mattress suture in the gingival margin and/or anchored to each tooth.

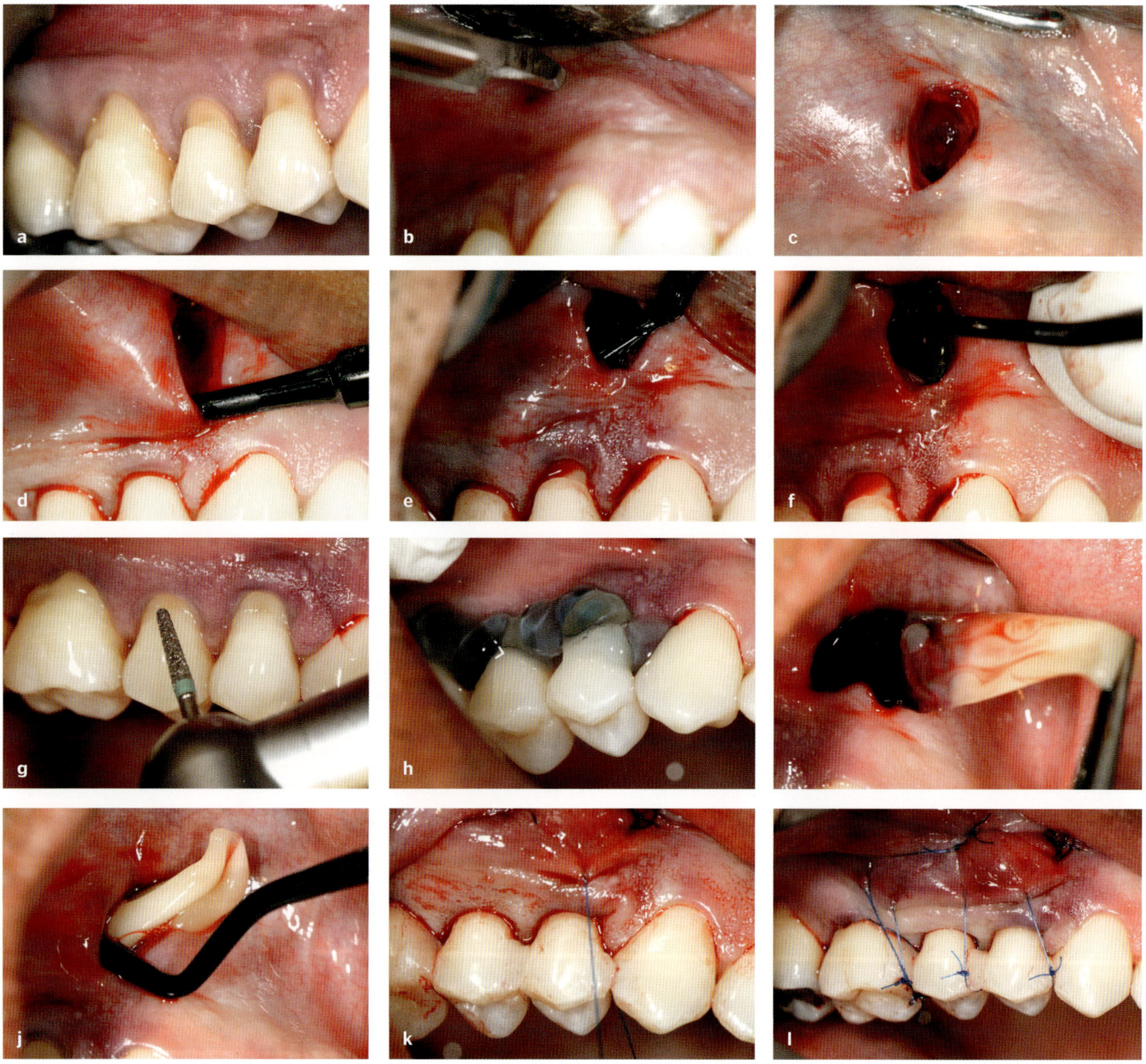

FIG 4-18 PRP membrane for gingival recession grafting. *(a)* Treatment of Miller Class I gingival recession. *(b)* Making a small vertical incision in the mucosal tissue. *(c)* Appearance of the "window" created. *(d)* Using specialized instruments to release the split-thickness mucosal tissue, the full-thickness attached gingiva, and the papillae. *(e)* Repositioning of the gingival flap to cover the receded areas. *(f)* Continued release of the flap for passive movement and coverage of the recession areas. *(g)* Recontouring the teeth if necessary after debridement with ultrasonics for thorough removal of plaque and tartar. *(h)* Conditioning of tooth structure with citric acid or ethylenediaminetetraacetic acid (EDTA). *(i)* Introduction of the first PRP membrane into the soft tissue "pocket." *(j)* Insertion of additional membranes into the prepared tunnel. *(k)* Suspending the soft tissues to allow for healing in the desired position to correct for the gingival recession preoperatively. *(l)* The sites are sutured and allowed to heal for approximately 3 weeks.

The PRP membrane eliminates the need for a donor site, the growth factors released from the PRP membrane stimulate faster revascularization and accelerate healing, and the minimally invasive approach requires no surface incisions and preserves the papilla. Most studies report 100% root coverage with this technique.[12-15] (See chapter 6 for more on soft tissue–related applications of PRP.)

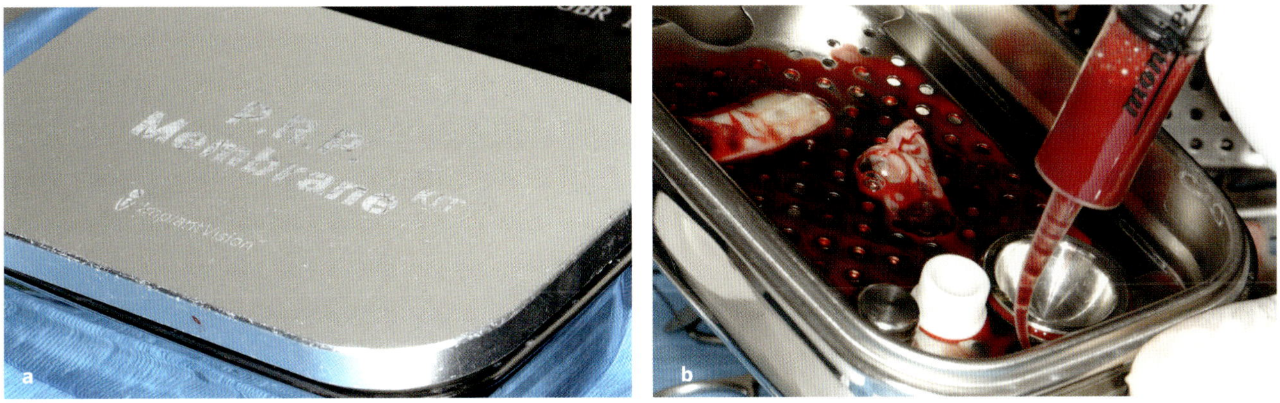

FIG 4-19 *(a and b)* Metal PRP box used to compress PRP gel into PRP membranes and obtain resultant PRP exudate.

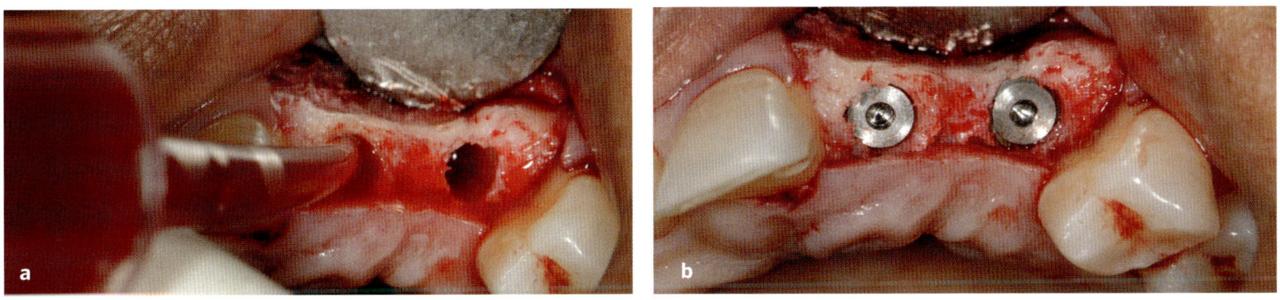

FIG 4-20 PRP exudate for irrigating an implant osteotomy prior to implant placement. *(a)* Irrigating the prepared osteotomies. *(b)* Placement of the implants.

FIG 4-21 PRP exudate to improve the handling qualities of sticky bone. *(a)* Sticky bone combined with a few drops of PRP exudate to improve clotting. *(b)* Observe the excellent clotting and handling qualities.

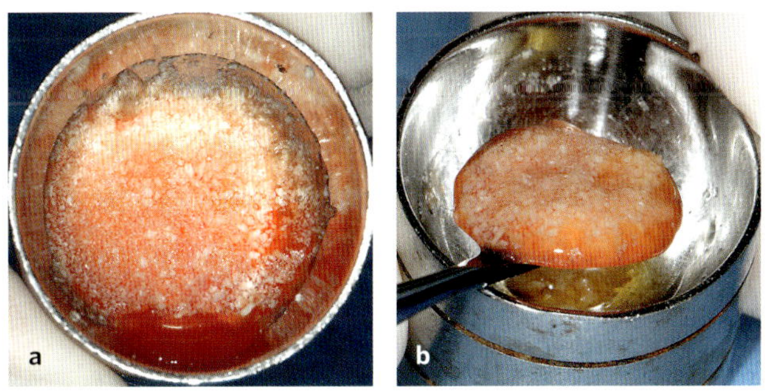

PRP Exudate

PRP exudate is the liquid residue remaining in the membrane box once the membrane has been removed (Fig 4-19). It is produced each time a PRP membrane is made. Comprised of just serum and thrombin, it can be used frequently but has few applications. Due to its lack of growth factors, PRP exudate should not be used in place of PRP liquid to hydrate particulate or freeze-dried bone or a collagen membrane. However, as a matter of convenience, it can be used in place of saline to irrigate an implant osteotomy site (Fig 4-20). Because of its high thrombin content, PRP exudate can be added to other configurations of PRP (Fig 4-21) to accelerate and improve clotting, as described below (see Fig 4-28c).

Formula for PRP exudate

Yield: 10 mL blood = 1–3 mL PRP exudate per test tube

Prep: 5–10 min | **Spin:** 12 min* | **Speed:** 2,000 rpm* **Varies based on machine used.*

Uses:

- Clotting of sticky bone
- Irrigation of implant osteotomy site

Needed:

- PRP centrifuge
- Phlebotomy kit
- Red-top test tubes
- Fine tweezers
- Glass dappen dish
- Scissors
- PRP membrane box
- Monoject syringe or 3-mL syringe with a 22-gauge blunt needle

Protocol:

1. Draw blood to fill two, four, six, or eight red-top vacutainer test tubes, depending on the volume of PRP exudate needed. Each test tube holds 10 mL of blood and will yield 1–3 mL of PRP exudate.
2. Place the test tubes in the centrifuge immediately.
3. Spin for approximately 12 minutes at about 2,000 rpm. (The exact time and speed will vary depending on your centrifuge device.)
4. Immediately after the spinning cycle is completed, remove the test tubes and place them in the test tube rack.
5. Move the rack to the clinical area if the centrifuge is located in another area of the office.
6. Using the fine long tweezers, after opening and removing the rubber stopper, reach into each of the red-top test tubes and grasp the clots in the middle portion.
7. If there is more than 1 mm of red portion attached, use the scissors to cut this portion off and discard it.
8. Place the gelatinous material into the PRP membrane box and cover it with the metal insert. Replace the lid and leave it undisturbed for 3–10 minutes or until you are ready to use it.
9. Open the lid of the PRP membrane box and carefully remove the membrane box insert.
10. Beneath the metal insert you will find the liquid PRP exudate, which contains the patient's own autologous thrombin.

Note:

PRP exudate can be used to improve the clotting of any of the other PRP configurations, or as an irrigating solution in an implant osteotomy.

PRP Plug

Like PRP membranes, PRP plugs are compressed and somewhat dehydrated forms of PRP gel. The difference is that the plug is compressed into a 3D configuration that has one-fourth the volume but four times the density of PRP gel (Fig 4-22).

Its primary value is as a space-occupying source of growth factors that can be used to avoid unnecessary expense in areas where a bone graft or collagen membrane is not needed (Fig 4-23).

Extraction socket defects

PRP plugs can be useful in the case of an extraction socket containing a large cyst that requires enucleation. Fortunately, a defect created by a cyst will naturally fill in as the socket heals. The actual extraction socket, however,

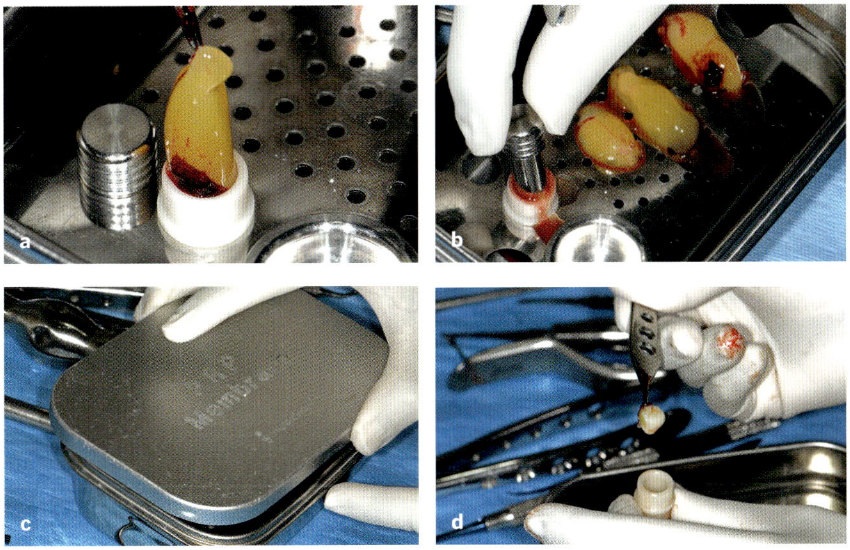

FIG 4-22 PRP plug. *(a)* Placement of a PRP gel clot into the custom plug compartment of the membrane box. *(b)* Compression of the gel clot into the compartment with a weighted plug press. *(c)* Replacing the lid on the membrane box. *(d)* Extraction of the PRP plug with tweezers.

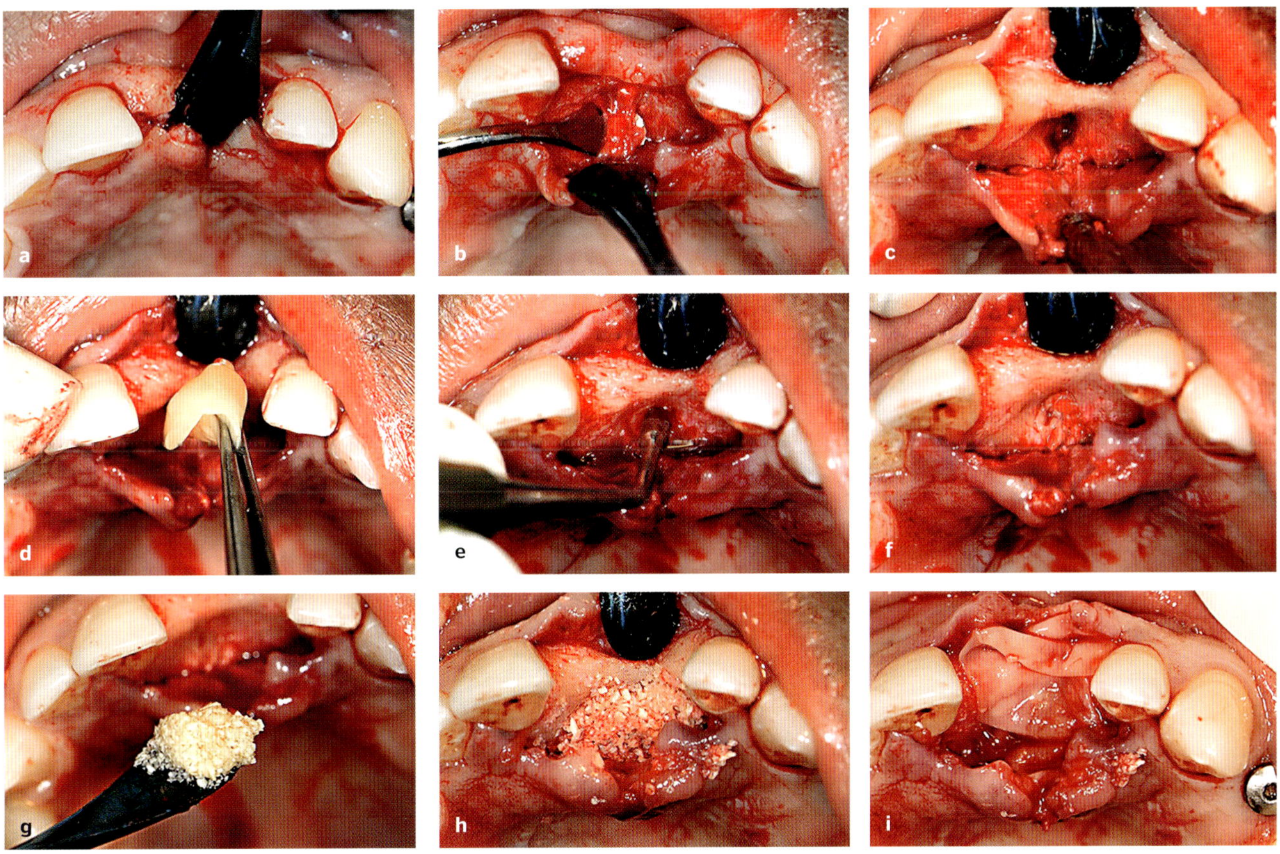

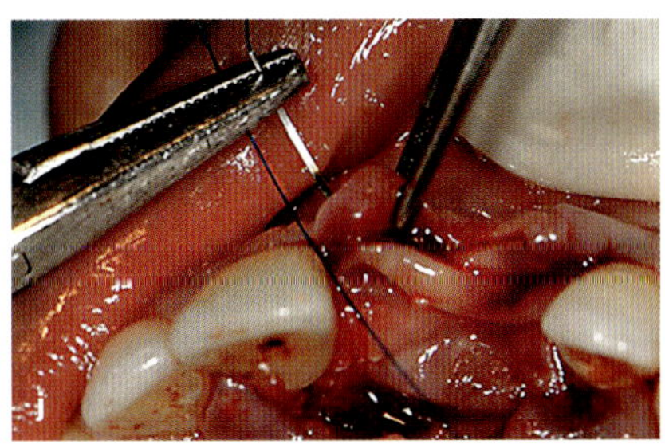

FIG 4-23 PRP plug to protect the neurovascular bundle. *(a)* Missing maxillary central incisor in the area of the nasopalatine canal with inadequate ridge width. *(b)* A crestal incision is made in preparation to expose the nasopalatine nerve. *(c)* The nasopalatine nerve and vessels are transected and removed from the canal in preparation for bone grafting of the canal. *(d)* A PRP plug is placed into the nasopalatine canal to protect the residual neurovascular bundle prior to augmentation of the canal and alveolar crest. *(e)* The PRP plug is condensed into the site. *(f)* Appearance of the PRP-filled canal. *(g)* The area is then bone grafted to allow for guided bone regeneration. *(h)* The graft is placed into the canal. *(i)* A PRP membrane is placed over the area to contain the graft particles and allow for bone regeneration. *(j)* Closure of the site.

Formula for PRP plug

Yield: 10 mL blood = 1 PRP plug per test tube

Prep: 5–10 min | **Spin:** ~10 min* | **Speed:** 2,000 rpm* **Varies based on machine used.*

Needed:

- PRP centrifuge
- Phlebotomy supplies
- Red-top test tubes
- Fine tweezers
- Scissors
- PRP membrane box

Protocol:

1. Draw blood to fill two, four, six, or eight red-top vacutainer test tubes, depending on the number of PRP plugs needed. Each test tube holds 10 mL of blood and will yield one PRP plug.
2. Place the test tubes in the centrifuge immediately.
3. Spin for approximately 10 minutes at about 2,000 rpm. (The exact time and speed will vary depending on your centrifuge device.)
4. Immediately after the spinning cycle is completed, remove the test tubes and place them in the test tube rack.
5. Move the rack to the clinical area if the centrifuge is located in another area of the office.
6. Using the fine long tweezers, after opening and removing the rubber stopper, reach into each of the red-top test tubes and grasp the clots in the middle portion.
7. If there is more than 1 mm of red portion attached, use the scissors to cut this portion off and discard it.
8. Place all of the clots in the plug compartment of a PRP membrane box and carefully lift the weighted plug press. Place the lid and leave the box undisturbed for 3–10 minutes or until ready to use.

Note:

This will give you the concentrated platelets, concentrated fibrin, and concentrated leukocytes (represented by the white line), as well as the youngest and most robust platelets contained in the top of the red portion,in a plug configuration.

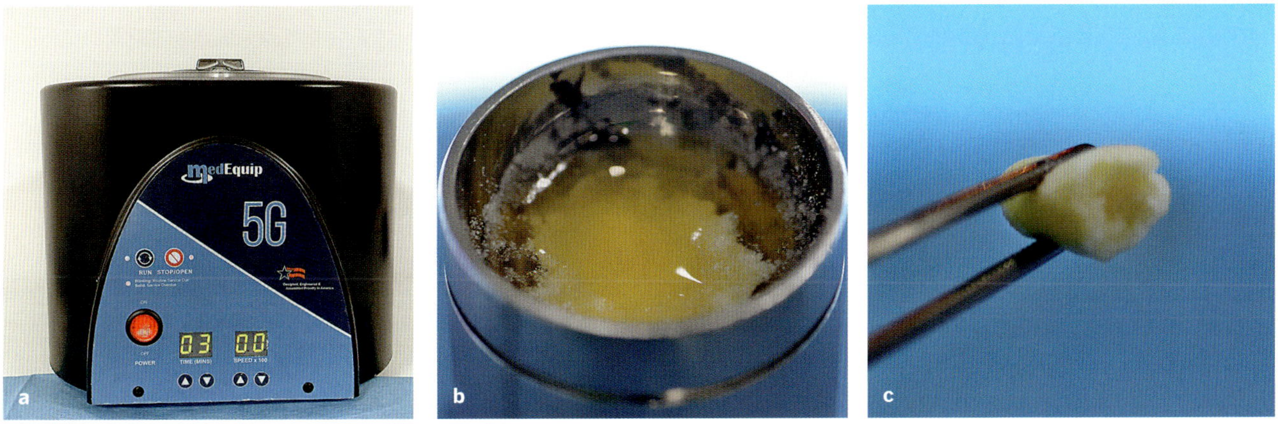

FIG 4-24 Sticky bone. *(a)* Centrifuge set for 3 minutes at 2,500 rpm. *(b)* PRP after one 2- to 4-minute spin, prior to clotting. This is added to bone grafting material in a small dish. *(c)* Appearance of rolled sticky bone prepared for socket grafting.

should be grafted to avoid bone resorption and/or to prepare for future implant placement. Because of the cyst defect and thus the expanded size of the socket, a greater quantity of grafting material would be required to fill it for predictable regeneration, thus increasing the cost of the surgery. At about $0.50 each, PRP plugs can be placed in the defect site prior to adding the bone grafting material into the extraction socket portion of the defect. As an extremely dense source of growth factors, they provide a simple, economical, and biologically compatible solution for filling the additional space.

Similarly, PRP plugs can be used when a sinus membrane has been overly reflected, or too far posteriorly reflected, above the third molar area during a sinus elevation procedure. When bone grafting material is being packed into the sinus cavity, it tends to fall into the posterior area where it may not be needed, increasing the amount that would normally be required. Typically, when this happens, collagen plugs are used to fill the space above the third molar area. However, PRP plugs are significantly less expensive than collagen plugs, which cost about $10 each, and are equally effective. It can also be used where a "cushion" is needed between the bone graft and some vital structure such as a nerve. (See chapter 7 for more on hard tissue–related applications of PRP.)

PRP Sticky Bone

Sticky or gummy bone (or Garg Sticky Bone) is made by putting the whole blood in a red-top test tube and spinning it for an abbreviated cycle (about 3–4 minutes), allowing the components to separate but not to fully clot. The result is a middle layer of concentrated platelets, leukocytes, and fibrin that is then used to hydrate particulate bone. After about 15 minutes, clotting will be complete, and the result will be "sticky bone" (Fig 4-24).

Excellent handling is the hallmark of sticky bone. Because it has the consistency of gummy bear candy, it can be easily molded into a surgical site and adapted to the shape of the bony defect. Furthermore, it maintains its 3D shape even after closure. The drawback of sticky bone is its relatively lower potency compared to PRP liquid. Whereas PRP liquid provides up to 10 times the concentration of growth factors found in whole blood, sticky bone provides about half that amount, or up to five times the baseline concentration. Nevertheless, the author has found that sticky bone made with an autogenous bone graft matures 25% faster than an autogenous bone graft without the benefit of PRP (3 months vs 4 months).

Formula for PRP sticky bone (aka Garg Sticky Bone)

Yield: 10 mL blood = 1–2 mL PRP sticky bone liquid per test tube

Prep: 5–10 min | **Spin:** ~2–4 min* | **Speed:** 2,500 rpm* **Varies based on machine used.*

Uses:

- Sinus grafting
- Ridge augmentation
- Socket preservation

Needed:

- PRP centrifuge
- Phlebotomy kit
- Red-top test tubes
- Fine tweezers
- Scissors
- Metal PRP membrane box
- Glass dappen dish
- Monoject syringe
- Graft material (freeze-dried bone allograft, autogenous bone, or synthetic particulate bone substitute)

Protocol:

1. Draw blood to fill two, four, six, or eight red-top test tubes, depending on the amount of bone graft material you are using. Each test tube holds 10 mL of blood and will yield 1–2 mL of PRP liquid for hydrating the particulate bone graft material.
2. Place the test tubes in the centrifuge immediately.
3. Spin for approximately 2–4 minutes at about 2,500 rpm. (The exact time and speed will vary depending on your centrifuge device.)
4. Immediately after the spinning cycle is completed, remove all of the test tubes except two and place them in the test tube rack. (The two remaining test tubes are reserved as a backup, as described in step 8 below.)
5. Move the rack to the clinical area if the centrifuge is located in another area of the office.
6. Draw up the top 3.5 mL from each test tube and discard it. This is the plasma or serum containing minimal or no growth factors.
7. Draw up the next 2–3 mL from each test tube and use it to hydrate the particulate bone in a glass dappen dish. (Avoid working under a direct air conditioning vent, as this will inhibit it from setting.) Cover the dish with 4 × 4 gauze and allow it to remain undisturbed for at least 15 minutes. This will create sticky bone.
8. In the event that your graft material remains too gelatinous for proper handling, use the two reserved test tubes of blood and follow the instructions below to prepare PRP exudate that will be added to your bone graft and give it better handling characteristics.

To create PRP exudate:

1. Place the two reserved test tubes in the centrifuge and spin for an additional 5–7 minutes at 1,200 rpm. When completed, take them to the clinical area.
2. Using the tweezers, reach into each test tube and grasp the middle portion of the gelatinous contents. If there is more than 1 mm of red portion attached, use the scissors to cut this portion off and discard it.
3. Place the gelatinous material into the PRP membrane box and cover it with the metal insert. Replace the lid and leave it undisturbed for 3–10 minutes.
4. Open the lid of the PRP membrane box and carefully remove the membrane box insert, which may require gentle force due to suction forces. Beneath the metal insert you will find the liquid PRP exudate, which contains the patient's own autologous thrombin.
5. Draw up this liquid with a syringe and add several drops to the dappen dish containing the sticky bone to thoroughly moisten it.
6. Leave the sticky bone undisturbed for 10 minutes or more before using it. The addition of the PRP exudate containing thrombin will greatly improve its handling characteristics.

Note:

PRP exudate can be used to solidify any of the other PRP configurations, as an irrigating solution in an implant osteotomy, or to hydrate a bone graft quickly and easily.

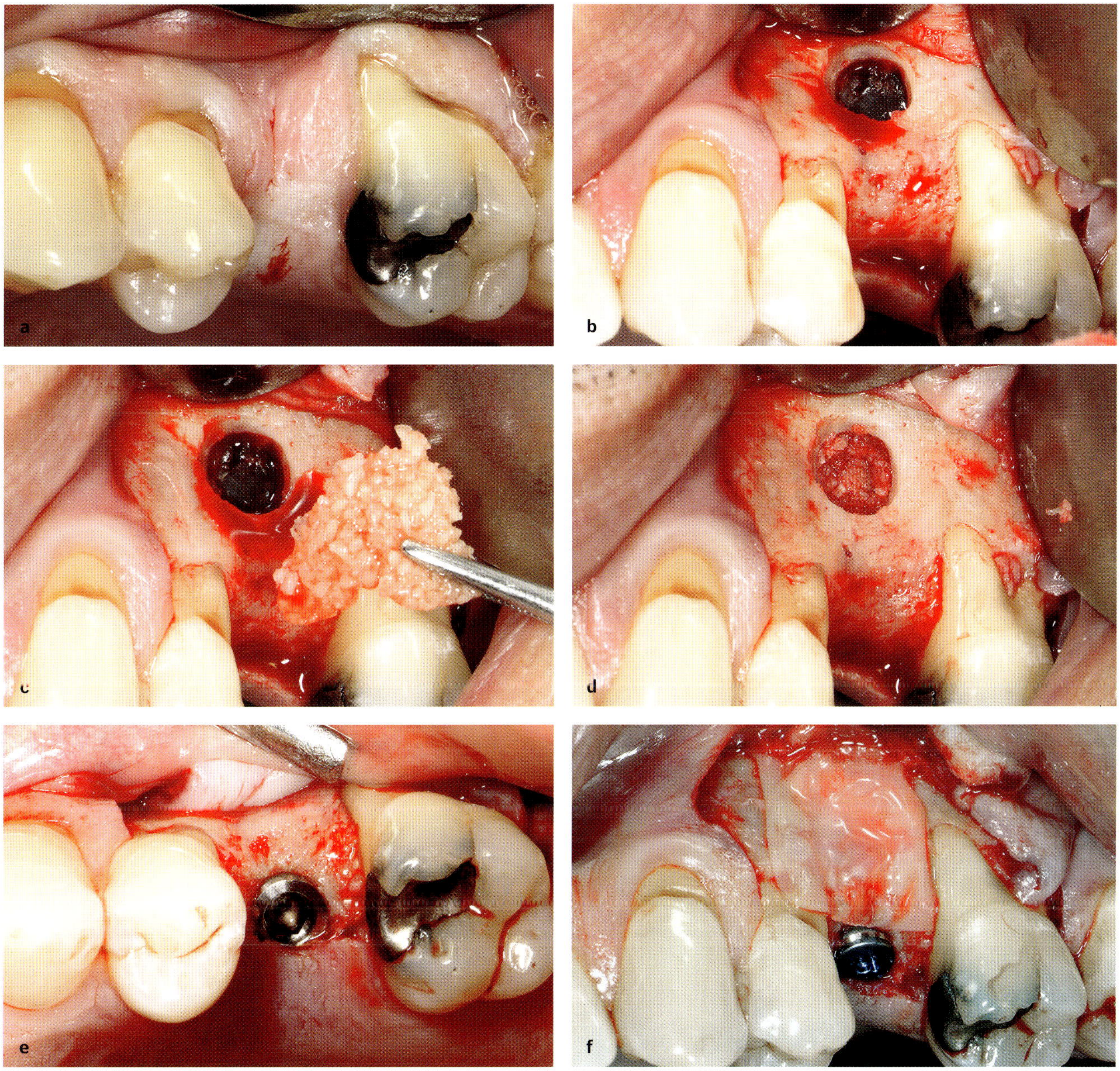

FIG 4-25 Sticky bone for sinus grafting. *(a)* Occlusal view of maxillary ridge at site of missing second premolar. *(b)* A flap has been elevated and a lateral window prepared for entry into the maxillary sinus cavity. *(c)* PRP sticky bone is introduced into the sinus to augment the alveolar ridge height. Note its excellent handling quality. *(d)* Appearance of the window packed with PRP sticky bone for bone regeneration. *(e)* The implant has been placed simultaneously. *(f)* A PRP membrane is placed over the lateral window prior to closure.

Horizontal ridge augmentation

Sticky bone can be used effectively in place of liquid PRP in sinus grafting (Fig 4-25) or sinus elevation procedures (Fig 4-26), but where this configuration provides the greatest benefit is in horizontal ridge augmentation surgery (Figs 4-27 and 4-28). For this procedure, a flap is reflected and the site is decorticated to obtain cells from the cancellous bone. The sticky bone, with its highly malleable consistency, is then adapted into the site, and the ridge is built out. Depending on the case, it can then be covered with a titanium mesh or collagen membrane or with a PRP membrane before a flap is advanced over it and sutured for tension-free primary closure. Using sticky bone in place of a standard bone graft for horizontal ridge augmentation makes the procedure easier and more predictable with accelerated healing. (See chapter 7 for more on hard tissue–related applications of PRP.)

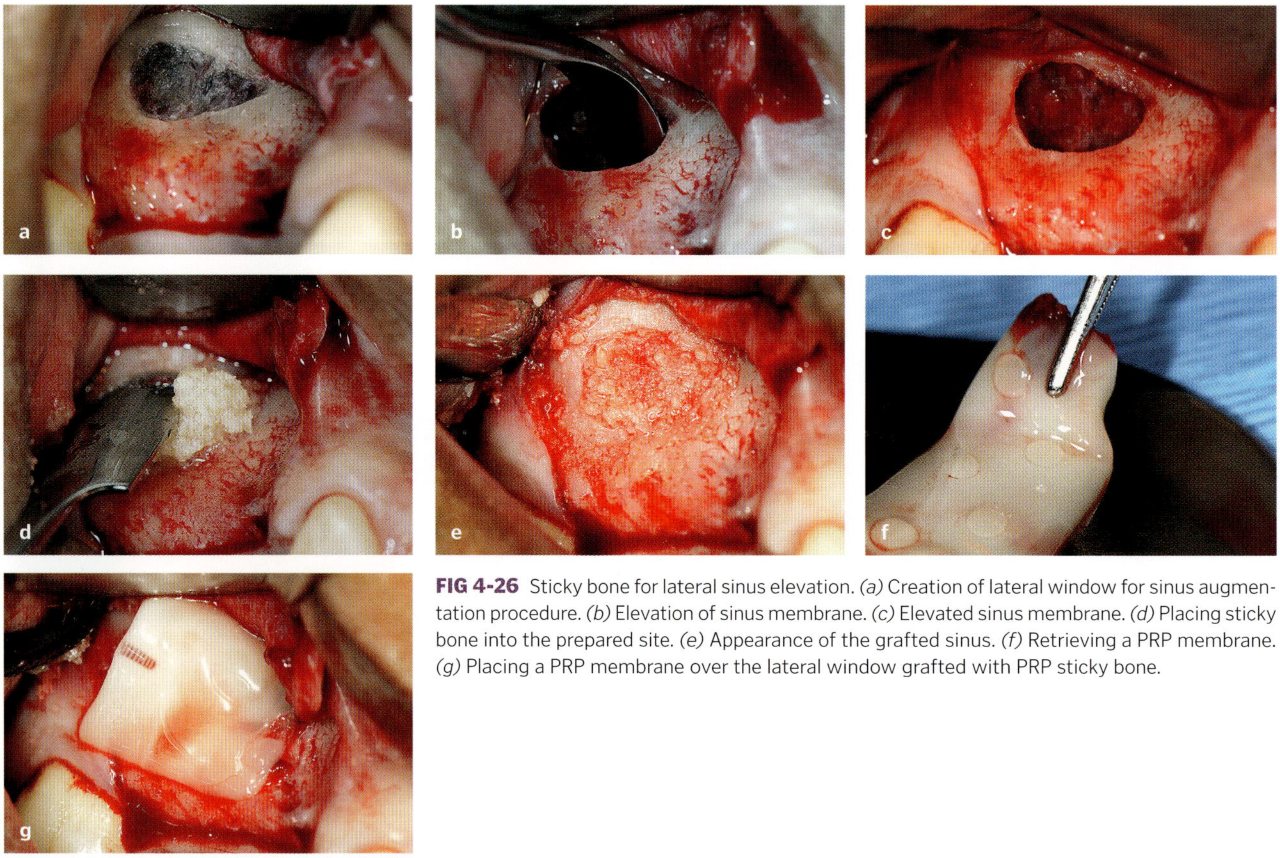

FIG 4-26 Sticky bone for lateral sinus elevation. *(a)* Creation of lateral window for sinus augmentation procedure. *(b)* Elevation of sinus membrane. *(c)* Elevated sinus membrane. *(d)* Placing sticky bone into the prepared site. *(e)* Appearance of the grafted sinus. *(f)* Retrieving a PRP membrane. *(g)* Placing a PRP membrane over the lateral window grafted with PRP sticky bone.

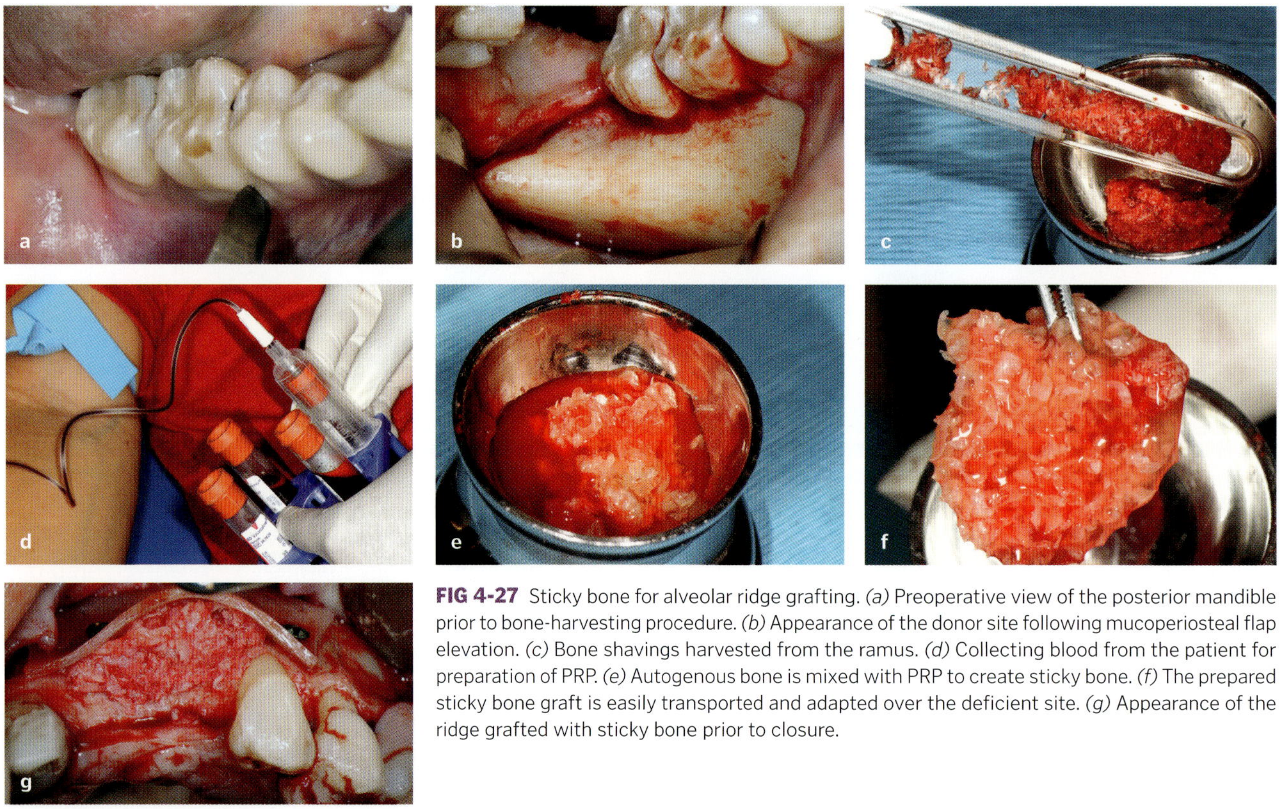

FIG 4-27 Sticky bone for alveolar ridge grafting. *(a)* Preoperative view of the posterior mandible prior to bone-harvesting procedure. *(b)* Appearance of the donor site following mucoperiosteal flap elevation. *(c)* Bone shavings harvested from the ramus. *(d)* Collecting blood from the patient for preparation of PRP. *(e)* Autogenous bone is mixed with PRP to create sticky bone. *(f)* The prepared sticky bone graft is easily transported and adapted over the deficient site. *(g)* Appearance of the ridge grafted with sticky bone prior to closure.

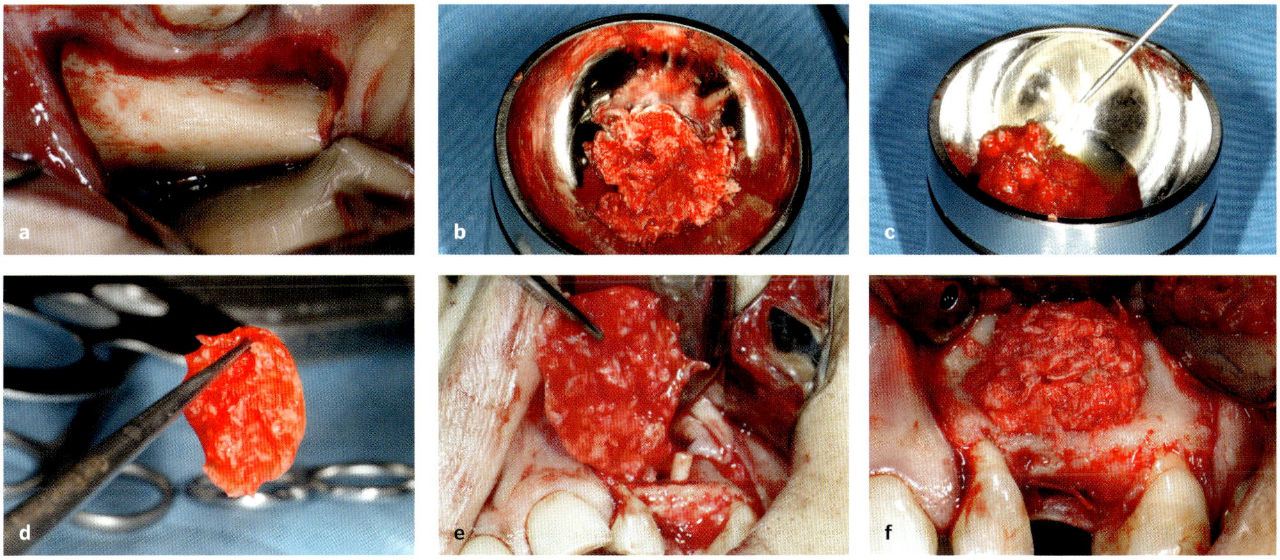

FIG 4-28 Sticky bone for alveolar ridge grafting. *(a)* Ramus is exposed for bone harvesting. *(b)* Bone is harvested from the ramus in a particulate configuration and combined with PRP to create sticky bone. *(c)* Addition of PRP exudate to improve the sticky bone's handleability. *(d)* Approximately 10 minutes later, the sticky bone graft demonstrates excellent handling characteristics. *(e)* The sticky bone graft is carried to the deficient ridge. *(f)* Appearance of the grafted site prior to closure.

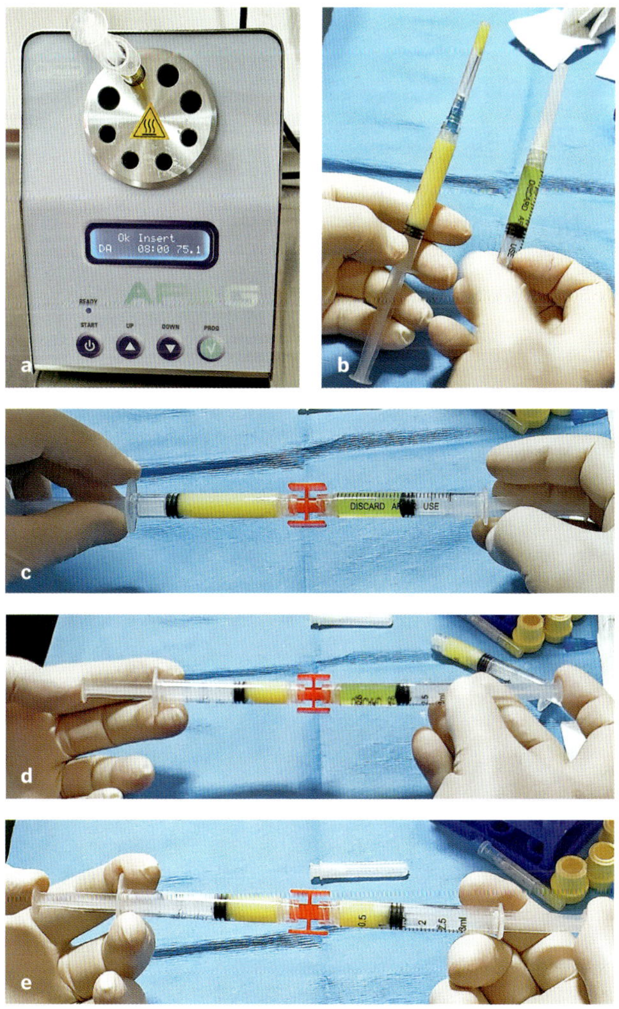

FIG 4-29 PRP with heated proteins. *(a)* Specialized "oven" used to heat and solidify the albumin proteins. *(b)* Syringes with Luer-lock tips containing solidified albumin proteins *(left)* and liquid buffy coat (PRP) layer of the test tube *(right)*. *(c to e)* Combining the contents of the two syringes using a Luer-lock attachment to create a filler with a creamy consistency.

PRP with Heated Proteins

Earlier in the chapter, the various applications for liquid PRP in facial esthetics were discussed, namely dermal microneedling and scalp injections for hair regrowth. PRP with heated proteins takes advantage of the plasma layer at the top of the test tube—which is usually discarded—to create a biologic filler, or biofiller, that performs as well as high-end dermal filler products like Juvéderm and Restylane but at a fraction of the cost.

The plasma layer consists mainly of proteins that, when heated, change from a liquid into a solid state (Fig 4-29). (Think of an egg, another type of liquid protein: when scrambled or boiled, it solidifies.) More importantly, heating the protein in the top layer also extends its longevity—from roughly 10 days (for PRP liquid) to 6 to 8 months.[16]

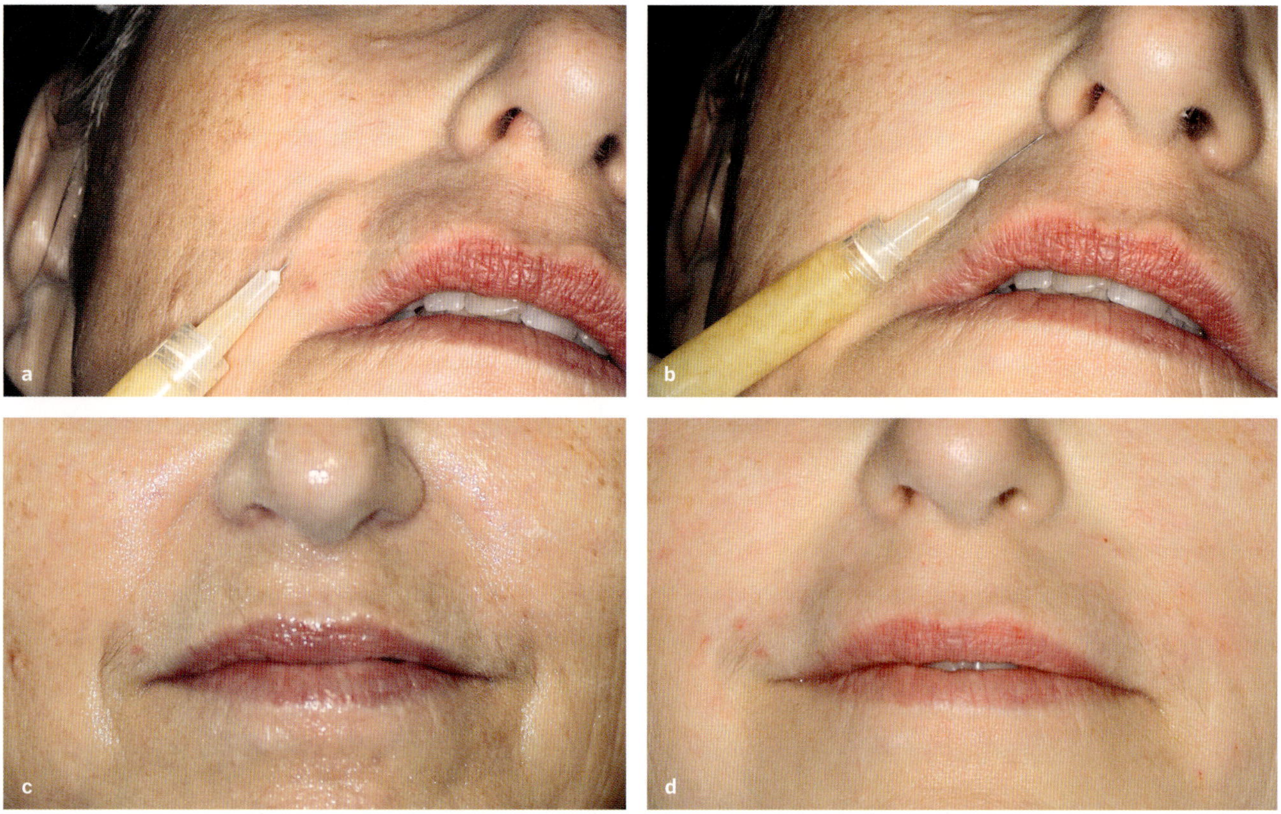

FIG 4-30 Facial rejuvenation treatment of nasolabial folds with PRP biofiller. *(a)* First injection into lower section of the nasolabial fold. *(b)* Second injection into upper section of the nasolabial fold. *(c and d)* Pretreatment view and immediate posttreatment view. The volumizing effect of the biofiller is already visible; the collagen-stimulating effect will develop over the next several weeks and will be maintained for up to 8 months.

The resulting heated platelet-poor plasma (PPP) can be used safely and effectively in place of commercial dermal fillers, which average $300 per syringe, for facial rejuvenation procedures. For example, it can be injected into nasolabial folds (Fig 4-30) to restore the lost volume, and the effects last for 6 to 8 months at a cost of about $0.50 per syringe. Because it contains no platelets, and therefore no growth factors, this plasma layer still has no stimulating effect.

However, when this PPP is mixed with PRP liquid, the result is a biofiller that combines the longevity and space-filling benefits of the former with the collagen-stimulating benefits of the latter.[17] When this biofiller is injected into a fold or wrinkle, its volumizing effect is visible immediately (see Fig 4-30). And, as noted earlier, the growth factors in the PRP liquid will stimulate the secretion of hyaluronic acid and type 1 collagen, which have been shown to improve skin texture, color homogeneity, elasticity, and firmness. Moreover, these benefits will continue for up to 8 months.[3,18,19]

PRP biofiller can be used not only to volumize and rejuvenate dynamic facial lines, folds, and wrinkles, but also to eliminate dark circles under the eyes, disrupt fibrotic adhesion in atrophic scars, and add plumpness to the lips and other areas of the face. The author has just published a new book on the step-by-step techniques for performing these procedures. *Dermal Fillers for Dental Professionals* (Quintessence, 2021) presents an experience-based framework for incorporating these procedures into a dental practice.[20]

Formula for PRP with heated proteins

Yield: 10 mL blood = 1–2 mL PRP with heated proteins per test tube

Prep: 5–10 min | **Spin:** ~2–4 min* | **Speed:** 3,500 rpm* *Varies based on machine used.*

Uses:

PRP facial rejuvenation with a dermal filler

Needed:

- PRP centrifuge
- Phlebotomy kit
- Yellow-top test tubes
- APAG (activated plasma albumin gel) device
- Cup of ice
- PRP membrane box
- Luer-lock female-to-female connector
- 3-mL syringes

Protocol:

1. Draw blood to fill six or eight yellow-top test tubes. Each test tube holds 10 mL of blood and will yield 1–2 mL PRP with heated proteins.
2. Place the test tubes in the centrifuge immediately.
3. Spin for approximately 4 minutes at about 3,500 rpm. (The exact time and speed will vary depending on your centrifuge device.)
4. Remove the test tubes and use a syringe to draw up the top 3-mL portion of each. Place the syringes in the APAG oven at 70°C for 10 minutes to heat and solidify the proteins.
5. Place the test tubes with the remaining PRP liquid into a cup of ice to keep it from naturally clotting over time.
6. Remove the syringes from the APAG oven and allow the cooked plasma protein to cool.
7. Remove the test tubes from the ice bath and use a 3-mL syringe to draw up the middle layer, being careful not to get the white line (leukocytes) or the red portion (erythrocytes).
8. Attach a Luer-lock female-to-female connector to this syringe. Attach a syringe with the cooled plasma protein to the other side of the Luer-lock connector. The ratio of the contents is 1:1. (This ratio can be varied depending on the consistency desired.)
9. Very gently and slowly mix the two substances back and forth 10 times until you have a creamy, uniform, and homogenous consistency. Repeat the process with all the syringes.

Note:

The amount of the liquid portion mixed with the amount of the cooked portion can be varied depending on the consistency desired.

PRP–Heated Protein Biomembrane

The heated protein biomembrane represents a natural step in the evolution of PRP, and yet only a handful of clinicians around the world are currently using it. For this configuration, the protocol for preparing PRP biofiller is modified to produce a membrane with delayed resorbability that will function both as a barrier and as a rich source of growth factors for use in procedures such as guided bone regeneration.[19,21]

As described earlier, a PRP membrane is useful in a variety of applications ranging from socket grafting to sinus

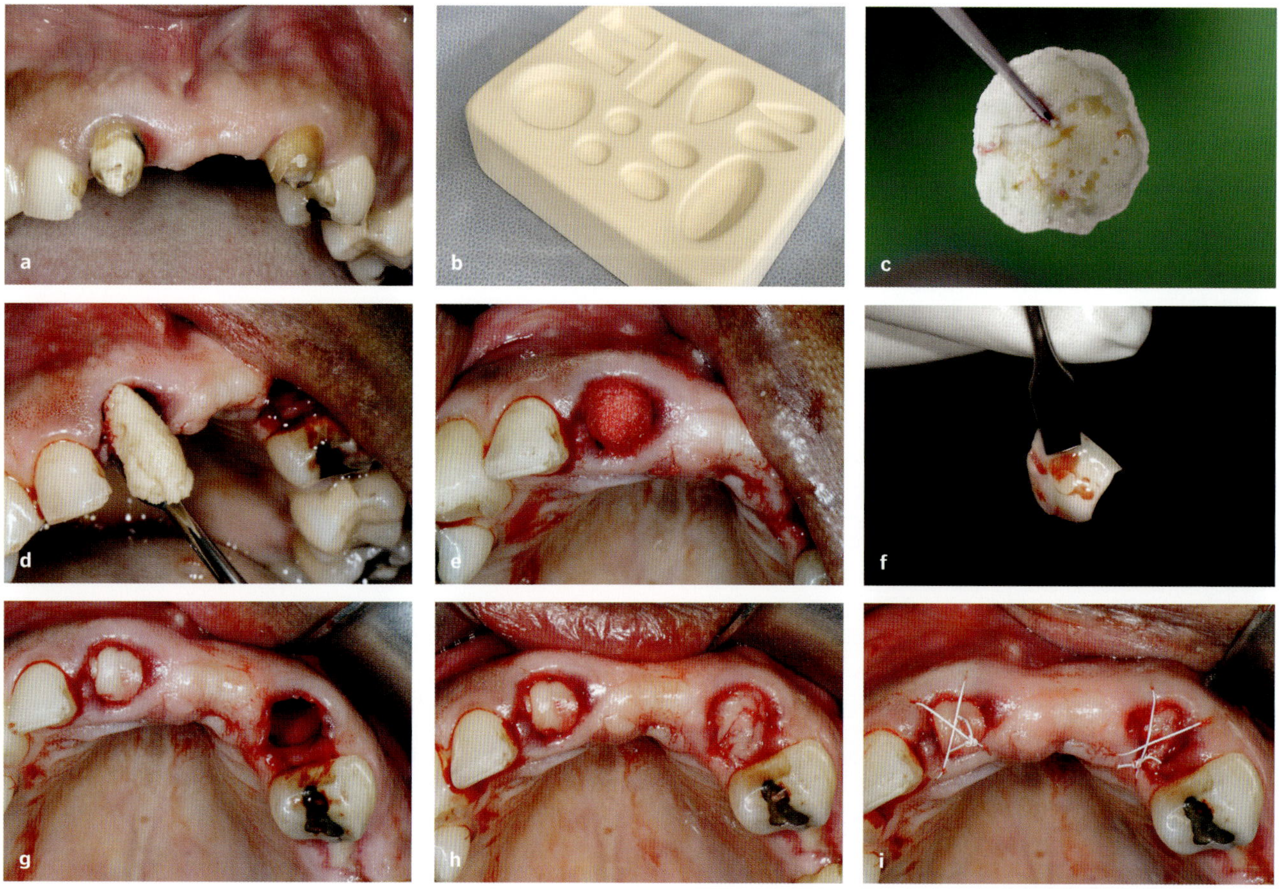

FIG 4-31 Heated biomembranes for socket preservation. *(a)* Extraction case with socket grafting followed by delayed implant placement. *(b)* Mold to create the best size and shape of membrane desired for a particular application. *(c)* Appearance of sticky bone before it is rolled for insertion into the socket. *(d and e)* Placement of the sticky bone into the extraction socket. *(f)* PRP heated biomembrane. *(g and h)* Placement of the PRP heated biomembrane over the grafted sockets. *(i)* Sutures placed.

membrane repair to gingival recession grafting, where it serves as a containment device and/or a source of growth factors to accelerate healing and enhance tissue regeneration. In these procedures, the rapid resorption of the membrane (about 10 days) is favorable to their success.

In guided bone regeneration, however, a membrane is placed over the regeneration site to prevent the invasion of connective tissue that would impede the formation of new bone over a period of 2 to 3 months. Collagen membranes—which are biocompatible, occlusive, durable, and resorbable—are the current gold standard for this purpose.

The heated protein biomembrane has all of the same characteristics as collagen membranes and additionally is capable of both immediate and prolonged release of growth factors that improve vascularity, accelerate healing, and increase the amount of new bone formed compared with non–PRP-treated sites (Fig 4-31). Given the autologous nature and nominal cost of the PRP-heated protein biomembrane, it is poised to become the new gold standard in bone regeneration procedures. (See chapter 7 for more on hard tissue–related applications of PRP.)

Formula for PRP-heated protein biomembrane

Yield: 3 heated biomembranes per test tube, depending on size

Prep: 5–10 min | **Spin:** ~8 min* | **Speed:** 2,500 rpm* **Varies based on machine used.*

Uses:

- Soft tissue procedures
- Bone grafting procedures
- Socket-preservation procedures
- Ridge augmentation

Needed:

- PRP centrifuge
- Phlebotomy kit
- Yellow-top test tubes
- Red-top test tubes
- APAG (activated plasma albumin gel) device
- Cup of ice
- PRP membrane box
- Membrane mold
- Syringe

Protocol:

1. Draw blood to fill two yellow-top vacutainer test tubes and two red-top vacutainer test tubes. Each test tube holds 10 mL of blood and will yield three heated biomembranes.
2. Place the test tubes in the centrifuge immediately.
3. Spin for approximately 8 minutes at about 2,500 rpm. (The exact time and speed will vary depending on your centrifuge device.)
4. Remove the yellow-top test tubes and use a syringe to draw up the top 2.5-mL portion of each. Place the syringes in the APAG oven at 74.5°C for 10 minutes to heat and solidify the proteins.
5. Place the yellow-top test tubes with the residual PRP liquid into a cup of ice to keep it from naturally clotting over time.
6. Using the tweezers, reach into the two red-top test tubes and grasp the clots in the middle portion. If there is more than 1 mm of red portion, use the scissors to cut off this portion and discard it. Place the clots in a PRP membrane box and cover them with the metal insert. Replace the lid and leave the box undisturbed for 3–10 minutes or until ready to use.
7. Open the lid of the membrane box and carefully remove the metal insert, which may require gentle force due to suction effects. The PRP membranes will be found on the reverse side of the metal insert or on the mesh grid at the bottom of the box. Remove them and set them aside. (See note below for suggested applications for PRP membranes.)
8. Beneath the mesh grid you will find the liquid PRP exudate, which contains the patient's own autologous thrombin. This will be used to expedite and enhance the clotting of the heated biomembrane.
9. Remove the syringes from the APAG oven and inject the cooked plasma protein into the molds according to the shape of the membranes needed. Gently pat them into the membrane thickness desired.
10. Remove the yellow-top test tubes from the ice bath and use a syringe to draw up the middle layer, which is liquid PRP. Add this to the cooked plasma protein that you just placed into the molds, using enough to completely cover the material.
11. From the PRP membrane box, draw up the PRP exudate, which contains the patient's thrombin, and add a few drops to the material in the molds. Leave the molds undisturbed for 10–15 minutes. This helps the clotting and improves the handling properties of your heated biomembranes.

Note:

Heated biomembranes are longer-lasting than regular PRP membranes, taking approximately 6 months to resorb. They also have excellent handling properties and deliver the healing and other growth factors found in the liquid PRP.

Conclusion

PRP can be configured in a variety of ways to enhance its efficacy in specific surgical and therapeutic procedures. Clinicians who are new to PRP should first focus on mastering the protocol for preparing PRP liquid, as all other forms depend on its potency. Once you and your patients experience its advantages firsthand—faster healing, less pain, lower rates of infection, more predictable outcomes—you can use the formulas and step-by-step instructions in this chapter to customize your PRP into forms that optimize its many clinical benefits.

References

1. Garg A. The platelet: Uses in dentistry. Dental Implantol Update 2011;22:57–62.
2. Wang S-Z, Fan W-M, Jia J, Ma L-Y, Yu JB, Wang C. Is exclusion of leukocytes from platelet-rich plasma (PRP) a better choice for early intervertebral disc regeneration? Stem Cell Res Ther 2018;9:199.
3. Matthews-Brzozowska T, Łącka M, Kawałkiewicz W, Mojs E, Kubisz L. Minimally invasive facial skin revitalization treatment: A case study. J Med Sci 2018;87:48–54.
4. Steigmann M, Garg AK. A comparative study of bilateral sinus lifts performed with platelet-rich plasma alone versus alloplastic graft material reconstituted with blood. Implant Dent 2005;14:261–266.
5. Gandevivala A, Sangle A, Shah D, et al. Autologous platelet-rich plasma after third molar surgery. Ann Maxillofac Surg 2017;7:245–249.
6. Cardarelli A, Garg AK. Effects of platelet-rich plasma and piezosurgery on impacted mandibular third molar surgery outcomes: Preliminary study. Adv J Otolaryngol 2020;1:1–7.
7. Kaul RP, Godhi SS, Singh A. Autologous platelet-rich plasma after third molar surgery: A comparative study. J Maxillofac Oral Surg 2012;11:200–205.
8. Barone A, Santini S, Sbordone L, Crespi R, Covani U. A clinical study of the outcomes and complications associated with maxillary sinus augmentation. Int J Oral Maxillofac Implants 2006;21:81–85.
9. Pikos MA. Maxillary sinus membrane repair: Update on technique for large and complete perforations. Implant Dent 2008;17:24–31.
10. Testori T, Wallace SS, Del Fabbro M, et al. Repair of large sinus membrane perforations using stabilized collagen barrier membranes: Surgical techniques with histologic and radiographic evidence of success. Int J Periodontics Restorative Dent 2008;28:9–17.
11. Eren G, Kantarci A, Sculean A, Atilla G. Vascularization after treatment of gingival recession defects with platelet-rich fibrin or connective tissue graft. Clin Oral Investig 2016;20:2045–2053.
12. Reddy S, MGS P, Bhowmik N, Singh S, Pandit HR, SK V. Vestibular incision subperiosteal tunnel access (VISTA) with platelet rich fibrin (PRF) and connective tissue graft (CTG) in the management of multiple gingival recession. A case series. Int J Appl Dent Sci 2016;2:34–37.
13. Tuttle D, Kurtzman GM, Bernotti AL. Gum drop technique: Minimally invasive soft-tissue platelet-rich fibrin grafting for marginal soft-tissue recession. Compend Contin Educ Dent 2018;39:e9–e12.
14. Jankovic S, Aleksic Z, Klokkevold P, et al. Use of platelet-rich fibrin membrane following treatment of gingival recession: A randomized clinical trial. Int J Periodontics Restorative Dent 2012;32:e31–e50.
15. Tunalı M, Özdemir H, Arabacı T, Gürbüzer B, Pikdöken L, Firatli E. Clinical evaluation of autologous PRF in the treatment of multiple adjacent gingival recession defects. A 12-month study. Int J Periodontics Restorative Dent 2015;35:105–114.
16. Orentreich N, Krajcik RA. Materials for soft tissue augmentation and methods of making and using same. New York: US Patent no. 7015198, 2006.
17. Del Pilar Rios M, Garg AK, Kurtzman GM, Rossi R. Warmed 5th generation PRP as a dermal filler for facial rejuvenation. J Acad Oral Implantol 2020;1:1–5.
18. Gheno E, Barros Mourão CF, de Mello-Machado R, et al. In vivo evaluation of the biocompatibility and biodegradation of a new denatured plasma membrane combined with liquid PRF (Alb-PRF) [epub ahead of print 12 June 2020]. Platelets doi:10.1080/09537104.2020.1775188.
19. Barros Mourão CF, Gheno E, Lourenço ES, et al. Characterization of a new membrane from concentrated growth factors associated with denaturized albumin (Alb-CGF) for clinical applications: A preliminary study. Int J Growth Factors Stem Cells Dent 2018;1:64–69.
20. Garg AK, Rossi R. Dermal Fillers for Dental Professionals. Chicago: Quintessence, 2021.
21. Kawase T, Kamiya M, Kobayashi M, et al. The heat compression technique for the conversion of platelet-rich fibrin preparation to a barrier membrane with a reduced rate of biodegradation. J Biomed Mater Res Part B 2015;103B:825–831.

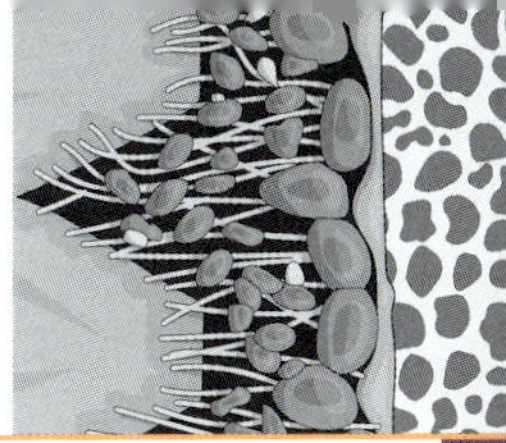

5

Dental Implants, Osseointegration, and Autologous Blood Concentrates

Endosseous dental implants have become increasingly popular and medically warranted to treat completely and partially edentulous patients.[1-3] The indispensable criterion for dental implant success is osseointegration, the process by which the dental implant becomes functional through firm anchorage in the alveolar bone.[4-7] Osseointegration is affected by a number of interrelated factors, including treatment planning, the condition of the host bone, the topography and biocompatibility of the implant material, the overall medical condition of the patient, and the timing and extent of functional load placed on the implant. Because the bonding of implant with bone involves the same kinds of wound healing processes seen in injured or damaged bone (including cell migration, adhesion, and proliferation), another important factor influencing osseointegration is the employment of biologic enhancements for wound healing, including the use of PRP. This chapter focuses on the process of wound healing involved in osseointegration and how the regeneration of bone cells and tissues—and thereby osseointegration—may be enhanced by the application of PRP at the wound/implant site.

Dental Implants and Osseointegration

Osseointegration in implant dentistry is the process by which a surgical implant becomes firmly anchored in the mandible or maxilla as bone grows around the implant, with no fibrous or connective tissue at the interface of the implant surface and bone, and with no loss of anchoring under implant function (load). Osseointegration is the primary measure of implant success.[8] A number of complementary factors determine successful osseointegration, including implant biocompatibility, implant surface and design, implant site health and bone quality, surgical method, site healing duration, and the functional loading process.[9-11]

The mechanism of osseointegration

Once an area of bone experiences damage (eg, fracture, lesion), direct bone healing begins in the bone matrix. Osseointegration occurs in much the same way (Fig 5-1). In fact, osseointegration closely resembles the healing process of bone over time. When damaged bone is exposed to blood, bone repair is activated via noncollagenous proteins and growth factors. Woven bone develops, followed by lamellar and parallel-fibered bone, which are then followed by remodeled bone.[12]

Osseointegration reflects implant stability in progressive ways, including primary and secondary stages. CAD/CAM software and radiographic treatment plans and surgical guides can assist the clinician in optimal placement in bone.[13,14] Primary stability of the implant must occur at implantation, via a mechanical connection between the implant and the host bone. This requires mechanical fixation of an implant with a biomimetic (roughened) surface placed at the optimal distance within the bone.[15] Optimal positioning of the implant in the host bone facilitates osteogenesis.[16]

Only when primary stability is established can the secondary stage of implant stability and osseointegration take place. This requires time as bone regeneration and remodeling create a long-lasting biologic connection

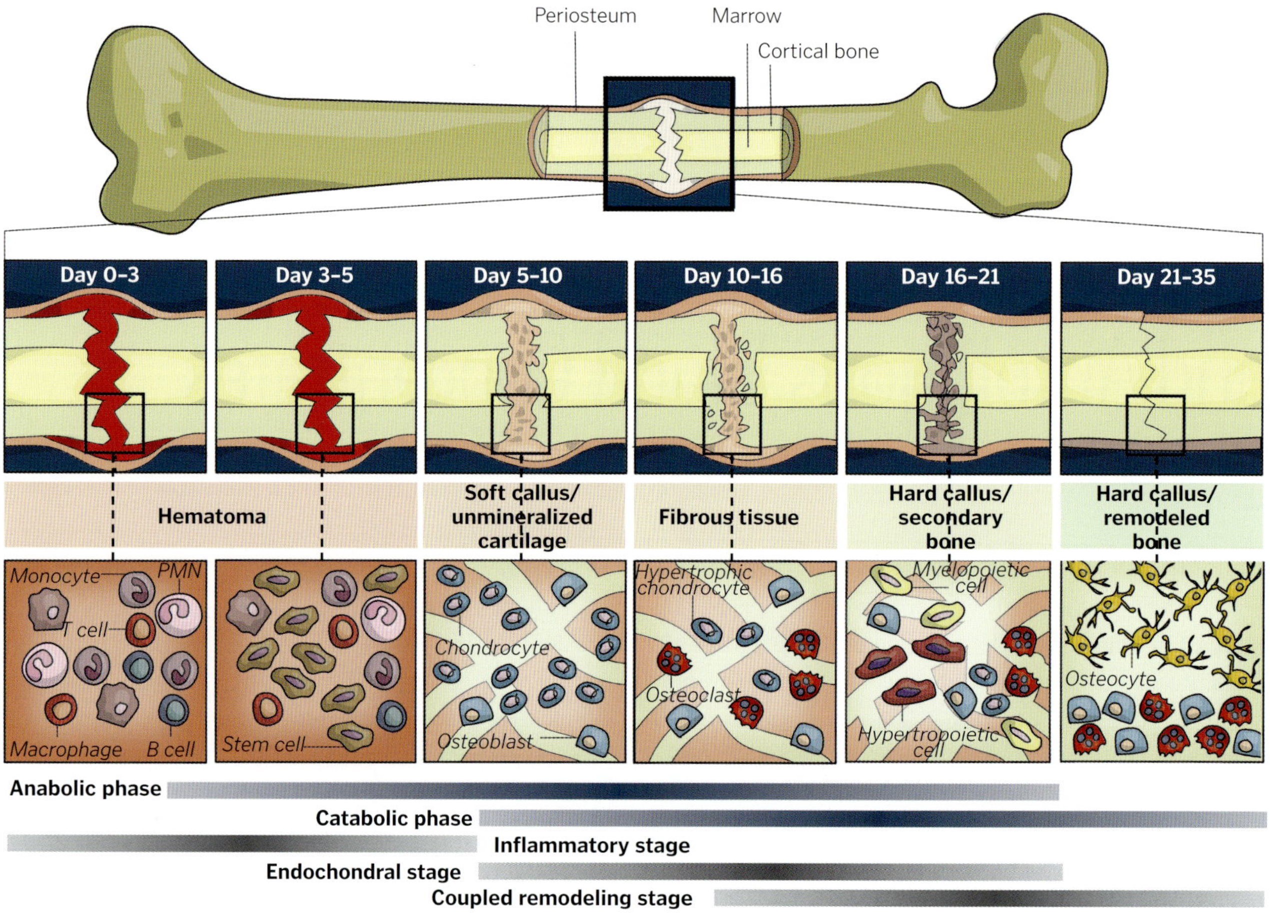

FIG 5-1 Osseointegration of a dental implant involves many of the same biologic processes that occur during direct bone healing. PMN, polymorphonuclear leukocyte.

between implant and bone. Once biologic stability is underway, loading of the implant in function can occur. The amount of bone originally surrounding the implant may add to the bond established between the two, as can physical contact via occlusal forces, which stimulate the bone after the implant is loaded for several years.[17] Eventually, a number of methods are used to determine if an implant has undergone osseointegration: implant immobility, radiographic evidence of direct contact between bone and the implant surface, and ultrasound devices with a metal instrument.[18-20]

Almost immediately after an implant is placed in the host bone, osteoblasts and mesenchymal cells in the blood move toward the surface of the implant for attachment (Figs 5-2a to 5-2d).[21,22] When they attach, they establish proteins and a noncollagenous network for controlling cell adhesion and mineral aggregation. The result is an implant surface covered with calcified afibrillar cells, a semimineralized bonelike cement comprised of calcium, phosphorous, osteopontin, and sialoprotein (Fig 5-2e).[22,23]

Clot formation is influenced by the interaction of blood cells with the implant surface, leading to both intercellular changes (spreading and aggregating) and intracellular changes (phosphotyrosine induction, calcium growth, and phospholipid hydrolysis). A fibrin matrix forms, providing the osteoconductive network by which osteogenic cells first migrate and then differentiate (Fig 5-3), inducing the formation of new bone in this order: osteoid tissue, trabecular bone, and then lamellar bone, which connects directly to the surface of the implant.[24]

New bone forms around the implant bidirectionally, moving outward from the implant surface via contact osteogenesis as well as inward from the host bone trabeculae via distance osteogenesis toward the implant's

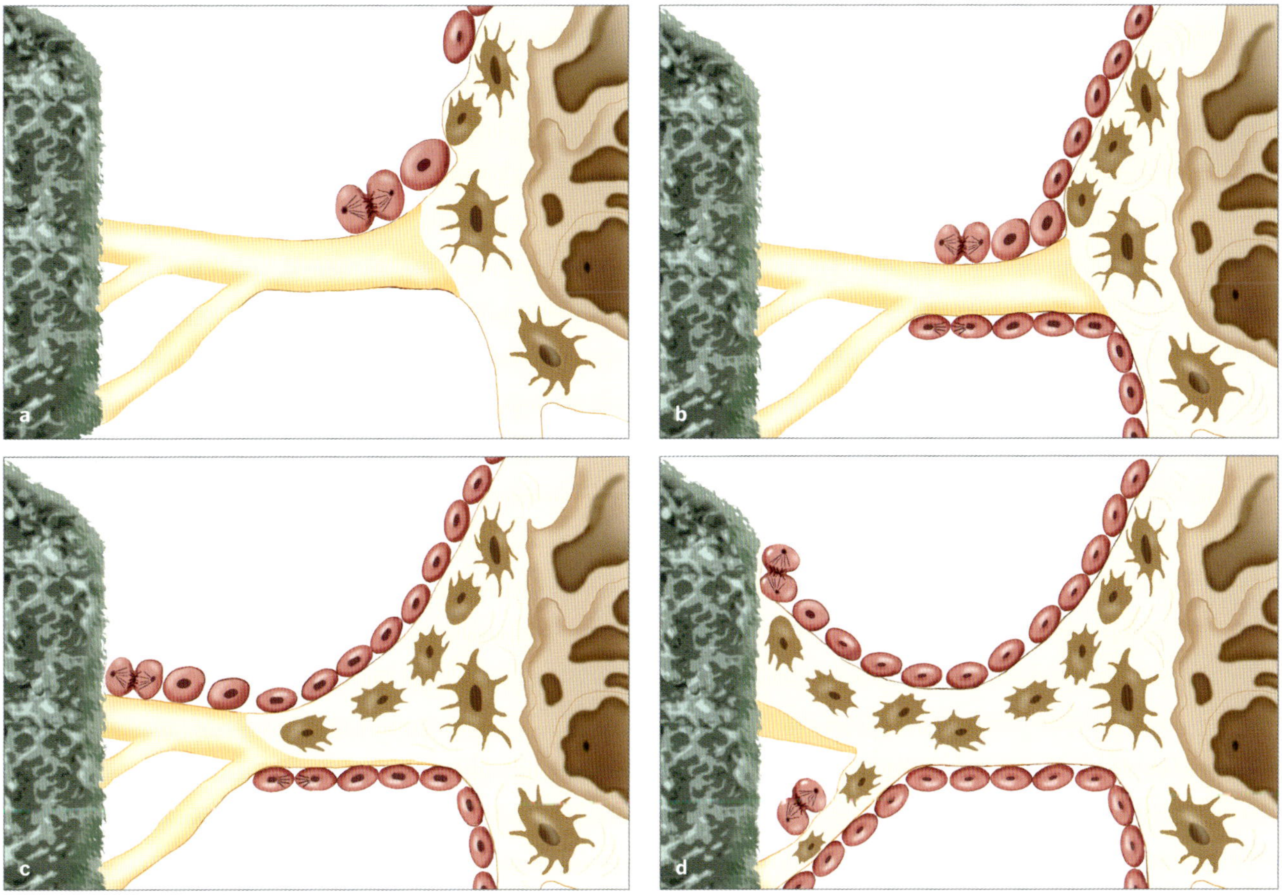

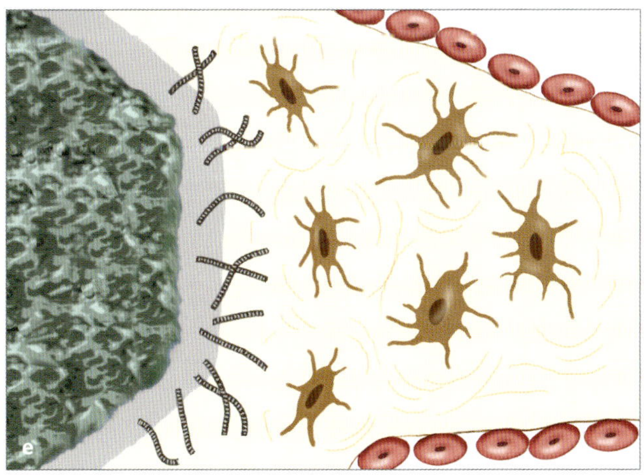

FIG 5-2 Osteoblasts and mesenchymal cells attach to the surface of an implant to achieve primary stability in the early stages of osseointegration. *(a)* Growth factor activation of osteoblasts causes cell division along the surface of the fibrin strand. *(b)* As cells divide along the fibrin strand, the lead daughter cell is advanced toward the implant surface, and the stationary daughter cell becomes an osteoblast that differentiates further and secretes osteoid. *(c)* As further cell divisions occur, the lead osteoblast daughter cells come into contact with the implant surface, and the trailing daughter cells that secrete osteoid become encased in their mineral matrix to become true osteocytes as a bridge of bone begins to develop toward the implant surface. *(d)* Still further cell divisions now advance the lead daughter cells along the implant surface, and more trailing osteoblasts mature into osteocytes and form the bridge of osseointegration across the microgap and now along the implant surface itself. *(e)* Osseointegration is actually bone anchored to the metal surface through the 5-μm-thick cementing substance. True bone is anchored in turn to the cementing substance when the bone embeds collagen fibers into the cementing substance. These collagen fibers perform a function similar to that of reinforced steel rods used in highway construction. (Reprinted with permission from Marx and Garg.[22])

surface. The implant becomes biologically attached to the host bone as a result of osteoblast production of a calcified/osteoid tissue layer on the implant surface and the presence of mesenchymal cells and blood vessels in the remaining spaces of the implant surface.

Bone chip remnants placed during the surgical procedure can assist in the early stages of osseointegration. They can facilitate trabecular bone formation via osteogenesis, osteoconduction, and osteoinduction when placed near the host bone and the implant.[15] After about 3 months, biologic fixation (osseointegration) is well underway as remodeled woven bone and then lamellar bone encase the implant and attach themselves to the implant surface.[25,26]

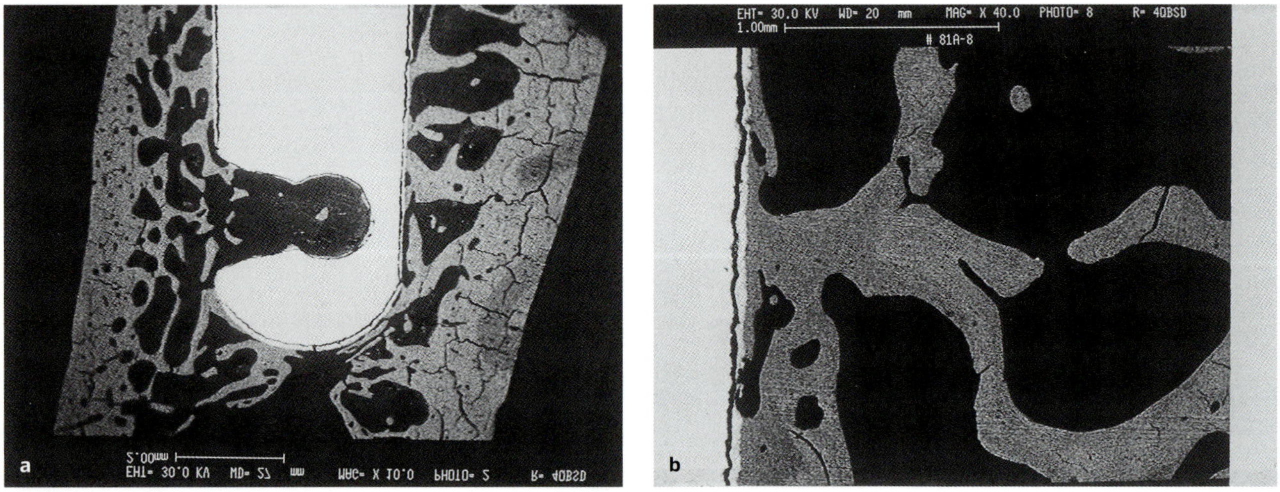

FIG 5-3 *(a)* Backscatter electron micrograph of an osseointegrated implant (in a dog sample) confirming this model of osseointegration and showing bony connections between the endosteal surface of a mandibular cortex to the implant surface and bone formation along the implant surface. *(b)* High-power view showing the migration of bone along the implant surface and a cement line between the implant surface and the actual bone. (Reprinted with permission from Marx and Garg.[22])

Table 5-1 Clinical factors influencing osseointegration positively and negatively

Factors	Enhancement of osseointegration	Inhibition of osseointegration
Implant-related factors	Design Chemical composition Coatings/surface topography	Inappropriate surface topography Inappropriate biomaterials
Status of host bone bed	Minimal surgical trauma Vascularity and cellularity of implantation site	Bone defect Osteoporosis Rheumatoid arthritis Smoking Advanced age
Mechanical stability	Primary implant stability No micromotion	Excessive implant mobility (interface motion)
Adjuvant therapies	Bone grafting Osteogenic coatings Biophysical stimulation Biostimulation	Irradiation Pharmacologic agents: cyclosporin A, methotrexate, cis-platinum, warfarin, indomethacin

Factors influencing osseointegration

Osseointegration can be compromised by implant movement,[27,28] faulty implant coating,[29] irradiated bone,[30] a variety of chemical elements (eg, cyclosporine, warfarin),[31-34] and poor patient health conditions and habits (eg, osteoporosis, smoking).[35] The preservation and formation of blood vessels around the implant are essential to osseointegration as angiogenesis helps to regulate the differentiation of osteogenic cells. Conditions that adversely affect angiogenesis, such as osteoporosis and a history of smoking, can thus restrict or inhibit osseointegration (Table 5-1).

Conversely, osseointegration can be facilitated by optimal implant design, chemical composition, surface topography, material, shape, length, diameter, and surface treatment; good host bone bed condition; favorable implant loading characteristics; adjuvant treatments that may be required (eg, bone grafting, biophysical stimulation); and any necessary pharmacologic agents (eg, simvastatin and bisphosphonates).[5]

The initial stability of the implant is affected by the design of the implant, the clinician's surgical technique, and the quality of bone at the host site. Implant characteristics have been shown to facilitate osseointegration, including material composition, length, diameter, and surface treatment (eg, roughened surface, hydroxyapatite, calcium phosphate).[9,36–38] Cortical bone usually provides better initial stability than cancellous bone, thus aiding initial osseointegration. Optimal functional loading can also facilitate integration of the implant with bone formation via occlusal forces.[39]

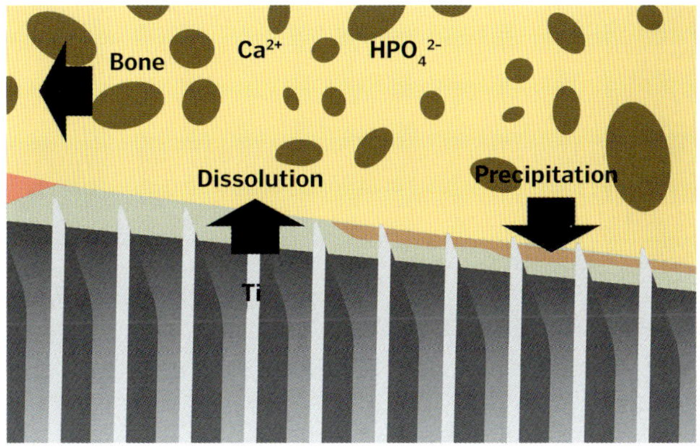

FIG 5-4 An anodized implant surface that incorporates calcium phosphate into the titanium oxide crystal structure improves the healing process, allowing a faster biologic integration of implants.

The surface of the implant can be particularly important in facilitating osseointegration. The interface between an alloplastic material (such as a dental implant) and living bone is a dynamic region of regeneration and remodeling. Titanium oxide, zirconia, and hydroxyapatite are implant surfaces that have been shown to readily establish osseointegration.[40] Titanium resists corrosion and does not damage macrophages or fibroblasts. An oxide layer is formed when the implant is exposed to air, and it can reoxidize if damaged. Calcium and phosphate ions in the oxide layers suggest an active ion interchange at the interface of implant and bone.[41]

Ionic exchanges are facilitated by porous surfaces on implants, helping to establish a chemical connection at the interface of implant and bone (Fig 5-4). Additionally, such surfaces can help bone grow on the implant's surface in three dimensions, leading to greater tensile strength. Most commercial implants have roughened surfaces.[42,43] The surface area exposure to bone growth reduces corrosion, increases load strengths of implants biomechanically, and facilitates bone formation through accelerated osteoblast differentiation. Bone patterns are more organized, and mineralization is higher at the junction of implant and bone.[44,45]

Physical characteristics of the implant's surface can aid osseointegration; for example, the lower the implant's susceptibility to mechanical stress, corrosion, and wear, the higher its likely osseointegration. A roughened surface and ion implantation can prevent wear. Additionally, an implant with a high surface energy and charge from heat and radiofrequency plasma glow discharge treatment can facilitate osseointegration.[46,47] Enhancing the wettability (hydrophilic characteristics) of implants at their interface with blood can also positively affect osseointegration by promoting osteoconduction.[48]

As noted above, a number of other factors affect osseointegration, including the alveolar site where the implant is placed. An irradiated host site or a site in a patient with osteoporosis could adversely affect the integration of bone and implant. Hyperbaric oxygen treatment can be used to help repair a site that has been irradiated for implantation.[49,50] Smoking is a contraindication for optimal osseointegration, as is diabetes; additionally, resorbed alveolar ridges may have to be treated with ridge augmentation via bone grafting or other therapies, such as ridge splitting with osteotomes or guided bone regeneration, to increase bone volume. Additionally, a gentle surgical technique enhances osseointegration, not only by preserving bone and soft tissues but also by keeping bone temperature low through low-speed drilling. High temperatures in bone can reduce its cells' capacity for repair.[10,51]

Yet another essential element of osseointegration concerns the timing and extent of functional load placed on the implant. Once initial stability is obtained, the clinician must prevent any motion in the implant, including micromotion, during early bone healing. Functional overloading of the implant can lead to implant failure due to compromised osseointegration. Careful placement of the provisional restoration can help the wound site heal before functional loading.

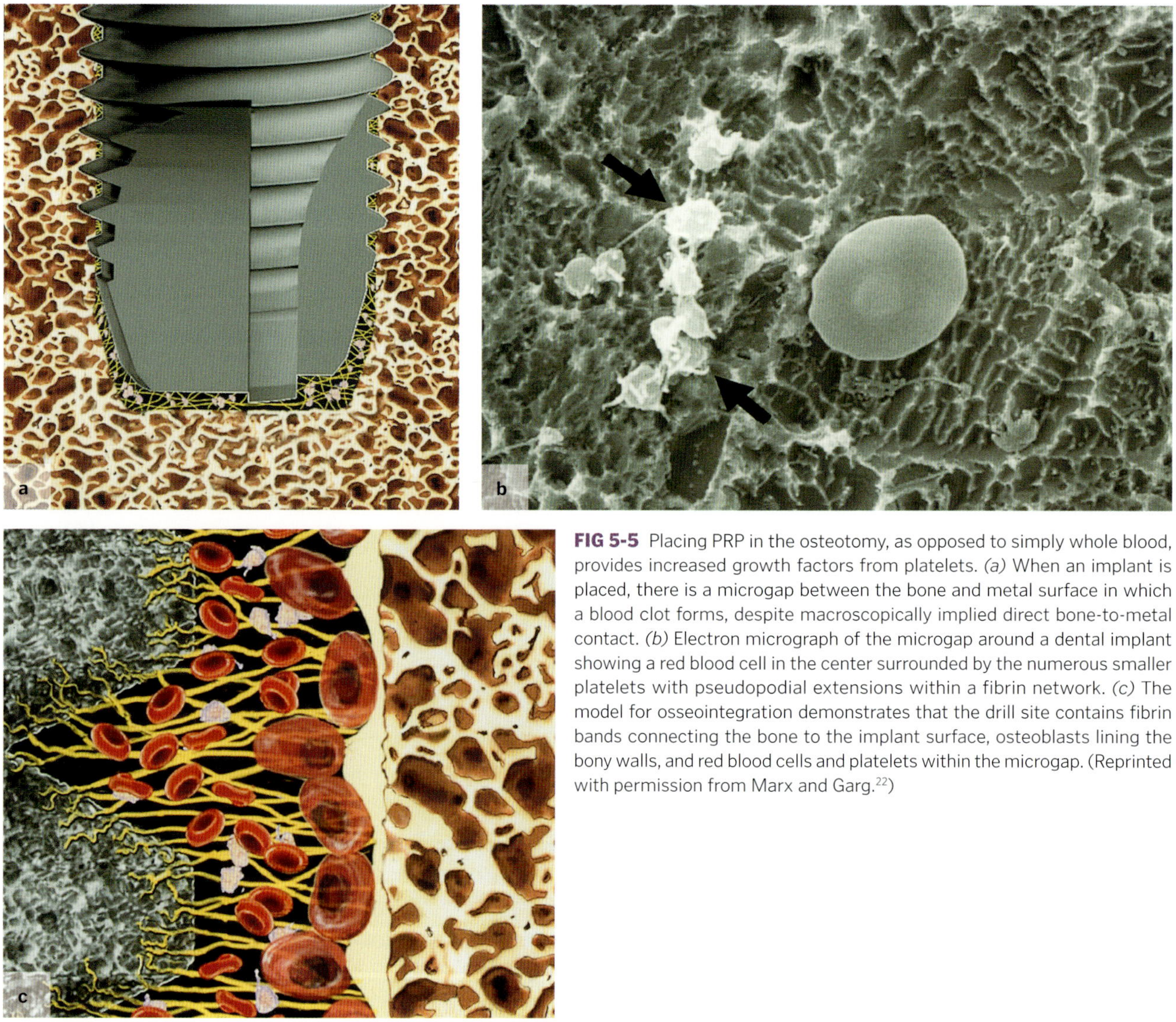

FIG 5-5 Placing PRP in the osteotomy, as opposed to simply whole blood, provides increased growth factors from platelets. *(a)* When an implant is placed, there is a microgap between the bone and metal surface in which a blood clot forms, despite macroscopically implied direct bone-to-metal contact. *(b)* Electron micrograph of the microgap around a dental implant showing a red blood cell in the center surrounded by the numerous smaller platelets with pseudopodial extensions within a fibrin network. *(c)* The model for osseointegration demonstrates that the drill site contains fibrin bands connecting the bone to the implant surface, osteoblasts lining the bony walls, and red blood cells and platelets within the microgap. (Reprinted with permission from Marx and Garg.[22])

Finally, growth factors and differentiation factors have been shown to enhance osseointegration under certain conditions.[52] In addition to PRP growth factors, other biologics in use include bone morphogenetic proteins 2 and 7 (BMP-2, BMP-7), osteogenic protein,[1,53] collagen, fibronectin, and vitronectin.[54–56]

PRP and Dental Implant Osseointegration

Dental implant osseointegration involves cell migration and differentiation as well as bone formation and remodeling, and PRP serves as a natural assistance to these processes.[22,57] Furthermore, PRP is often used to enhance wound healing for osseointegration of dental implants in compromised sites, including in patients with osteoporosis and diabetes. Healing in the posterior maxilla also benefits from PRP assistance. Placing PRP in the site prepared by drilling encases the surface of the implant while bridging the microgap (approximately 25 μm) between the implant and the native bone and adds molecular adhesives (fibrin, fibronectin, and vitronectin) to the mix of cells in the wound (ie, red and white blood cells and platelets). Via granulation, the platelets secrete growth factors, which induce proliferation and migration of osteoblasts and marrow stem cells across the molecularly adhesion-filled microgap (Fig 5-5). By this means, the marrow cells actually "knit" the fibrin strands from

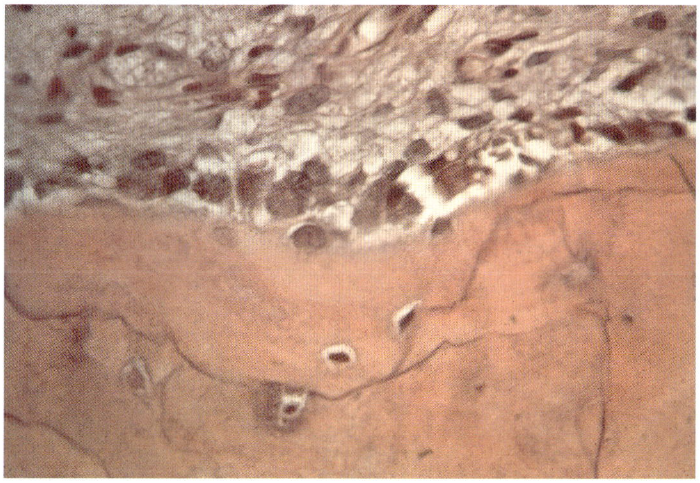

FIG 5-6 The cement line that connects the implant surface to bone is identical to the reversal or resting lines of von Ebner, which are commonly seen in normal bone histology and function physiologically to anchor new bone to old bone and prevent their separation. (Reprinted with permission from Marx and Garg.[22])

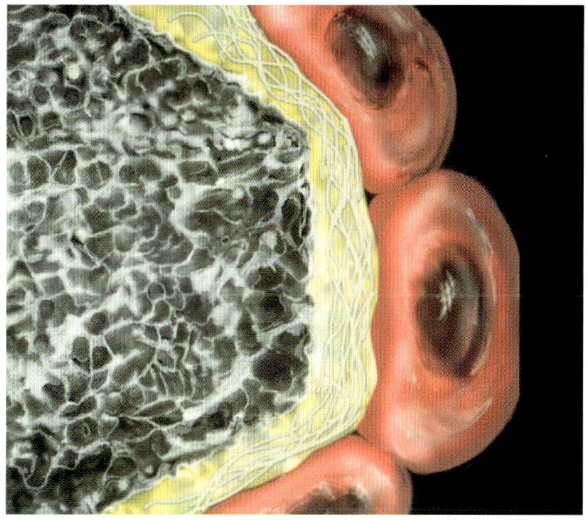

FIG 5-7 The cementing substance that is secreted onto the metal surface is not true bone but the reversal line of von Ebner. This cement is composed of osteopontin and sialoprotein in which crystals of calcium hypophosphate form as the mineral rather than calcium hydroxyapatite. (Reprinted with permission from Marx and Garg.[22])

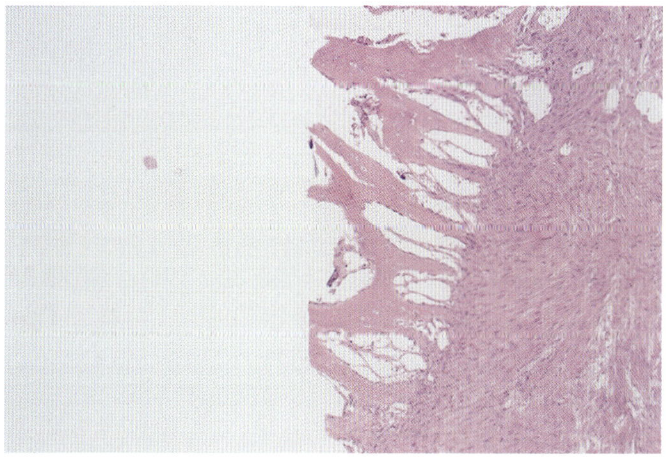

FIG 5-8 Histology of osseointegration confirms the model shown in Fig 5-2. Note the stem cells to the right forming projections of osteoid toward the now-vacated implant surface on the left. There is osteoblastic rimming on the osteoid projections and encasement of osteoblasts into true osteocytes. Note the remnants of the fibrin bands between the osteoid projections. (Reprinted with permission from Marx and Garg.[22])

the surface of the implant to the bone (especially if it is rough-surfaced and not simply machined) as the stem cells simultaneously differentiate and create osteoid.[21]

Osseointegration, like the remodeling of new bone, requires a biologic bone cement to seal the new bone to the surface of the (ideally) rough-surfaced implant.[58] Platelets play a key role because their growth factors induce cells not only to proliferate, migrate, and differentiate but also to help produce a cement-osteoid secretion comprised of sialoprotein, osteopontin, and calcium hypophosphate crystals (Fig 5-6). This substance knits the bone of the drill site (type 1 collagen, calcium hydroxyapatite crystals, BMP, and insulinlike growth factor) to the rough surface of the implant, leading to what we know as osseointegration. The hypophosphate crystals harden the cement, allowing the osteoblast to weave the collagen fibers in its osteoid to knit itself to the cement substance, binding the new bone to the cement (Fig 5-7). In short, the cement (5-μm-thick) joins (osseointegrates) the implant surface to the bone (Fig 5-8).

An autogenous product itself, PRP was naturally assumed to work best with autogenous bone and tissues as well as grafts, yet it has also been shown to be effective in wound healing with bone substitutes for grafts.[59,60] Researchers learned that while autogenous bone from a donor site transplants osteoprogenitor cells to the recipient site, the bone substitute graft acts as a conduit, or conductor, of those migrating cells into the substitute bone graft's

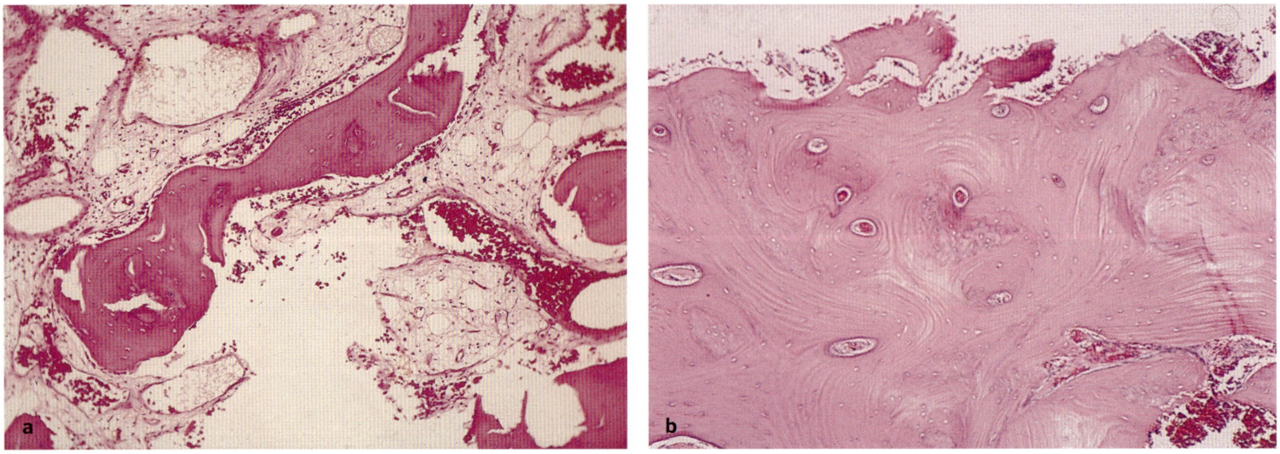

FIG 5-9 Histomorphometry shows that PRP-enhanced grafts exhibit increased density of bone as well as an advanced rate of maturity. *(a)* Histomorphometry of a native posterior mandible serves as a baseline. It contains 38% trabecular bone area consisting of mature bone within a fibrovascular stroma. *(b)* In contrast to *a*, histomorphometry of an autogenous bone graft enhanced with PRP at 4 months shows that the graft has an 80% bone density, consists almost entirely of mature bone with lamellar architecture, and features mature Haversian systems with limited bone remodeling. (Reprinted with permission from Marx and Garg.[22])

"nooks and crannies." Just as new bone bonds to a dental implant (which itself is a kind of bone substitute) via osseointegration, so does new bone bond to substitute bone grafts via osteoconduction, particularly when fibrin is readily absorbed by the substitute material and when the material's size, particular arrangement, and physical configuration (eg, porosity) lend themselves to the interstitial atmosphere conducive for osteoconduction. Though the sheer volume of osteoprogenitor cells transplanted by autogenous bone grafts is far superior to those conducted by bone substitutes (Fig 5-9), the latter can be a desirable grafting material depending on a number of conditions. As a result, PRP's promotion of cell migration makes it a valuable upregulator of osteoprogenitor cells when bone substitutes are used for grafting. In reality, no type of bone other than autogenous bone or freeze-dried bone allograft is actually osteoconductive in humans.[61]

Osteoprogenitor cells from a host site are necessary for bone grafts to provide osteoconduction for new bone. For example, bone graft particles used in oral surgery can be placed in a blood clot, accompanied by fibrin, fibronectin, vitronectin, red and white blood cells, and platelets. Applying PRP to the mixture multiplies platelet presence by 400% to 700%; then, after degranulation, platelets release their abundant growth factors, promoting capillary ingrowth in the graft site's native bone as well as inducing cell migration and differentiation. This growth factor cascade leads to bone production similar to that involved in osseointegration and osteoconduction. A fibrin network is established, knitting (cementing) the distance between and among native bone and bone substitute particles via osteoprogenitor cell migration, similar to the way osseointegration between a drill site and dental implants with hydroxyapatite coating is facilitated. While autogenous bone grafts work much faster to form more new bone in the graft site, PRP helps to narrow this time and volume gap when any bone substitutes are used as grafts because PRP provides the growth factors and resultant migration and differentiation of cells necessary to cement the spaces between native bone and substitute bone.[62–65]

Conclusion

Dental implant success is measured by implant osseointegration—the implant's functional anchorage in alveolar bone. Many factors influence osseointegration, including dental implant treatment planning, the condition of the host bone, the topography and biocompatibility of the implant material, the overall medical condition of the patient, and the timing and extent of functional load placed on the implant. Because the bonding of implant with bone involves the same wound healing processes seen in injured or damaged bone, biologic enhancements—including PRP—have been used to boost osseointegration. PRP is an important adjunct in helping completely and partially edentulous patients regain dental function.

References

1. White GS. Treatment of the edentulous patient. Oral Maxillofac Surg Clin North Am 2015;27:265–272.
2. Boyce RA, Klemons G. Treatment planning for restorative implantology. Dent Clin North Am 2015;59:291–304.
3. Moraschini V, Velloso G, Luz D, Porto Barboza E. Implant survival rates, marginal bone level changes, and complications in full-mouth rehabilitation with flapless computer-guided surgery: A systematic review and meta-analysis. Int J Oral Maxillofac Surg 2015;44:892–901.
4. Heitz-Mayfield LJ, Mombelli A. The therapy of peri-implantitis: A systematic review. Int J Oral Maxillofac Implants 2014;29(suppl):325–345.
5. Mavrogenis AF, Dimitriou R, Parvizi J, Babis GC. Biology of implant osseointegration. J Musculoskelet Neuronal Interact 2009;9:61–71.
6. Wróbel E, Witkowska-Zimny M, Przybylski J. Biological mechanisms of implant osseointegration. Orthop Traumatol Rehabil 2010;12:401–409.
7. Parithimarkalaignan S, Padmanabhan TV. Osseointegration: An update. J Indian Prosthodont Soc 2013;13:2–6.
8. Albrektsson T, Brånemark PI, Hansson HA, Lindström J. Osseointegrated titanium implants. Requirements for ensuring a long-lasting, direct bone-to-implant anchorage in man. Acta Orthop Scand 1981;52:155–170.
9. Ogle OE. Implant surface material, design, and osseointegration. Dent Clin North Am 2015;59:505–520.
10. Möhlhenrich SC, Modabber A, Steiner T, Mitchell DA, Hölzle F. Heat generation and drill wear during dental implant site preparation: Systematic review. Br J Oral Maxillofac Surg 2015;53:679–689.
11. Heinemann F, Hasan I, Bourauel C, Biffar R, Mundt T. Bone stability around dental implants: Treatment-related factors. Ann Anat 2015;199:3–8.
12. Jayesh RS, Dhinakarsamy V. Osseointegration. J Pharm Bioallied Sci 2015;7(suppl 1):S226–S229.
13. Lewis RC, Harris BT, Sarno R, Morton D, Llop DR, Lin WS. Maxillary and mandibular immediately loaded implant supported interim complete fixed dental prostheses on immediately placed dental implants with a digital approach: A clinical report. J Prosthet Dent 2015;114:315–322.
14. Proussaefs P, Proussaefs P. Use of CAD/CAM healing abutment immediately after dental implant placement for the non-esthetic zone. A guided soft tissue healing technique. J Oral Implantol 2016;42:189–193.
15. Franchi M, Fini M, Martini D, et al. Biological fixation of endosseous implants. Micron 2005;36:665–671.
16. Okamatsu K, Kido H, Sato A, Watazu A, Matsuura M. Ultrastructure of the interface between titanium and surrounding tissue in rat tibiae—A comparison study on titanium-coated and -uncoated plastic implants. Clin Implant Dent Relat Res 2007;9:100–111.
17. Albrektsson T, Jansson T, Lekholm U. Osseointegrated dental implants. Dent Clin North Am 1986;30:151–174.
18. Wataha JC. Materials for endosseous dental implants. J Oral Rehabil 1996;23:79–90.
19. Gupta RK, Padmanabhan TV. Resonance frequency analysis. Indian J Dent Res 2011;22:567–573.
20. Cooper LF. Biologic determinants of bone formation for osseointegration: Clues for future clinical improvements. J Prosthet Dent 1998;80:439–449.
21. Puleo DA, Nanci A. Understanding and controlling the bone-implant interface. Biomaterials 1999;20:2311–2321.
22. Marx RE, Garg A. Dental and Craniofacial Applications of Platelet-Rich Plasma. Chicago: Quintessence, 2005.
23. Colombo JS, Satoshi S, Okazaki J, Crean SJ, Sloan AJ, Waddington RJ. In vivo monitoring of the bone healing process around different titanium alloy implant surfaces placed into fresh extraction sockets. J Dent 2012;40:338–346.
24. Stoppie N, Van Oosterwyck H, Jansen J, Wolke J, Wevers M, Naert I. The influence of Young's modulus of loaded implants on bone remodeling: An experimental and numerical study in the goat knee. J Biomed Mater Res A 2009;90:792–803.
25. Craig RG, LeGeros RZ. Early events associated with periodontal connective tissue attachment formation on titanium and hydroxyapatite surfaces. J Biomed Mater Res 1999;47:585–594.
26. Yamanishi Y, Yamaguchi S, Imazato S, Nakano T, Yatani H. Influences of implant neck design and implant-abutment joint type on peri-implant bone stress and abutment micromovement: Three-dimensional finite element analysis. Dent Mater 2012;28:1126–1133.
27. Freitas AC Jr, Bonfante EA, Giro G, Janal MN, Coelho PG. The effect of implant design on insertion torque and immediate micromotion. Clin Oral Implants Res 2012;23:113–118.
28. Baril E, Lefebvre LP, Hacking SA. Direct visualization and quantification of bone growth into porous titanium implants using micro-computed tomography. J Mater Sci Mater Med 2011;22:1321–1332.
29. Zen Filho EV, Tolentino ES, Silva Santos PS. Viability of dental implants in head and neck irradiated patients: A systematic review. Head Neck 2016;38(suppl 1):E2229–E2240.
30. Sakakura CE, Marcantonio E Jr, Wenzel A, Scaf G. Influence of cyclosporine A on quality of bone around integrated dental implants: A radiographic study in rabbits. Clin Oral Implants Res 2007;18:34–39.
31. Calciolari E, Donos N, Park JC, Petrie A, Mardas N. A systematic review on the correlation between skeletal and jawbone mineral density in osteoporotic subjects. Clin Oral Implants Res 2016;27:433–442.
32. Clementini M, Rossetti PH, Penarrocha D, Micarelli C, Bonachela WC, Canullo L. Systemic risk factors for peri-implant bone loss: A systematic review and meta-analysis. Int J Oral Maxillofac Surg 2014;43:323–334.
33. Bacci C, Berengo M, Favero L, Zanon E. Safety of dental implant surgery in patients undergoing anticoagulation therapy: A prospective case-control study. Clin Oral Implants Res 2011;22:151–156.
34. Chrcanovic BR, Albrektsson T, Wennerberg A. Smoking and dental implants: A systematic review and meta-analysis. J Dent. 2015;43:487–498.

35. Barfeie A, Wilson J, Rees J. Implant surface characteristics and their effect on osseointegration. Br Dent J 2015;218:E9.
36. Shibata Y, Tanimoto Y. A review of improved fixation methods for dental implants. Part I: Surface optimization for rapid osseointegration. J Prosthodont Res 2015;59:20–33.
37. Feller L, Chandran R, Khammissa RA, et al. Osseointegration: Biological events in relation to characteristics of the implant surface. SADJ 2014;69:112,114–117.
38. Lin D, Li Q, Li W, Swain M. Dental implant induced bone remodeling and associated algorithms. J Mech Behav Biomed Mater 2009;2:410–432.
39. Frydman A, Simonian K. Review of models for titanium as a foreign body. J Calif Dent Assoc 2014;42:829–833.
40. Suba C, Velich N, Turi C, Szabó G. Surface analysis methods of biomaterials used in oral surgery: Literature review. J Craniofac Surg 2005;16:31–36.
41. Aljateeli M, Wang HL. Implant microdesigns and their impact on osseointegration. Implant Dent 2013;22:127–132.
42. Altman EI, Schwarz UD. Mechanisms, kinetics, and dynamics of oxidation and reactions on oxide surfaces investigated by scanning probe microscopy. Adv Mater 2010;22:2854–2869.
43. Xuereb M, Camilleri J, Attard NJ. Systematic review of current dental implant coating materials and novel coating techniques. Int J Prosthodont 2015;28:51–59.
44. Dunne CF, Gibbons J, FitzPatrick DP, Mulhall KJ, Stanton KT. On the fate of particles liberated from hydroxyapatite coatings in vivo. Ir J Med Sci 2015;184:125–133.
45. De Maeztu MA, Braceras I, Álava JI, Recio C, Piñera M, Gay-Escoda C. Human study of ion implantation as a surface treatment for dental implants. Int J Oral Maxillofac Surg 2013;42:891–896.
46. Rapuano BE, Singh H, Boskey AL, Doty SB, MacDonald DE. Heat and radiofrequency plasma glow discharge pretreatment of a titanium alloy: Evidence [corrected] for enhanced osteoinductive properties. J Cell Biochem 2013;114:1917–1927.
47. Gittens RA, Scheideler L, Rupp F, et al. A review on the wettability of dental implant surfaces II: Biological and clinical aspects. Acta Biomater 2014;10:2907–2918.
48. MacInnes A, Lamont T. Radiotherapy associated with higher rates of dental implant loss. Evid Based Dent 2014;15:27–28.
49. Chambrone L, Mandia J Jr, Shibli JA, Romito GA, Abrahao M. Dental implants installed in irradiated jaws: A systematic review. J Dent Res 2013;92(12 suppl):119S–130S.
50. Mishra SK, Chowdhary R. Heat generated by dental implant drills during osteotomy—A review: Heat generated by dental implant drills. J Indian Prosthodont Soc 2014;14:131–143.
51. Wikesjö UM, Qahash M, Huang YH, Xiropaidis A, Polimeni G, Susin C. Bone morphogenetic proteins for periodontal and alveolar indications; Biological observations, clinical implications. Orthod Craniofac Res 2009;12:263–270.
52. Baas J. Adjuvant therapies of bone graft around non-cemented experimental orthopedic implants: Stereological methods and experiments in dogs. Acta Orthop Suppl 2008;79:1–43.
53. Rammelt S, Heck C, Bernhardt R, et al. In vivo effects of coating loaded and unloaded Ti implants with collagen, chondroitin sulfate, and hydroxyapatite in the sheep tibia. J Orthop Res 2007;25:1052–1061.
54. Ku Y, Chung CP, Jang JH. The effect of the surface modification of titanium using a recombinant fragment of fibronectin and vitronectin on cell behavior. Biomaterials 2005;26:5153–5157.
55. Zeifang F, Grunze M, Delling G, et al. Improved osseointegration of PTFEP-coated titanium implants. Med Sci Monit 2008;14:BR35–BR40.
56. Roffi A, Filardo G, Kon E, Marcacci M. Does PRP enhance bone integration with grafts, graft substitutes, or implants? A systematic review. BMC Musculoskelet Disord 2013;14:330.
57. Kundu R, Rathee M. Effect of platelet-rich-plasma (PRP) and implant surface topography on implant stability and bone. J Clin Diagn Res 2014;8:ZC26–ZC30.
58. Davies JE. In vitro modeling of the bone/implant interface. Anat Rec 1996;245:426–445.
59. Aghaloo TL, Moy PK. Which hard tissue augmentation techniques are the most successful in furnishing bony support for implant placement? Int J Oral Maxillofac Implants 2007;22(suppl):49–70 [erratum 2008;23:56].
60. Aichelmann-Reidy ME, Reynolds MA. Predictability of clinical outcomes following regenerative therapy in intrabony defects. J Periodontol 2008;79:387–393.
61. Garg AK. Grafting materials in repair and restoration. In: Lynch SE, Genco RJ, Marx RE (eds). Tissue Engineering: Applications in Maxillofacial Surgery and Periodontics. Chicago: Quintessence, 1999:83–101.
62. Öncü E, Alaaddinoglu EE. The effect of platelet-rich fibrin on implant stability. Int J Oral Maxillofac Implants 2015;30:578–582.
63. Simonpieri A, Del Corso M, Vervelle A, et al. Current knowledge and perspectives for the use of platelet-rich plasma (PRP) and platelet-rich fibrin (PRF) in oral and maxillofacial surgery, part 2: Bone graft, implant and reconstructive surgery. Curr Pharm Biotechnol 2012;13:1231–1256.
64. Marrelli M, Tatullo M. Influence of PRF in the healing of bone and gingival tissues. Clinical and histological evaluations. Eur Rev Med Pharmacol Sci 2013;17:1958–1962.
65. Boora P, Rathee M, Bhoria M. Effect of platelet rich fibrin (PRF) on peri-implant soft tissue and crestal bone in one-stage implant placement: A randomized controlled trial. J Clin Diagn Res 2015;9:ZC18–ZC21.

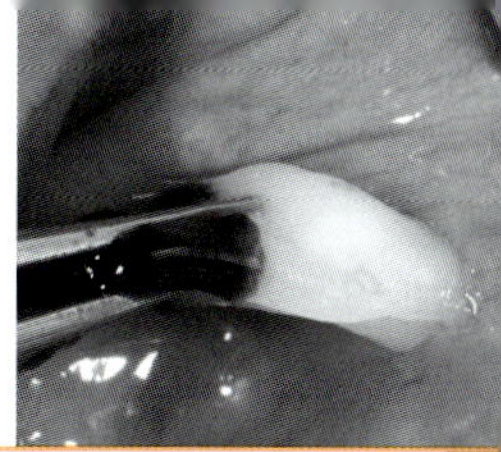

6

Oral Cavity Soft Tissue Healing and Autologous Blood Concentrates

Soft tissue healing has become the focus of a number of medical disciplines in the last 20 years in fields such as plastic and reconstructive surgery, maxillofacial surgery, dermatology, cosmetic surgery, periodontics, sports medicine, orthopedics, and the treatment of chronic wounds. Significant developments in biologics and other growth factor–driven clinical practices have led to improvements across the entire wound healing landscape. As noted in previous chapters, PRP is often the biologic of choice to enhance soft tissue healing in a number of medical disciplines. The focus of this chapter is soft tissue healing in oral and maxillofacial surgery, periodontology, and implantology, and specifically on how PRP can be used to enhance such healing.[1-4]

Implant Surgery, Soft Tissue Flaps, and PRP

Soft tissue flaps that are raised to place dental implants heal best when they have an uninterrupted blood supply and the vascular connective tissue surrounding the bone readily promotes osteogenesis.[5,6] However, many implant patients present with one or more common conditions that can impede such healing, such as persistent inflammation from chronic periodontitis, tooth loss with mucoperiosteal flap reflection, abscesses, apicoectomies, multiple periodontal surgeries, and failed grafting procedures. The use of PRP can significantly improve healing under such conditions. PRP is also recommended to counter the impairment of soft tissue healing resulting from tobacco use, steroids, radiation therapy, diabetes, peripheral vascular disease, systemic lupus erythematosus, and advanced age.[7-9]

The usual method for placing dental implants is a crestal incision to elevate a full-thickness mucoperiosteal flap.[10,11] Other methods include elevating flaps from buccal or palatal incisions or using a vascularized periosteal-connective tissue flap. Because a high proportion of patients have conditions that require PRP enhancement when implants are placed, PRP is recommended regardless of the flap technique used for placement.[1,12,13]

When a flap is reflected and subsequently sutured, it contracts, making it challenging for the surgeon to maintain tension-free closure during healing. Primary closure of the flap by flap advancement eversion can be accomplished by the use of an incision to undermine the mucosa of the periosteum superficially to muscle attachments. Separating the flap from attachments to facial muscles in this way prevents microseparations during closure as well as consequent leakage of saliva when the lips and cheeks move.[12]

PRP liquid can be applied to the flap's undersurface or over bone and implant before closure to obtain a suture line seal that will prevent leakage (Fig 6-1). Alternatively, PRP membranes can be placed over the implant or cover screw, or between the implant and the gingival healing abutment for the same purpose (Fig 6-2). PRP's growth factors establish flap stabilization through molecular adhesion. Over time, the growth factors increase the rate of vascular ingrowth for osteoinduction to prevent loss of crestal bone. During stage-two implant surgery, PRP can also be applied to flaps, bone, and implants.

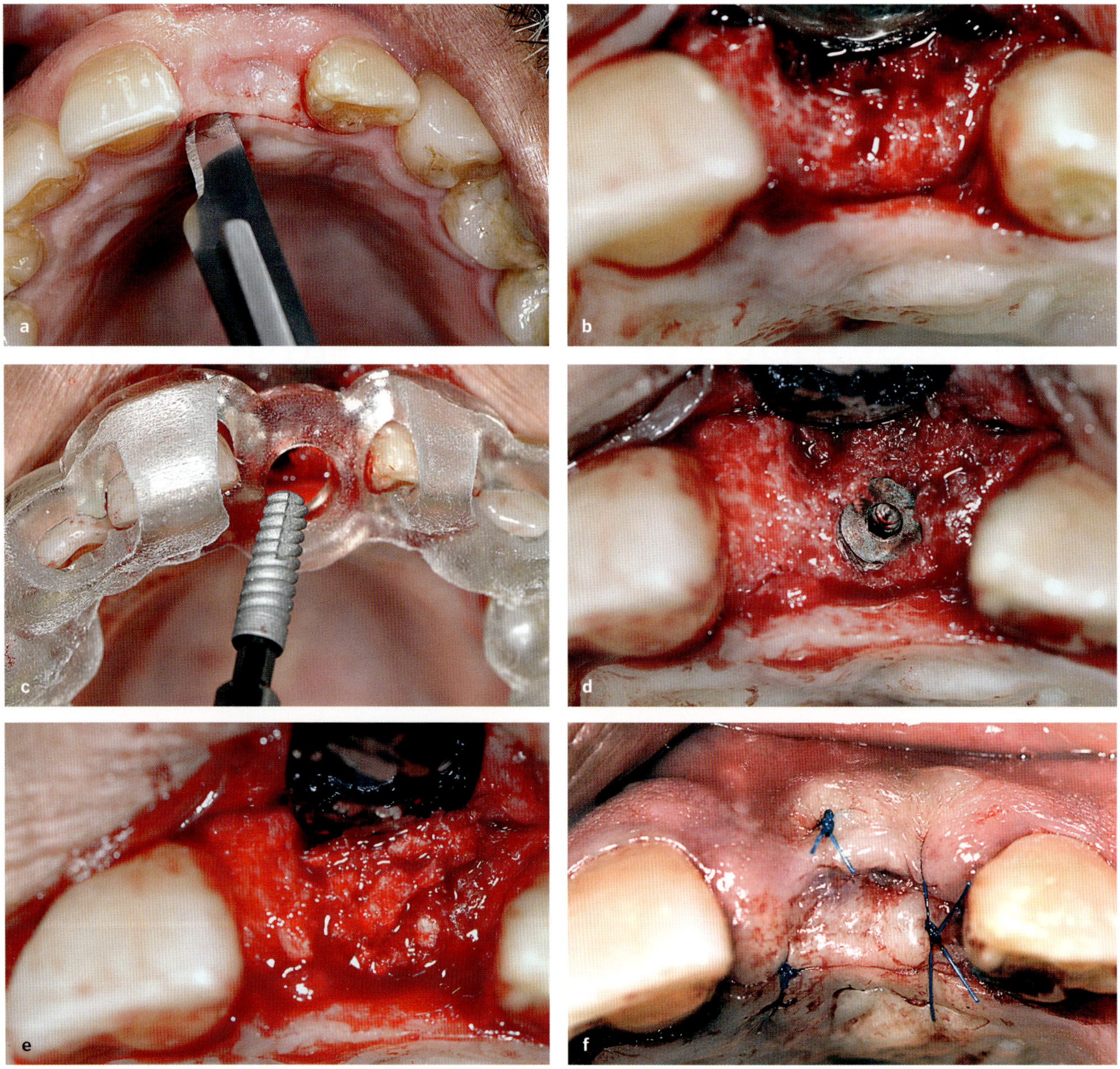

FIG 6-1 Implant placement in the esthetic zone. *(a)* Initial flap for implant placement in an area of limited bone and osteoid following tooth extraction and healing. *(b)* Appearance of the ridge after flap reflection. *(c)* Placement of implant. *(d)* Implant in situ. *(e)* PRP liquid and bone chips are placed over the cover screw and buccal aspect of the ridge due to immaturity of the osteoid and to help protect the site from microleakage at the incision line. *(f)* Final suturing.

Gingival Recession and PRP

Whether located in a specific area of the gingiva or generally in the oral cavity, gingival recession can lead to problems related to esthetics and/or function.[14,15] Traditional root coverage procedures involve the use of flaps (eg, laterally sliding and double papilla), grafts (eg, subepithelial connective tissue, coronally advanced, and free gingival), and guided tissue regeneration techniques.[16,17]

More recently, innovative treatments have been developed for root coverage, including the use of an acellular dermal matrix as graft material, enamel matrix derivative, and recombinant human growth factors, including PRP.[18] Minimally invasive techniques such as the pinhole surgical

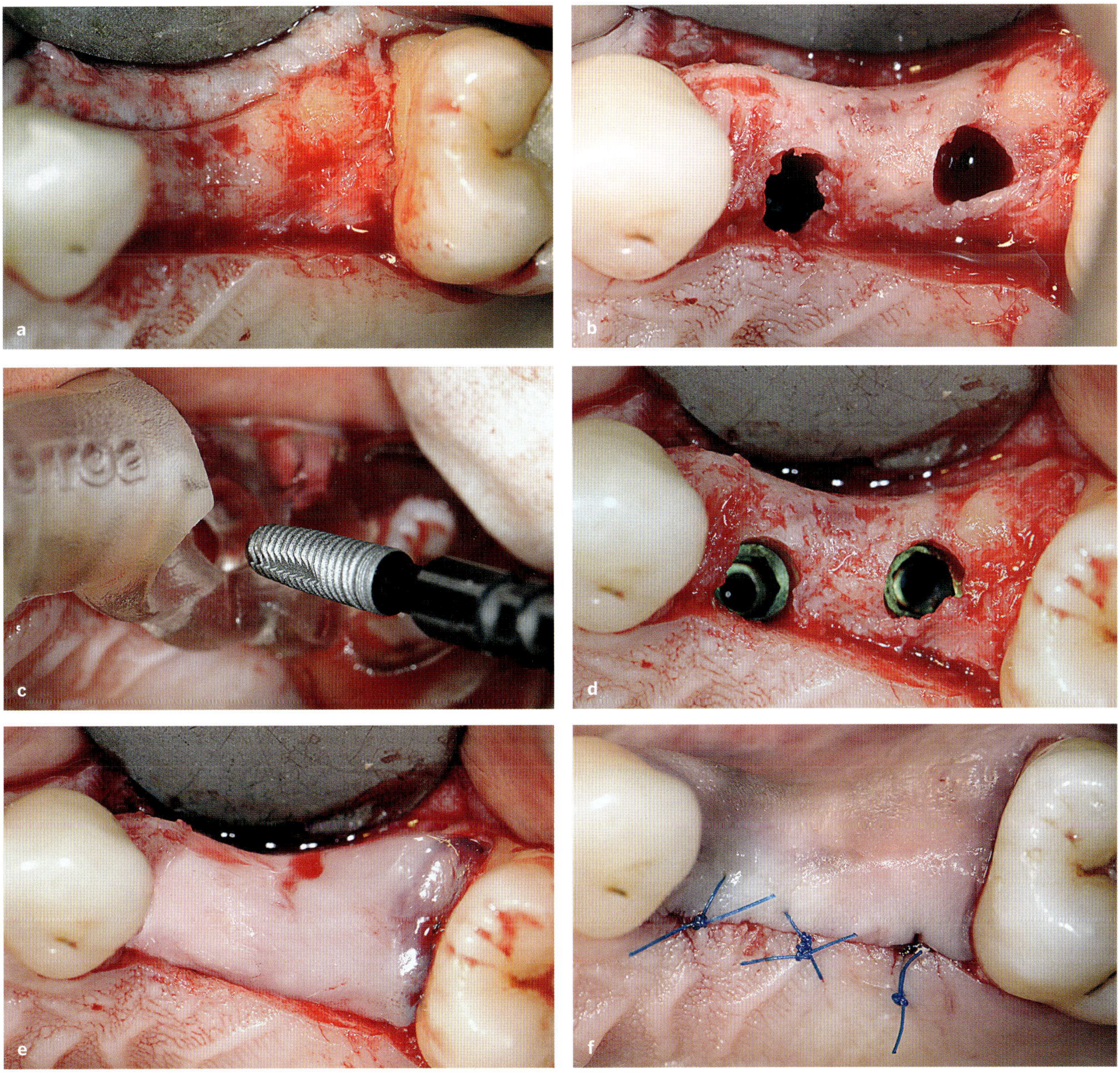

FIG 6-2 Implant placement in the posterior mandible. *(a)* Flap reflection. *(b)* Implant osteotomies. *(c)* Implant placement. *(d)* Implants in situ prior to placement of cover screws. *(e)* PRP membrane placement prior to closure in order to minimize microleakage from the incision line to the implant site. *(f)* Primary closure of the flap.

technique and FASTAB are also being performed successfully and becoming increasingly popularized.

Free gingival grafts

Free gingival grafting, a relatively minor but essential oral soft tissue procedure, is often used to cover an exposed root or implant and to enhance esthetics. It is most commonly used, however, to restore attached tissue after gingival recession or a high mucogingival attachment (Fig 6-3).[12]

The graft, consisting of keratinized mucosa, is harvested from the lateral palatal shelf opposite molars and premolars. Because of the abundant vascular and nerve tissue

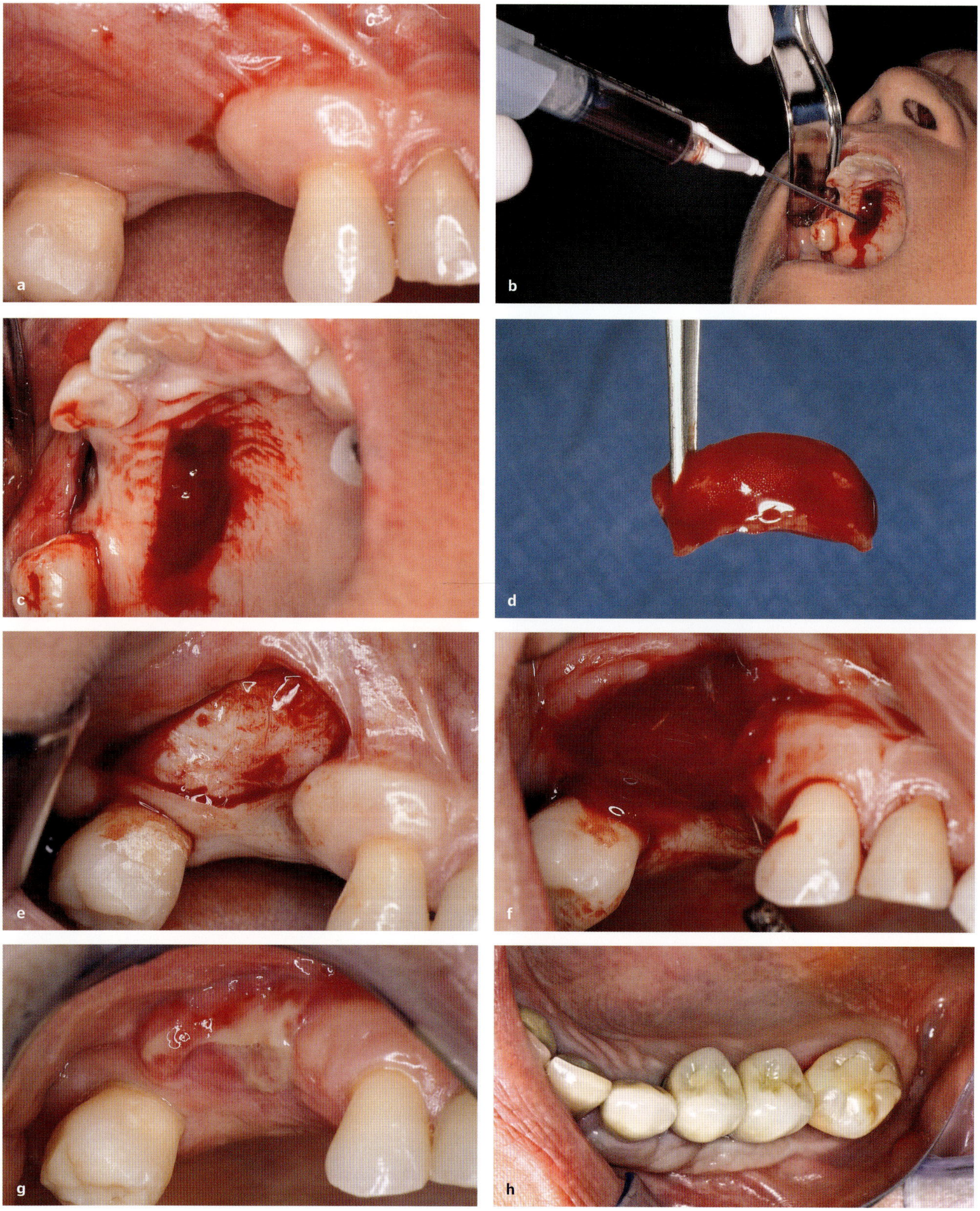

FIG 6-3 Free gingival graft. *(a)* Missing attached gingiva in an edentulous area can be restored with an autologous soft tissue graft of keratinized mucosa. *(b)* Keratinized mucosa is usually obtained from the palate. PRP is applied to control the bleeding that can result from this tissue harvest as well as to enhance its healing. *(c)* Coating the donor area with PRP helps achieve hemostasis and enhances healing time and comfort. *(d)* Before placement, the gingival tissue graft is coated with PRP on the internal surface. Because PRP growth factors are released quickly, the graft should be coated just prior to placement. *(e)* Once the recipient site is prepared to the periosteal level, a coating of PRP can be applied. *(f)* After the free gingival graft is placed, another coating of PRP is applied. *(g)* An initial graft acceptance is revealed at suture removal 7 days later. Note the correction of the mucogingival banding. *(h)* The matured gingival graft developed a normal-appearing attached gingiva. (Reprinted with permission from Marx and Garg.[12])

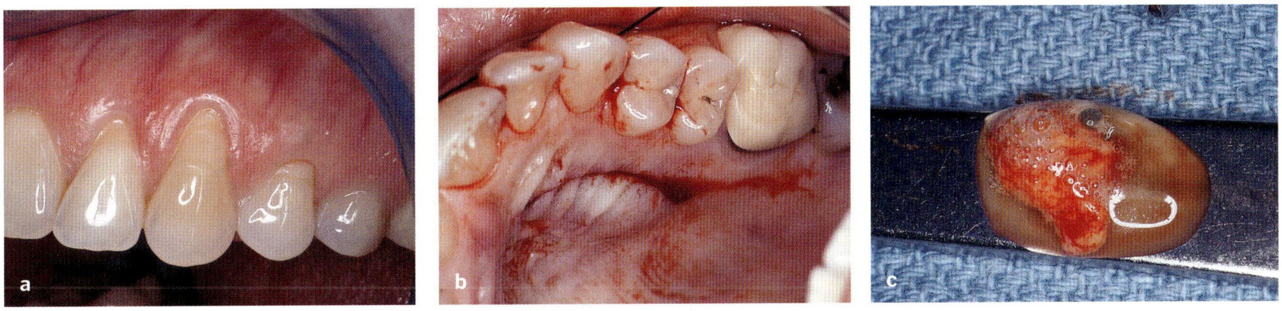

FIG 6-4 Connective tissue harvest and graft for root coverage. *(a)* Abrasive brushing habits can cause recession of the vulnerable thin mucosa over the canine prominence. *(b)* A pocket technique is used to remove connective tissue from under the surface epithelium, between the alveolar and palatal processes, leaving them intact for suturing and avoiding a denuded wound. *(c)* The palatal connective tissue is covered with and stored in PRP liquid just before placement at the recipient site. (Reprinted with permission from Marx and Garg.[12])

in this area of the palate, platelet-poor plasma (PPP) or PRP is often used to control bleeding and to enhance healing after the graft is harvested.[19] Electrocautery and suturing a small compression bolster can also help control donor site bleeding. Activated PRP can also be used for immersion of the full-thickness graft during preparation of the recipient site so that growth factors can attach to the graft's cell membranes and then cover the graft's collagen fibrils with cell adhesion molecules to facilitate graft success via recipient site adhesion and capillary and connective tissue growth.[3,20]

Connective tissue grafts

Connective tissue grafting has been performed in esthetic dentistry for root coverage or to thicken and shape the gingiva surrounding a tooth or implant-retained crown for a more natural-looking interface of tooth and gingiva (Fig 6-4).[12] Generally, the harvest site for the connective tissue graft is the same lateral area of the palate accessed for the free gingival graft, but the primary closure of the donor site incision for the connective tissue graft prevents the kind of bleeding and pain normally associated with a free gingival graft. As a result, application of PRP at the harvest site is unnecessary but will certainly assist in healing if used; like the free gingival graft, the harvested connective tissue graft should be immersed in PRP during recipient site preparation to help ensure graft success mediated by the growth factors' biologic action in the graft before and after attachment to the recipient site.[3,20]

To remove proteinaceous deposits from the root surface, the clinician should apply a saturated solution of citric acid, tetracycline hydrochloride, or ethylenediaminetetraacetic acid (EDTA). These solutions also etch the dentin, which opens the dentinal tubules to promote graft ingrowth as well as sound fibrous attachment to the root. The root surface should then be coated with activated PRP.

To gain space for the graft, the mucosa is elevated from the periosteum using an interproximal sulcular incision and dissection, sparing the papilla. Next, 3-0 silk suture material is used to pull the connective tissue graft through the tunnel created by the incision. The mucosa is then repositioned coronally and sutured to the palatal mucosa, resulting in a stable, integrated graft. The gingival contour is thus improved esthetically while the exposed root is covered. The application of PRP provides optimal revascularization and reduces graft shrinkage. The clinician should note that PRP enhances tissue repair and wound healing through angiogenesis and collagen formation only when graft trauma and mobility are controlled through sound surgical principles.[12]

PRP membranes serve as an effective substitute for connective tissue grafts in the treatment of gingival recession defects. In addition to avoiding the painful surgery required to harvest connective tissue from the palate, PRP membranes promote earlier blood vessel and tissue formation compared with connective tissue grafts.

PRP membranes also can be used in place of connective tissue or collagen grafts in minimally invasive root coverage procedures such as the vestibular incision

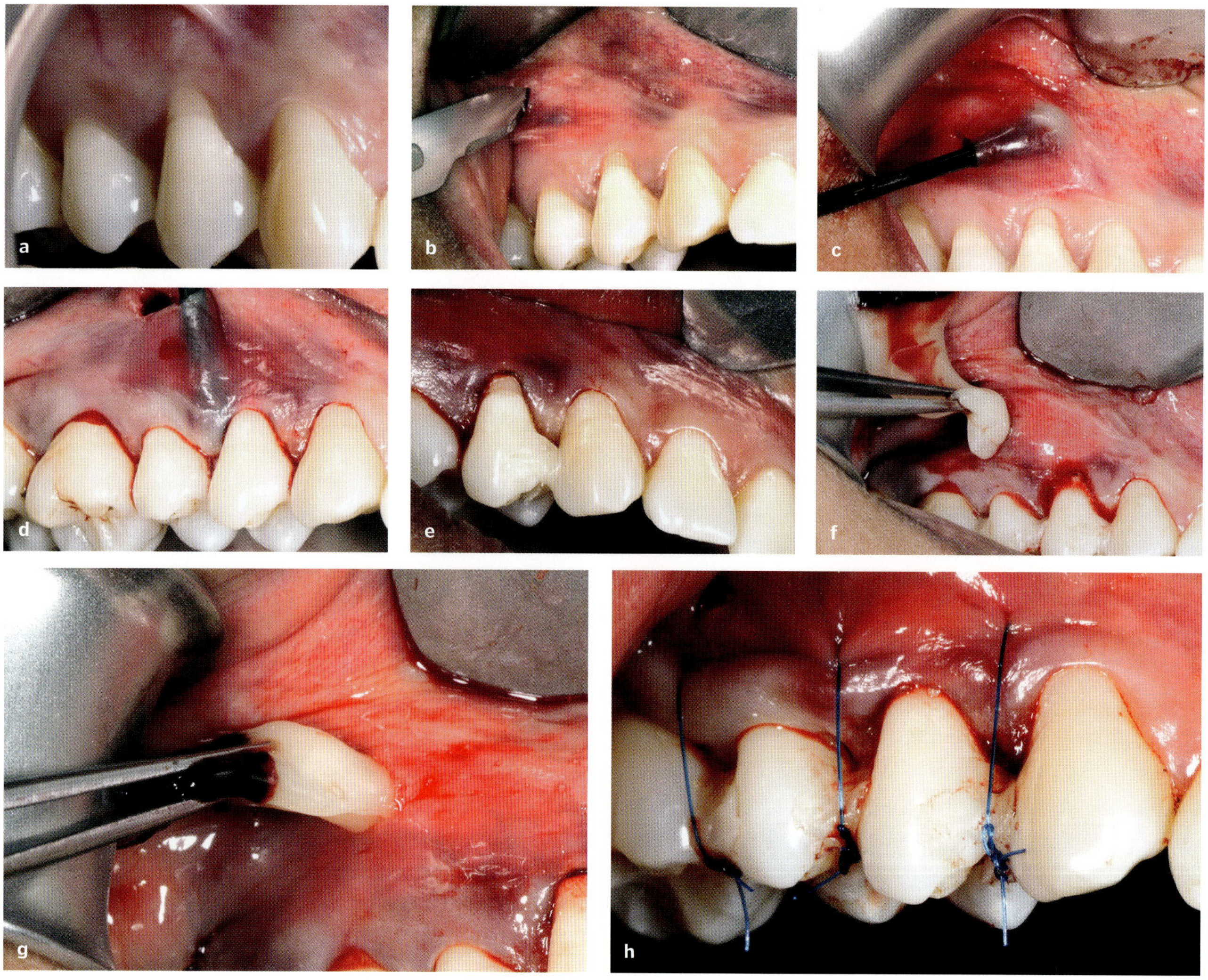

FIG 6-5 Minimally invasive technique with PRP membrane for gingival recession. *(a)* Preoperative view of gingival recession at the maxillary right first premolar. *(b)* Small incision created remotely in the mucosa. *(c)* Split-thickness dissection of the mucosa surrounding and overlying the area of gingival recession. *(d)* Dissection with a full-thickness flap in the attached gingiva surrounding and overlying the recession as well as the papillary areas mesial and distal to it. *(e)* A drop of composite is placed in each contact area (without etching the tooth) to suspend the sutures. *(f)* Placement of PRP membranes into the tunnel created after prophylaxis of the tooth and etching and rinsing of the tooth surface. *(g)* Placement of additional PRP membranes into the tunnel and areas surrounding the gingival recession. *(h)* Suspension of sutures placed to hold the gingiva in its more coronal position during the first 3 weeks of healing.

subperiosteal tunnel access (VISTA) technique (Fig 6-5). The PRP membrane eliminates the need for a donor site, the growth factors released from the PRP membrane stimulate faster revascularization and accelerate healing, and the minimally invasive approach requires no surface incisions and preserves the papilla.

With this approach, a small access hole is made in a remote area of the unattached gingiva, and a subperiosteal tunnel is created into the areas surrounding the gingival defect as well as the papillary areas mesial and distal to it. PRP membranes are then introduced to cover the root dehiscence and bulk up the soft tissue. Finally, the mucogingival complex is advanced coronally and stabilized in the new position with a horizontal mattress suture in the gingival margin and/or anchored to each tooth.

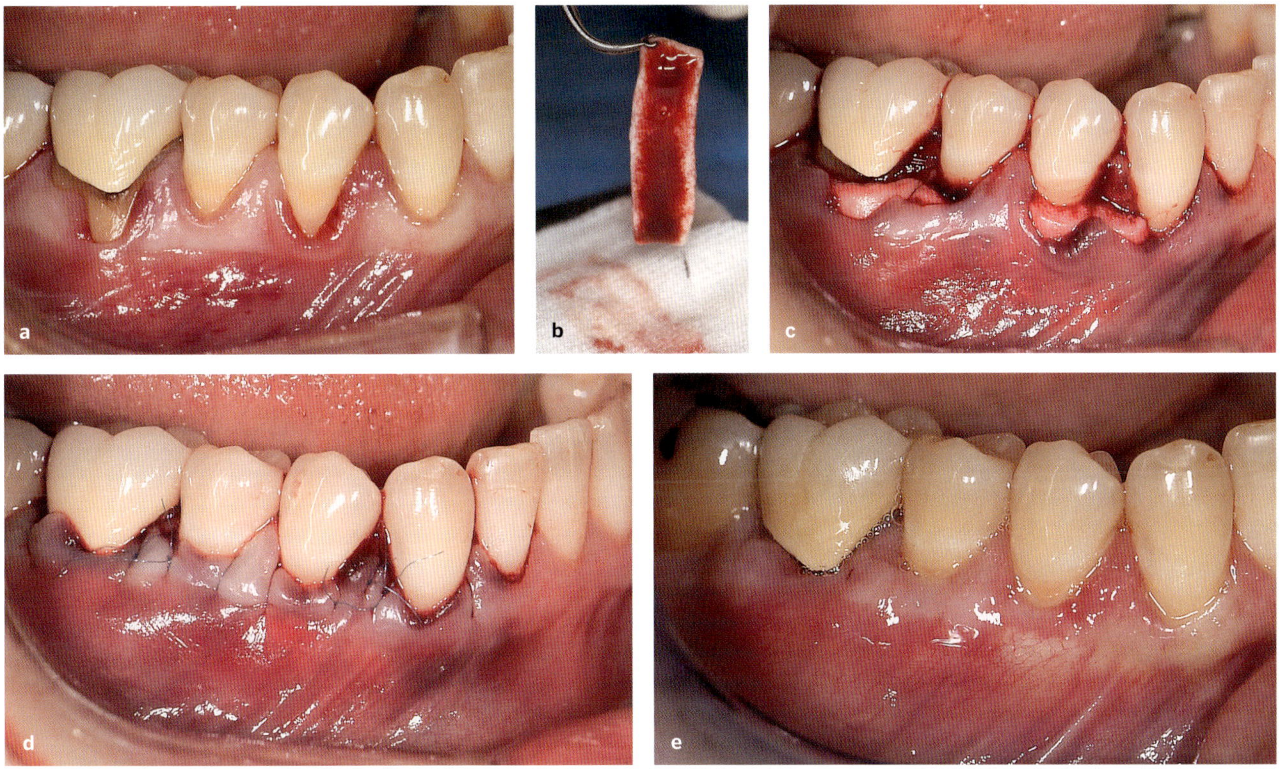

FIG 6-6 Allogeneic dermis and full-thickness flap for coverage of root exposure. *(a)* The roots are exposed in the right mandibular quadrant, including the canine, the two premolars, and, more severely, the mesial root of the first molar. *(b)* The allogeneic dermis is shaped to completely fill in the soft tissue defect. A comparable amount of soft tissue graft harvested from the oral cavity would cause significant morbidity. *(c)* The soft tissue graft is evenly distributed under the reflected flap to cover the roots. *(d)* The surgical area is covered with PRP immediately before it is sutured to ensure immobility of the graft and overlying flap. *(e)* As shown here, PRP used with an allogeneic human dermis graft can produce excellent results in as little as 3 weeks. (Reprinted with permission from Marx and Garg.[12])

Excessive Root Exposure: Repositioned Flaps, Allogeneic Dermis, and PRP

A coronally repositioned flap can be used predictably in conjunction with allogeneic human dermis and activated PRP to cover excessive root exposure of (usually) several teeth and bone (Fig 6-6).[12] The allogeneic dermis is rehydrated in PRP while a full-thickness flap is elevated, and a saturated citric acid solution or other agent is used to prepare the root surface. The allogeneic dermis covers the interproximal spaces to the lingual or palatal gingiva. Tacks that are designed to secure barrier membranes are here used to attach the dermis apically to the bone. Before suturing, the clinician applies PRP to the attached allogeneic dermis and beneath the mucoperiosteal flap. The technique works well even for advanced cases of root exposure, as neither free gingival grafts nor connective tissue grafts can adequately cover such cases. Additionally, rehydrating the dermis with PRP instead of the standard saline solution has been shown to yield superior results.[21,22]

The allogeneic dermis acts as a temporary periosteum to prevent apical retraction and promote healing of the coronally repositioned flap. The cell adhesion molecules released by PRP (fibrin, fibronectin, and vitronectin) adhere to the graft and help it integrate with the bone adjacent to the exposed roots.

PRP and Other Forms of Oral Soft Tissue Wound Healing

PRP has also been shown to increase soft tissue wound healing in the management of medication-related osteonecrosis of the jaw and avascular necrosis.[23] A 2004 study notes that while its results (as well as the results from a number of similar studies) are not controlled or randomized, it is nevertheless superb for enhancing postsurgical wound healing. In fact, a PRP gel is particularly effective in countering the lack of vascularization typical of osteonecrosis of the jaw. Additionally, a 2013 study reported the positive effects of using PRP to treat the medication-related osteonecrosis of the jaws in a patient suffering from extreme pain and hemimandibular paresthesia. While hard tissues were treated through a resection of necrotic bone and PRP application, the researchers noted marked success in closing an ulcer in the soft tissue accompanied by less pain and paresthesia. A follow-up after 1 year revealed complete absence of pain and paresthesia.[24]

The mechanism by which PRP enhances wound healing in oral soft tissues is readily observable in a skin graft. To demonstrate the effects of PRP on wound healing, a randomized prospective study was performed on identical adjacent skin graft donor site wounds. One wound was treated with PRP activated with bovine thrombin, and the adjacent wound received only bovine thrombin as a hemostatic agent (Fig 6-7). The differences in the rate and esthetic outcome of healing were dramatic. As early as day 6, accelerated healing in the PRP-treated site was apparent clinically as well as histologically when compared with the control site (see Figs 6-7d to 6-7h). At 6 months, the PRP-treated site showed a significant reduction in the scar and wound contraction and improved pigment regeneration, and patients reported a 40% reduction in pain at the PRP site (see Figs 6-7k and 6-7l). These effects of PRP on dermal wound healing, while less apparent, are equally impressive when used in oral soft tissue harvest and graft sites.

PRP has also been shown to have positive effects on the healing of the hard and soft tissues of extraction sockets.[25,26] A 2014 systematic review of the literature concluded that using PRP in retained third molar surgery lacks sufficient scientific evidence to recommend it in the absence of randomized clinical trials.[27] Nevertheless, there is enough evidence from individual studies to suggest that PRP's biocompatibility and growth factors can markedly improve soft tissue healing and lead to bone regeneration and density within extraction sockets while also reducing patient pain.[28] For example, a 2012 study used PRP in mandibular third molar extraction sockets to enhance soft tissue healing.[29] In the absence of grafting material, the researchers evaluated not only soft tissue healing but also gingival attachment. Analysis of PRP's effectiveness was based on wound dehiscence and probing depth and revealed an improvement in the socket where PRP was used. A similar study also reported a decrease in alveolar osteitis as well as improvements in soft tissue flap healing, including less edema in the week following socket treatment where PRP was used as a biologic adjunct.[30] Another extraction-socket study reported improved epithelialization when PRP was used.[31]

A 2015 randomized controlled clinical trial assessing the efficacy of PRP in the treatment of extraction socket healing found that its use can enhance the healing of extraction sockets by reducing postoperative complications and stimulating soft tissue healing. Analysis by CBCT demonstrated more than twice the amount of socket regeneration for PRP-treated patients. Additionally, those receiving PRP experienced significantly less pain and inflammation during the first week of healing. Finally, soft tissue healing was greatly increased in the PRP group, along with thicker keratinized epithelium, as revealed by histologic analysis.[32]

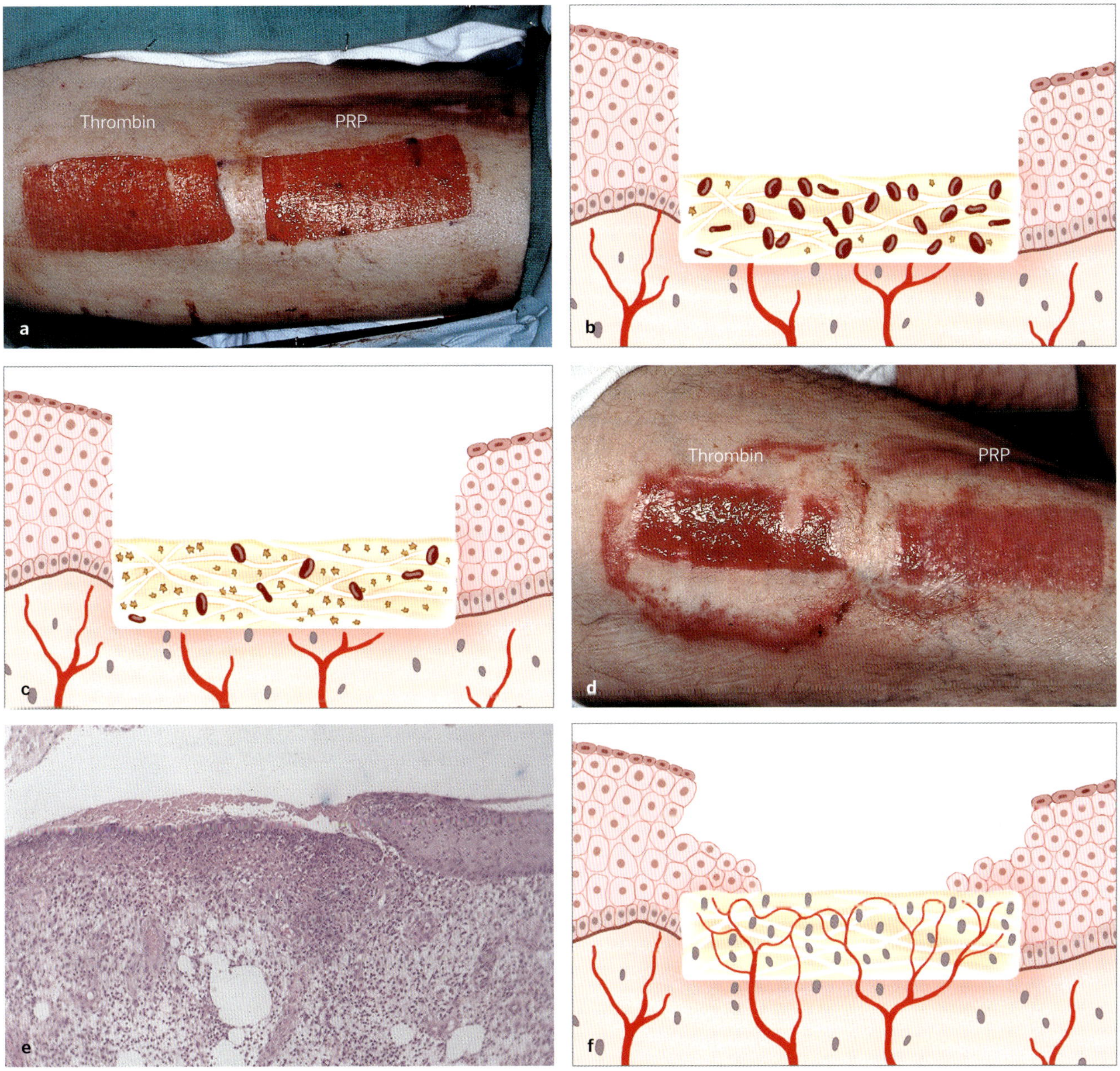

FIG 6-7 Comparison of identical adjacent donor site wounds treated with and without PRP. *(a)* Adjacent split-thickness skin graft donor sites at time of harvest. One site has a thrombin-activated normal blood clot, and the other has a thrombin-activated PRP clot. Note that above the PRP clot is an older skin graft donor site that shows significant scarring, contraction, and pigment alteration. *(b)* A split-thickness skin graft without PRP forms a surface clot containing 5% platelets and 95% red blood cells within a fibrin network. *(c)* A split-thickness skin graft enhanced with PRP forms a surface clot containing 95% platelets and 5% red blood cells within a fibrin network. *(d)* The same donor sites at 6 days. The normal clot donor site has a peripheral erythema and exuberant granulation tissue with minimal epithelial ingrowth. The PRP-enhanced donor site has no peripheral erythema and a nearly complete epithelial covering at this early time frame. *(e)* Histology of the split-thickness skin graft donor site not enhanced with PRP at 6 days shows no epithelial budding and numerous plump immature fibroblasts and macrophages within a granulation tissue base. *(f)* Illustration of the histology shown in *e*. →

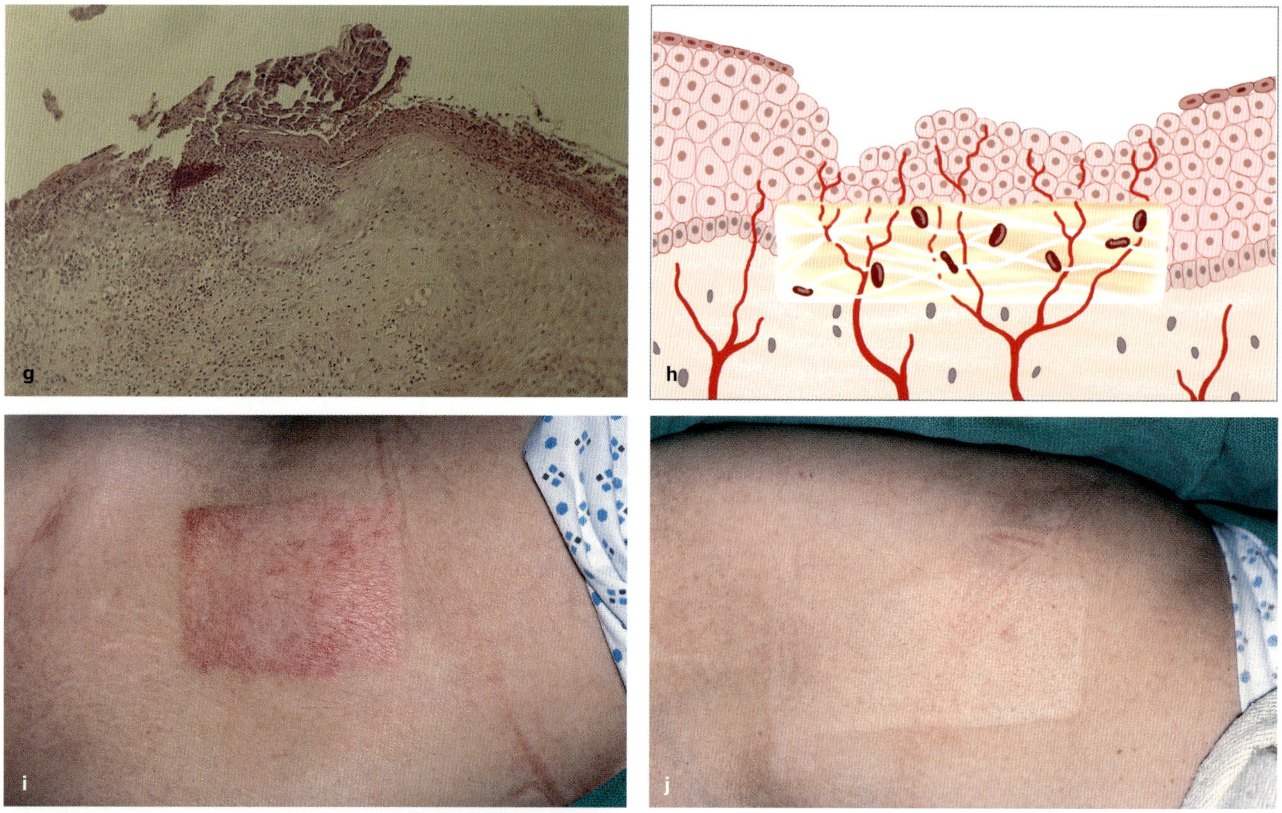

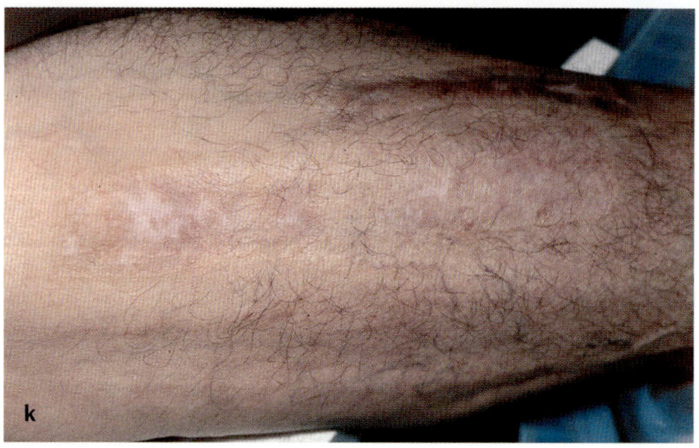

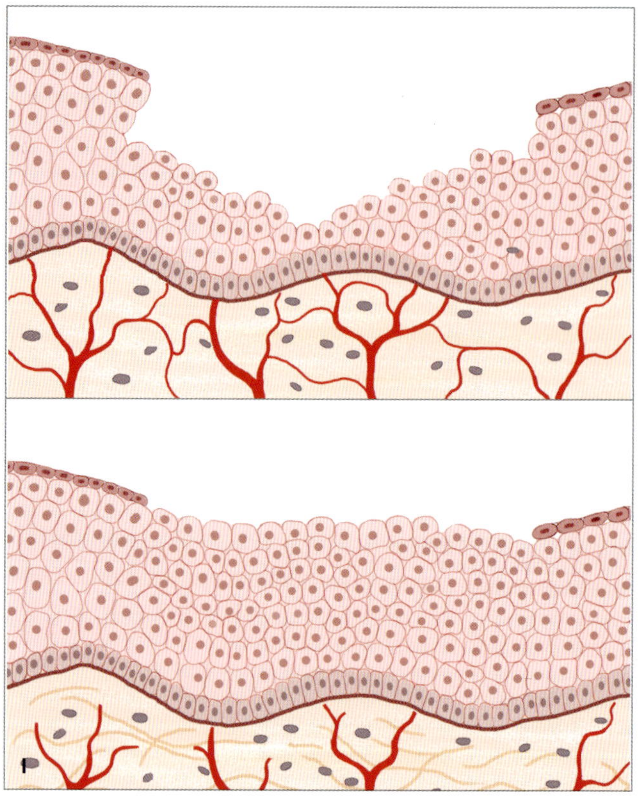

FIG 6-7 *(cont) (g)* Histology of the split-thickness skin graft donor site enhanced with PRP at 6 days shows obvious epithelial budding and mature connective tissue development exemplified by spindle-shaped fibroblasts and collagen bundles. *(h)* Illustration of the histology shown in *g*. *(i)* At 45 days, a split-thickness skin graft donor site without PRP enhancement features only a thin epithelium layer over a hypervascular connective tissue indicative of immature healing. *(j)* At 45 days, a split-thickness skin graft donor site (on the opposite leg of the same individual) enhanced with PRP is more flesh-colored, indicating a thicker epithelial covering, and shows regression of the hypervascular phase, indicating wound maturation. *(k)* The split-thickness skin graft donor sites shown in *a* and *d* now at 6 months. Note the increased scar development and pigment alterations in the site not enhanced with PRP *(left)*. *(l)* Cellular representations of split-thickness skin graft donor sites without *(top)* and with *(bottom)* PRP enhancement at 6 months. (Reprinted with permission from Marx and Garg.[12])

PRP's Antimicrobial Properties and Oral Soft Tissue Wound Healing

Just as certain conditions in the patient's medical and dental history can affect soft tissue healing negatively (eg, chronic periodontitis and other oral health issues, tobacco use, diabetes, and so on) or positively (eg, a rich vascular network, youth, and good overall health), so can conditions associated with PRP itself (especially how it is prepared) impede or improve wound healing. In particular, the percentage of leukocytes in the composition of PRP could positively or negatively affect PRP's biologic and mechanical properties, affecting inflammation and wound healing. These compositional outcomes are particularly important with regard to oral soft tissues and hard tissues, as well as periodontal ligaments and cementum, because periodontal disease is one of the most widespread chronic inflammatory conditions affecting humans.

The gram-negative anaerobic bacterium *Porphyromonas gingivalis* is a known cause of periodontitis, and lipopolysaccharide—considered by many to be another foundation for the inflammation associated with periodontitis—is a significant part of the composition of the outer membrane of gram-negative bacteria. It should be noted that the exact nature and function of the antimicrobial factors in PRP are a matter of debate.[33-36] Researchers have not reached consensus concerning the effect of leukocytes in the composition of PRP. For example, some believe that including them can improve scaffold stability, antimicrobial activity, and inflammatory response.[37] Others, however, believe that leukocytes can adversely affect wound healing due to an inflammatory response.[38]

A 2014 in vitro study attempted to determine the bacteria-inhibiting components in leukocyte-poor PRP, and the authors concluded that plasma components and platelet-derived factors work in unison and that coagulation is an essential initial element in the process for fighting *Enterococcus faecalis*, *Streptococcus agalactiae*, *Streptococcus oralis*, and *Staphylococcus aureus*. The researchers believe that their conclusions have fundamental impact on the clinical application of PRP. Their tests included before-and-after activation with calcium chloride on pure PRP, platelets resuspended in phosphate-buffered saline, and PPP.[39]

Furthermore, a 2015 study comparing the effect of PRP under normal and inflammatory conditions concluded that PRP containing leukocytes negatively affected fibrin scaffolds; additionally, it stimulated proinflammatory conditions, thus inhibiting cell proliferation. The study noted that these outcomes could be detrimental to tissue repair and cell regeneration.[40] These conclusions appear to confirm that leukocyte-free plasma rich in growth factors has more predictable properties (including the release of growth factors and increased strength of the fibrin scaffold). These properties appear to survive the transition from noninflammatory to inflammatory conditions.

Conclusion

PRP can be used for soft tissue healing in a number of treatments in the oral cavity, including implant placement, endodontic surgery, and for free gingival and connective tissue grafts for gingival recession. PRP has also been shown to have positive effects in wound (mucosa) healing in the management of medication-related osteonecrosis of the jaw, avascular necrosis, and healing of the hard and soft tissues of extraction sockets. Finally, PRP's biologic action has been shown to improve scaffold stability, antimicrobial activity, and inflammatory response in oral cavity procedures. Just as PRP is often the biologic of choice to enhance soft tissue healing in cosmetic surgery, sports medicine, orthopedics, and the treatment of chronic wounds, it has also been shown to improve the entire wound healing landscape in a variety of procedures in oral and maxillofacial surgery, implantology, and periodontology.

References

1. Albanese A, Licata ME, Polizzi B, Campisi G. Platelet-rich plasma (PRP) in dental and oral surgery: From the wound healing to bone regeneration. Immun Ageing 2013;10:23.
2. Nikolidakis D, Jansen JA. The biology of platelet-rich plasma and its application in oral surgery: Literature review. Tissue Eng Part B Rev 2008;14:249–258.
3. Petrungaro PS. Using platelet-rich plasma to accelerate soft tissue maturation in esthetic periodontal surgery. Compend Contin Educ Dent 2001;22:729–732,734,736.
4. Anitua E. The use of plasma-rich growth factors (PRGF) in oral surgery. Pract Proced Aesthet Dent 2001;13:487–493.
5. Cebrian-Carretero JL, Guiñales-Díaz de Cevallos J, Sobrino JA, Yu T, Burgueño-García M. Predictable dental rehabilitation in maxillomandibular reconstruction with free flaps. The role of implant guided surgery. Med Oral Patol Oral Cir Bucal 2014;19:e605–e611.
6. Okay DJ, Buchbinder D, Urken M, Jacobson A, Lazarus C, Persky M. Computer-assisted implant rehabilitation of maxillomandibular defects reconstructed with vascularized bone free flaps. JAMA Otolaryngol Head Neck Surg 2013;139:371–381 [erratum 2013;139:771].

7. Tagliareni JM, Clarkson E. Basic concepts and techniques of dental implants. Dent Clin North Am 2015;59:255–264.
8. Gómez-de Diego R, Mang-de la Rosa Mdel R, Romero-Pérez MJ, Cutando-Soriano A, López-Valverde-Centeno A. Indications and contraindications of dental implants in medically compromised patients: Update. Med Oral Patol Oral Cir Bucal 2014;19:e483–e489.
9. Diz P, Scully C, Sanz M. Dental implants in the medically compromised patient. J Dent 2013;41:195–206.
10. Vohra F, Al-Kheraif AA, Almas K, Javed F. Comparison of crestal bone loss around dental implants placed in healed sites using flapped and flapless techniques: A systematic review. J Periodontol 2015;86:185–191.
11. Mounir M, Beheiri G, El-Beialy W. Assessment of marginal bone loss using full thickness versus partial thickness flaps for alveolar ridge splitting and immediate implant placement in the anterior maxilla. Int J Oral Maxillofac Surg 2014;43:1373–1380.
12. Marx RE, Garg A. Dental and Craniofacial Applications of Platelet-Rich Plasma. Chicago: Quintessence, 2005.
13. Simonpieri A, Del Corso M, Vervelle A, et al. Current knowledge and perspectives for the use of platelet-rich plasma (PRP) and platelet-rich fibrin (PRF) in oral and maxillofacial surgery, part 2: Bone graft, implant and reconstructive surgery. Curr Pharm Biotechnol 2012;13:1231–1256.
14. Caplanis N, Romanos G, Rosen P, Bickert G, Sharma A, Lozada J. Teeth versus implants: Mucogingival considerations and management of soft tissue complications. J Calif Dent Assoc 2014;42:841–858.
15. Chambrone L, Tatakis DN. Periodontal soft tissue root coverage procedures: A systematic review from the AAP Regeneration Workshop. J Periodontol 2015;86(2 suppl):S8–S51.
16. Cairo F, Nieri M, Pagliaro U. Efficacy of periodontal plastic surgery procedures in the treatment of localized facial gingival recessions. A systematic review. J Clin Periodontol 2014;41(suppl 15):S44–S62.
17. Pini-Prato G, Nieri M, Pagliaro U, et al. Surgical treatment of single gingival recessions: Clinical guidelines. Eur J Oral Implantol 2014;7:9–43.
18. Naik AR, Ramesh AV, Dwarkanath CD, Naik MS, Chinnappa AB. Use of autologous platelet-rich plasma to treat gingival recession in esthetic periodontal surgery. J Indian Soc Periodontol 2013;17:345–353.
19. Del Corso M, Vervelle A, Simonpieri A, et al. Current knowledge and perspectives for the use of platelet-rich plasma (PRP) and platelet-rich fibrin (PRF) in oral and maxillofacial surgery, part 1: Periodontal and dentoalveolar surgery. Curr Pharm Biotechnol 2012;13:1207–1230.
20. Griffin TJ, Cheung WS. Treatment of gingival recession with a platelet concentrate graft: A report of two cases. Int J Periodontics Restorative Dent 2004;24:589–595.
21. Weiss SR, Tenney JM, Thomson JL, et al. The effect of AlloDerm on the initiation and growth of human neovessels. Laryngoscope 2010;120:443–449.
22. Cairo F, Pagliaro U, Nieri M. Treatment of gingival recession with coronally advanced flap procedures: A systematic review. J Clin Periodontol 2008;35(8 suppl):136–162.
23. Longo F, Guida A, Aversa C, et al. Platelet-rich plasma in the treatment of bisphosphonate-related osteonecrosis of the jaw: Personal experience and review of the literature. Int J Dent 2014;2014:298945.
24. Anitua E, Begoña L, Orive G. Treatment of hemimandibular paresthesia in a patient with bisphosphonate-related osteonecrosis of the jaw (BRONJ) by combining surgical resection and PRGF-Endoret. Br J Oral Maxillofac Surg 2013;51:e272–e274.
25. Alissa R, Esposito M, Horner K, Oliver R. The influence of platelet-rich plasma on the healing of extraction sockets: An explorative randomised clinical trial. Eur J Oral Implantol 2010;3:121–134.
26. Del Fabbro M, Corbella S, Taschieri S, Francetti L, Weinstein R. Autologous platelet concentrate for post-extraction socket healing: A systematic review. Eur J Oral Implantol 2014;7:333–344.
27. Barona-Dorado C, González-Regueiro I, Martín-Ares M, Arias-Irimia O, Martínez-González JM. Efficacy of platelet-rich plasma applied to post-extraction retained lower third molar alveoli. A systematic review. Med Oral Patol Oral Cir Bucal 2014;19:e142–e148.
28. Vivek GK, Sripathi Rao BH. Potential for osseous regeneration of platelet-rich plasma: A comparative study in mandibular third molar sockets. J Maxillofac Oral Surg 2009;8:308–311.
29. Kaul RP, Godhi SS, Singh A. Autologous platelet-rich plasma after third molar surgery: A comparative study. J Maxillofac Oral Surg 2012;11:200–205.
30. Mancuso JD, Bennion JW, Hull MJ, Winterholler BW. Platelet-rich plasma: A preliminary report in routine impacted mandibular third molar surgery and the prevention of alveolar osteitis. J Oral Maxillofac Surg 2003;61:40.
31. Anitua E. Plasma rich in growth factors: Preliminary results of use in the preparation of future sites for implants. Int J Oral Maxillofac Implants 1999;14:529–535.
32. Anitua E, Murias-Freijo A, Alkhraisat MH, Orive G. Clinical, radiographical, and histological outcomes of plasma rich in growth factors in extraction socket: A randomized controlled clinical trial. Clin Oral Investig 2015;19:589–600.
33. Tohidnezhad M, Varoga D, Wruck CJ, et al. Platelets display potent antimicrobial activity and release human beta-defensin 2. Platelets 2012;23:217–223.
34. Yeaman MR. Platelets: At the nexus of antimicrobial defence. Nat Rev Microbiol 2014;12:426–437.
35. Speth C, Löffler J, Krappmann S, Lass-Flörl C, Rambach G. Platelets as immune cells in infectious diseases. Future Microbiol 2013;8:1431–1451.
36. Morrell CN, Aggrey AA, Chapman LM, Modjeski KL. Emerging roles for platelets as immune and inflammatory cells. Blood 2014;123:2759–2767.
37. Dohan Ehrenfest DM, Bielecki T, Mishra A, et al. In search of a consensus terminology in the field of platelet concentrates for surgical use: Platelet-rich plasma (PRP), platelet-rich fibrin (PRF), fibrin gel polymerization and leukocytes. Curr Pharm Biotechnol 2012;13:1131–1137.
38. Boswell SG, Schnabel LV, Mohammed HO, Sundman EA, Minas T, Fortier LA. Increasing platelet concentrations in leukocyte-reduced platelet-rich plasma decrease collagen gene synthesis in tendons. Am J Sports Med 2014;42:42–49.
39. Drago L, Bortolin M, Vassena C, Romanò CL, Taschieri S, Del Fabbro M. Plasma components and platelet activation are essential for the antimicrobial properties of autologous platelet-rich plasma: An in vitro study. PLoS One 2014;18;9:e107813.
40. Anitua E, Zalduendo M, Troya M, Padilla S, Orive G. Leukocyte inclusion within a platelet-rich plasma–derived fibrin scaffold stimulates a more pro-inflammatory environment and alters fibrin properties. PLoS One 2015;10:e0121713.

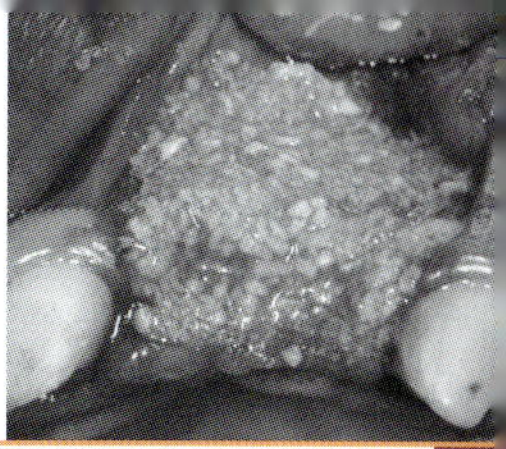

7

Oral Cavity Hard Tissue Healing and Autologous Blood Concentrates

When bone is damaged or injured, or in situations where bone grafting is required, healing is possible only through the processes of osteogenesis, osteoconduction, and osteoinduction.[1-4] The growth factors released by PRP are believed to facilitate osteogenesis, and cell adhesion molecules released by PRP can enhance osteoconduction.[5] Therefore, the biologic assistance provided by PRP can facilitate bone healing under a range of implant-related procedures involving autogenous bone, allogeneic bone, synthetic bone, and bone graft composites. This chapter describes the efficacy of using PRP in a variety of implant-related procedures, from simple to complex, to preserve and regenerate bone as well as to facilitate implant placement.

Periodontal Disease and Periodontal Defects

PRP can help prevent and treat periodontal disease, which is the leading cause of tooth loss around the world.[6-10] The underlying cause of periodontal disease is the presence of subgingival microorganisms, which causes inflammation that leads to alveolar bone resorption and tooth loss.[11-13] Debriding the defect and planing the exposed root surfaces can facilitate the regeneration of bone and interrupt the cycle of tooth loss.

Afterward, the use of an appropriate grafting material and technique can restore both the bone and the soft tissue surrounding it in preparation for implant placement. Combining PRP with bone grafting materials (including porous bovine bone or other bone or bone substitutes or composites) can facilitate bone regeneration in intrabony defects.[14-16] Graft material should be hydrated in PRP liquid while the clinician debrides and irrigates the defect, allowing cell adhesion molecules and clotting to create a malleable composite. The PRP membranes can then be placed in layers over the graft's surface, facilitating tissue regeneration (specifically, osteoconduction) not only in the graft via adjacent bony walls but also in the closing soft tissue flap to reduce the chances of dehiscence and loss of grafting material. Using autogenous bone as part of the composite adds osteogenesis to the regeneration process, particularly in one-wall and two-wall bony defects, which have reduced quantities of osteoprogenitor cells.[5]

Alveolar Ridge Preservation

For optimal implant placement following tooth removal, steps should be taken to preserve the alveolar bone, especially in the anterior maxilla, also known as the *esthetic zone*.[17-19] Complications to the tooth socket bone include tooth fracture below the alveolar crest, brittle teeth that have received root canal treatment, ankylosed roots, a thin buccal cortical plate, fenestrations, and buccal plate fistulae[5] (Fig 7-1).

An intrasocket approach should be taken to minimize trauma to the site via proper rotational or elliptical force and digital support, as well as whatever adjunctive measures are needed for atraumatic tooth removal (Fig 7-2). Following implant placement, socket gaps of 2 mm or less can be treated with PRP alone to facilitate bone

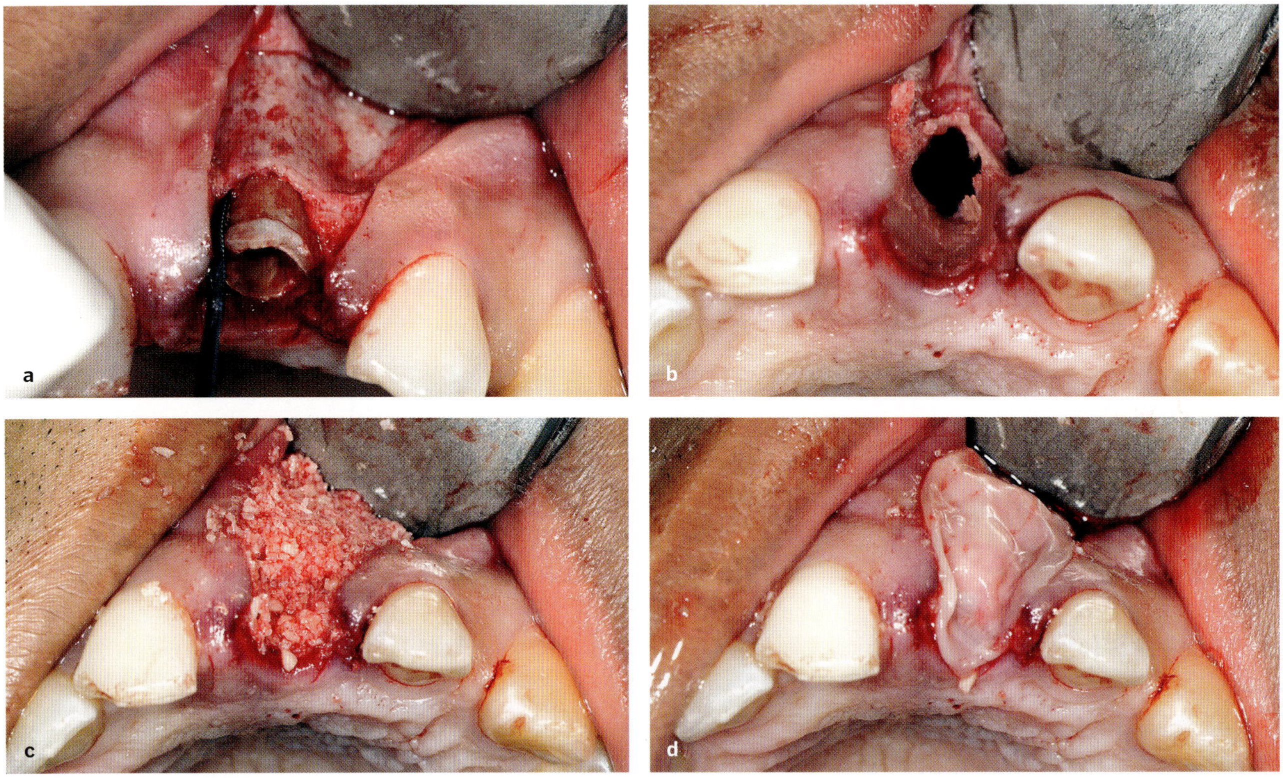

FIG 7-1 Use of PRP-hydrated particulate graft material (sticky bone) in the restoration of a five-wall defect. *(a)* Extraction of maxillary left central incisor. *(b)* Residual bony plate and extraction socket exhibiting an apical defect. *(c)* Socket preservation with freeze-dried bone allograft (FDBA) hydrated with PRP liquid to create sticky bone. *(d)* Coverage of the graft material and extraction socket with a PRP membrane before flap advancement and primary closure.

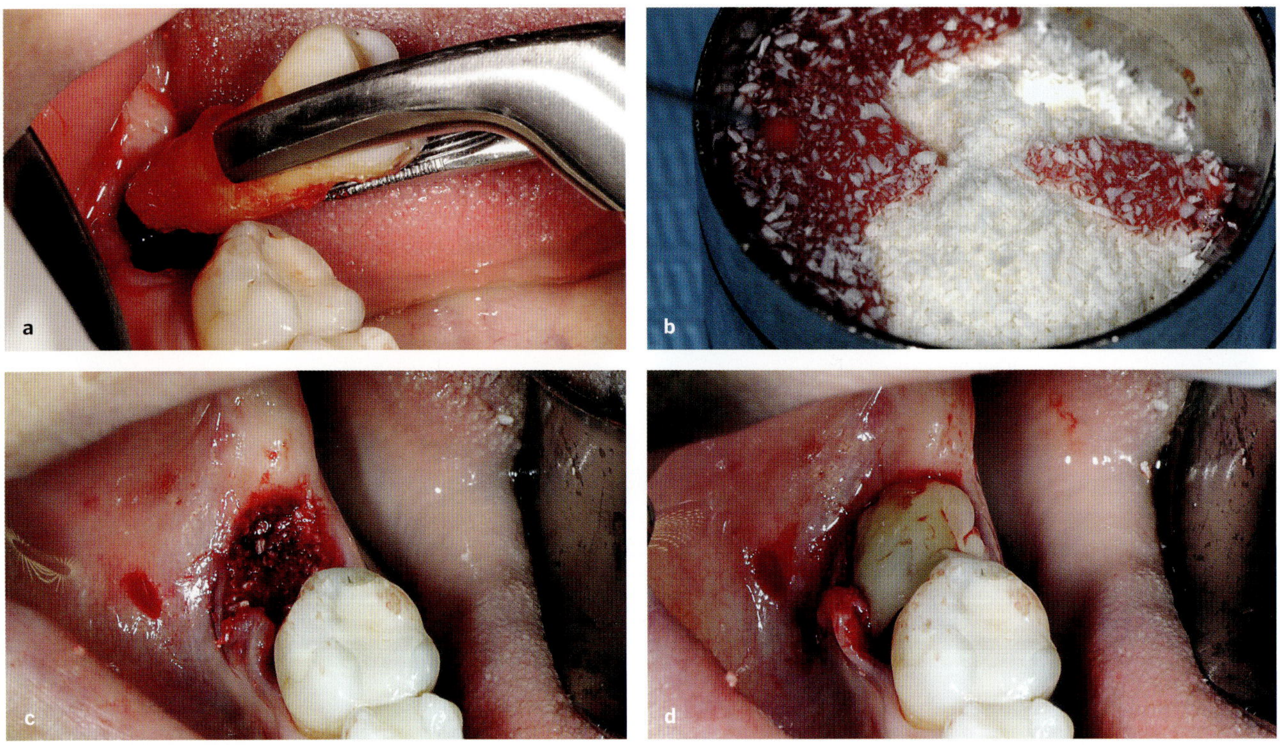

FIG 7-2 Use of PRP in an extraction socket preservation procedure. *(a)* Extraction of mandibular right second molar. *(b)* FDBA hydrated with PRP liquid to create sticky bone. *(c)* Placement of the sticky bone grafting material into the socket. *(d)* Coverage of the grafted socket with a PRP membrane prior to suturing.

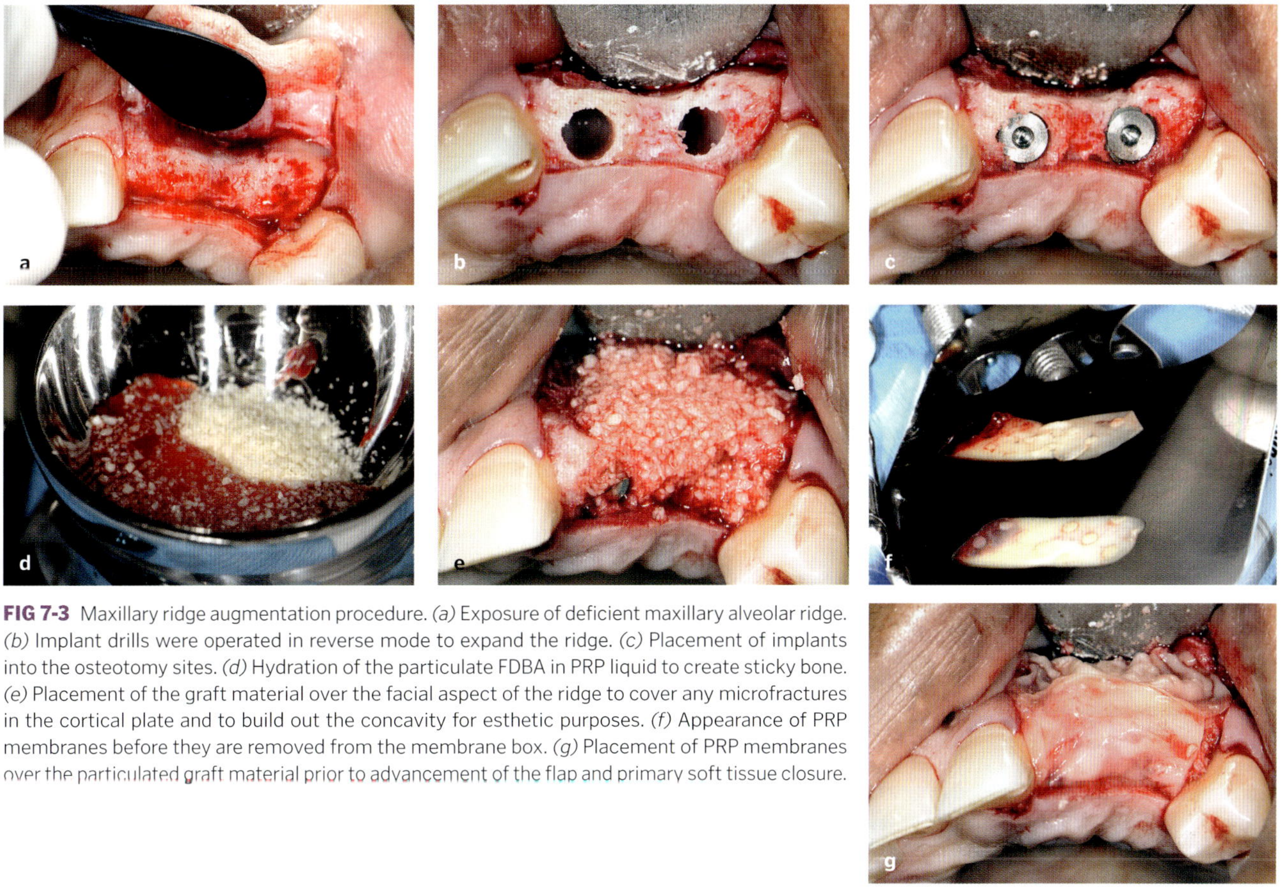

FIG 7-3 Maxillary ridge augmentation procedure. *(a)* Exposure of deficient maxillary alveolar ridge. *(b)* Implant drills were operated in reverse mode to expand the ridge. *(c)* Placement of implants into the osteotomy sites. *(d)* Hydration of the particulate FDBA in PRP liquid to create sticky bone. *(e)* Placement of the graft material over the facial aspect of the ridge to cover any microfractures in the cortical plate and to build out the concavity for esthetic purposes. *(f)* Appearance of PRP membranes before they are removed from the membrane box. *(g)* Placement of PRP membranes over the particulated graft material prior to advancement of the flap and primary soft tissue closure.

regeneration by osteoconduction via the marrow stem cells and endosteal osteoblasts of the socket's bony wall (Fig 7-3). For gaps of more than 2 mm, a composite graft of allogeneic bone or bone substitute combined with PRP can be used to facilitate bone regeneration via osteoinduction, paving the way for eventual osseointegration of bone and implant. PRP can also be used following the loss of a thin buccal plate. Immediate grafting is needed in such cases, especially in the esthetic zone, with a combination of PRP and particulate allogeneic nondemineralized bone. The allogeneic bone should be completely saturated in PRP so that the malleable PRP-graft composite can be positioned in the socket to conform to the buccal spaces. The socket and buccal plate should also be treated with a small amount of PRP. Soft tissue primary closure should be used to enclose the PRP-graft composite, which is tightly packed into the defect, leaving a dense fill for an osteoconductive matrix to form.[5]

Ridge Augmentation Grafting

Maxillary and mandibular ridges that have deficient bone for dental implants can be augmented both horizontally and vertically (Fig 7-4). Afterward, a minimum of 3 weeks is needed to protect the ridges from masticatory forces (eg, via a provisional appliance) during healing.

Horizontal ridge augmentation begins with an osteotomy that can be augmented horizontally (Fig 7-5) by ridge splitting with a bur, chisel, and mallet or ridge-splitting osteotomes. An implant can then be placed immediately with graft material in the buccal and lingual cortices.[20,21] PRP can be applied between the split cortices and within the grafting material, and a PRP membrane and/or a commercially available collagen membrane can be used to cover the split ridge to limit fibrous ingrowth and deliver growth factors to both the soft tissue flaps and the ridge itself.

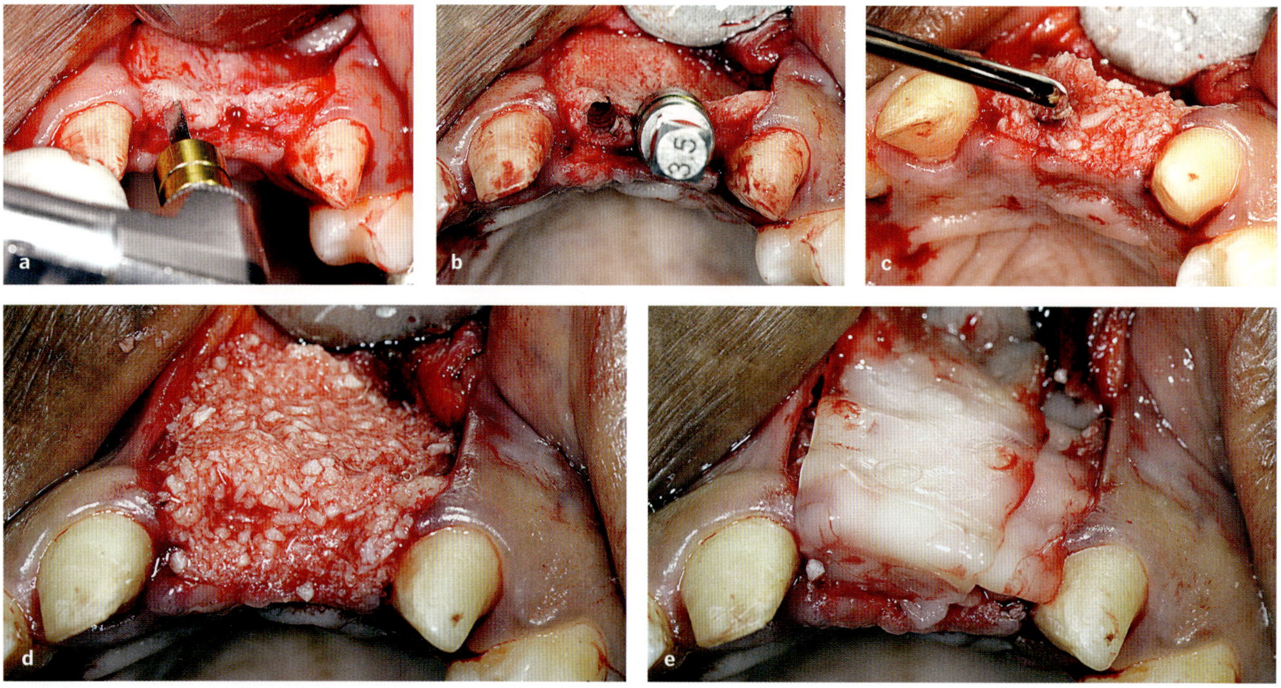

FIG 7-4 Use of PRP-hydrated particulate graft material (sticky bone) in a ridge expansion procedure. *(a)* Pilot hole of the alveolar ridge for ridge expansion at the site of the maxillary left central and lateral incisors following removal of the provisional four-unit maxillary fixed partial denture that replaced teeth nos. 8–11. *(b)* Continuation of ridge expansion procedure using a ridge expansion kit. *(c)* Placement of sticky bone into the osteotomy. *(d)* Placement of additional graft material over the facial aspect of the maxillary ridge to eliminate the buccal concavity for esthetic purposes. *(e)* Placement of PRP membranes over the entire site prior to primary closure and replacement of the provisional fixed partial denture.

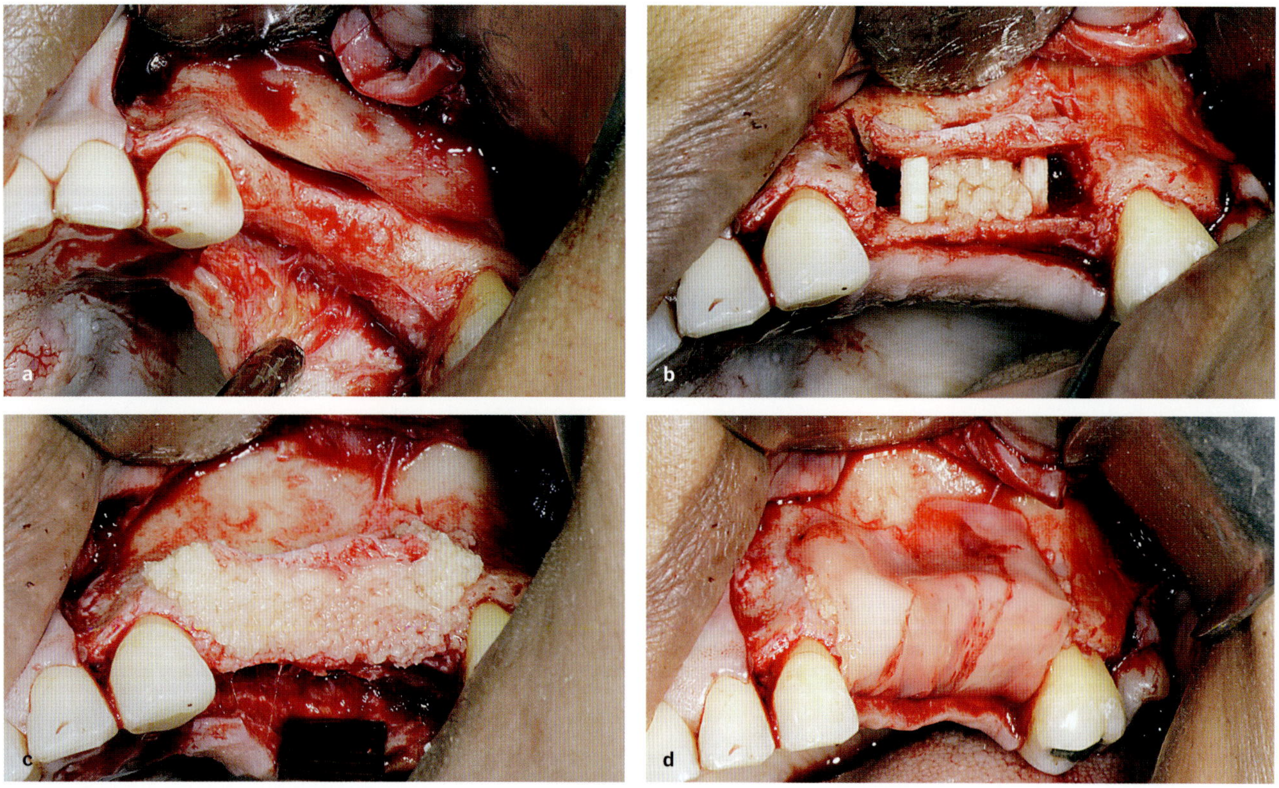

FIG 7-5 Ridge split with sticky bone. *(a)* Preoperative view of a ridge with inadequate width for implant placement. *(b)* Ridge split with cortical bone dowels placed interpositionally to hold the expanded portion more buccal and facilitate the introduction of sticky bone into the expanded portion. *(c)* Continued placement of sticky bone into the expanded ridge and around the sharp edges of the mobilized segment. *(d)* Site covered with PRP membranes prior to flap advancement and primary closure.

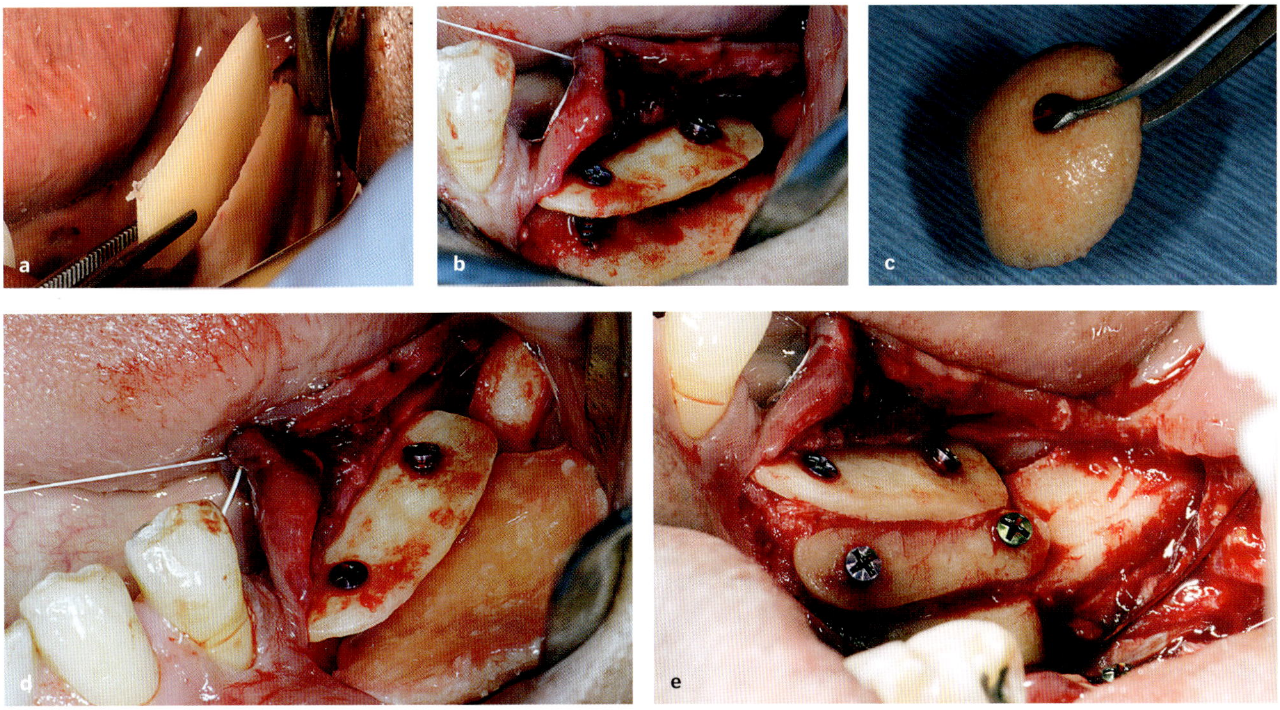

FIG 7-6 Mandibular ridge augmentation with bone harvested from the ramus. *(a)* Harvesting of two bone plates—one thin and the other thicker—from the ramus. An Ebner bone grafter was used to thin the plates and obtain bone scrapings. *(b)* Placement of thin bone plate above the residual crest to increase the height of the mandible. *(c)* Autogenous bone scrapings combined with PRP liquid to create sticky bone. *(d)* Placement of sticky bone beneath the thin crestal bone plate. *(e)* Placement of the second bone plate on the facial aspect to increase the width of the mandible. Fixation screws are used to secure the plates. The site is then covered with PRP membranes, and the flap is advanced over it to obtain primary closure.

For knife-edge ridges and other conditions that limit or preclude ridge splitting, the ridges can be augmented with an onlay graft harvested from the patient's ramus, chin, calvaria, or anterior/posterior ilium, depending on the defect size (Fig 7-6). The addition of compacted cancellous marrow and PRP can help prevent graft reduction and immobility (Fig 7-7). PRP should be placed between the graft and host bone to facilitate their union by lag screws. Compacted cancellous marrow is then placed around the grafts and into any voids during the procedure. The composite block onlay graft and cancellous marrow should be covered with PRP to facilitate healing. Placement of a PRP membrane on the surface of the grafted area reduces flap dehiscence and membrane exposure, allowing implants to be placed in approximately 3 months.[5]

Vertical ridge augmentation is problematic due to soft tissue restriction, limitations on vertical distraction from an osteotomy, and the need to prevent masticatory forces from damaging the graft. Onlay grafts from the same sources used for horizontal grafting are usually the best option: a rigid lag screw in conjunction with cancellous marrow/PRP composite and PRP membrane (Fig 7-8).[5] Of added importance in vertical ridge augmentation is the need for complete soft tissue undermining and advancement over the graft to gain an absolutely tension-free closure (Fig 7-9). PRP can be used on the edges of the flap and the bone graft surface to prevent dehiscence. Once the graft has matured to a sufficient height, implants can be placed. As an alternative, large defects in the ridge can be repaired with a titanium basket crib holding an all-cancellous marrow graft (or half-cancellous graft and half-allogeneic bone/bone substitute) (Fig 7-10).[5,22,23]

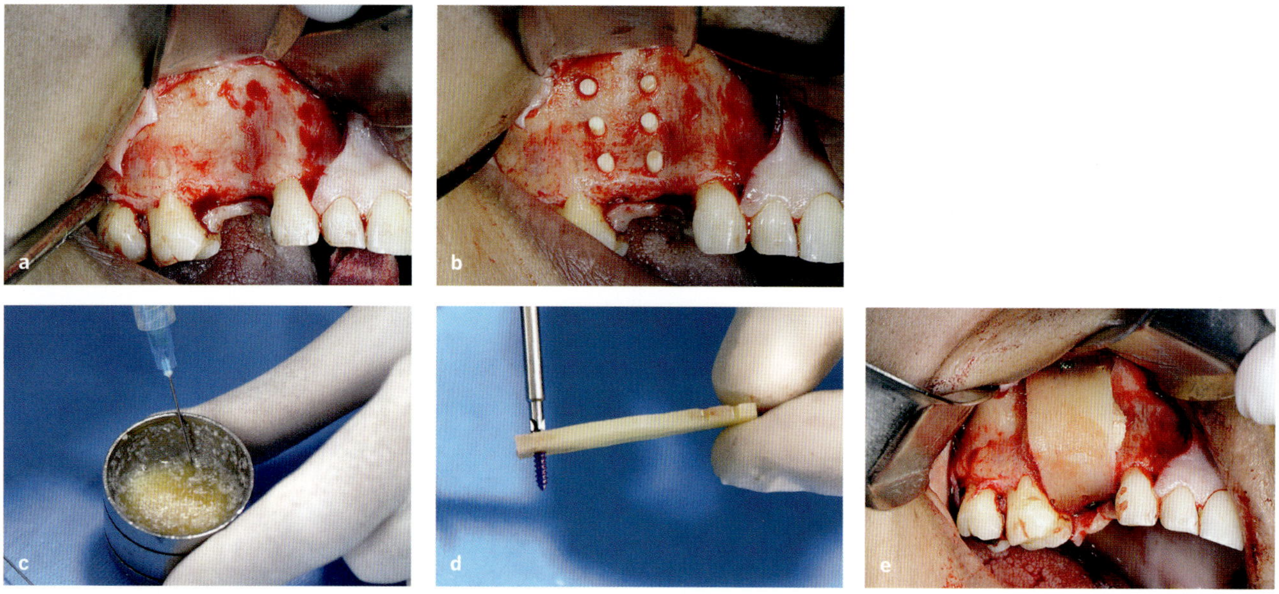

FIG 7-7 Bone graft for increased bone width using bone dowels and bone plates with sticky bone. *(a)* Preoperative view after flap reflection. *(b)* Placement of six bone dowels. *(c)* Preparation of sticky bone. *(d)* Demineralized FDBA bone plate. *(e)* Placement of sticky bone around and over the six bone dowels and covered with the bone plate, which is fixated at each end apically and palatally prior to flap advancement and primary closure.

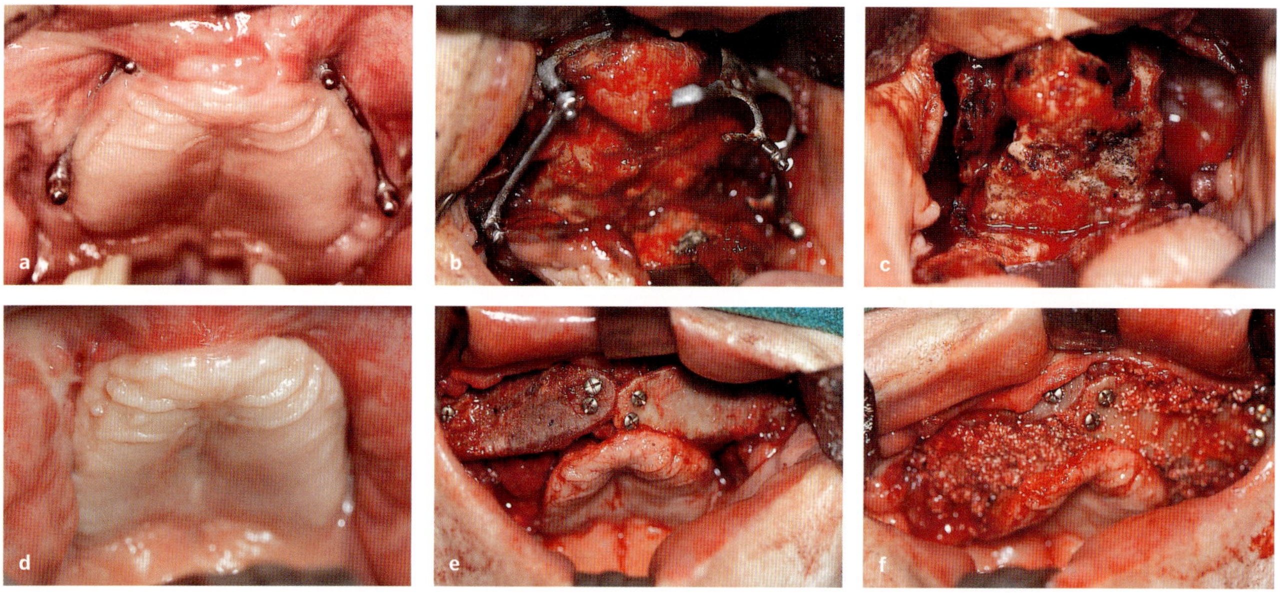

FIG 7-8 Use of PRP in the reconstruction of a maxilla mutilated by chronically infected subperiosteal implants. *(a)* Failing maxillary subperiosteal implant with chronic long-term infection. *(b)* Infected subperiosteal implants such as these must be removed to resolve the infection. *(c)* Nearly all of the maxilla is lost due to the subperiosteal implants and the related trauma and infection. *(d)* The maxilla can be reconstructed once the infection resolves. However, the soft tissue loss becomes evident as the wound contracts. *(e)* Autogenous corticocancellous block grafts lag-screwed to the host maxilla with a layer of PRP placed in between. *(f)* Additional autogenous cancellous marrow is placed between and around the block grafts. (Reprinted with permission from Marx and Garg.[5])

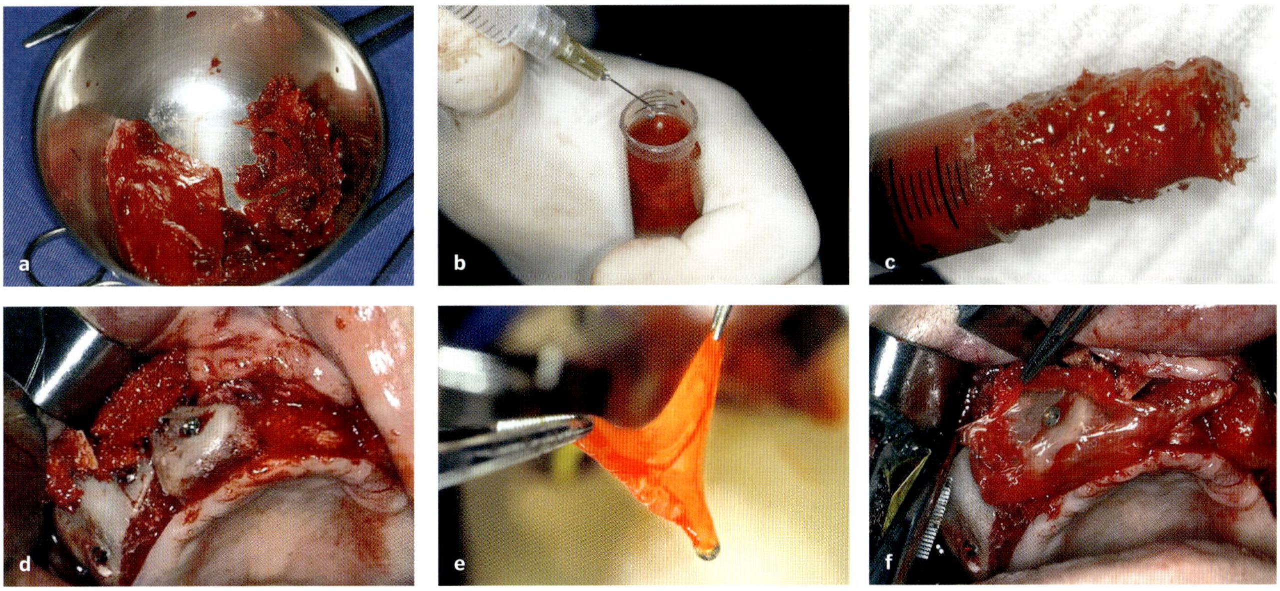

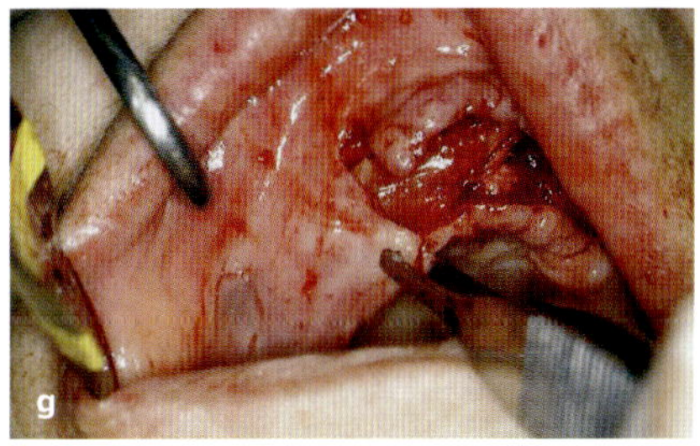

FIG 7-9 Use of PRP with particulate and block grafts for rehabilitation of severe alveolar ridge resorption. The amount of bone loss in this patient requires not only particulate bone but also bone blocks. *(a)* Because of the volume and configurations required, bone was harvested from the iliac crest in an operating room under general anesthesia. *(b)* Once the nonactivated PRP has saturated the graft material, the activating agent is injected. This step is performed immediately prior to placement of the graft in the recipient site. *(c)* For demonstration purposes, the graft material is pushed out onto a sterile towel. Clinically, it is dispensed directly to the recipient site in this manner. *(d)* Reshaping of the blocks continues as they are placed to fill the spaces. *(e)* PRP membrane. *(f)* This membrane is placed over the blocks of bone and particulate graft material. *(g)* The flap is advanced for primary passive closure.

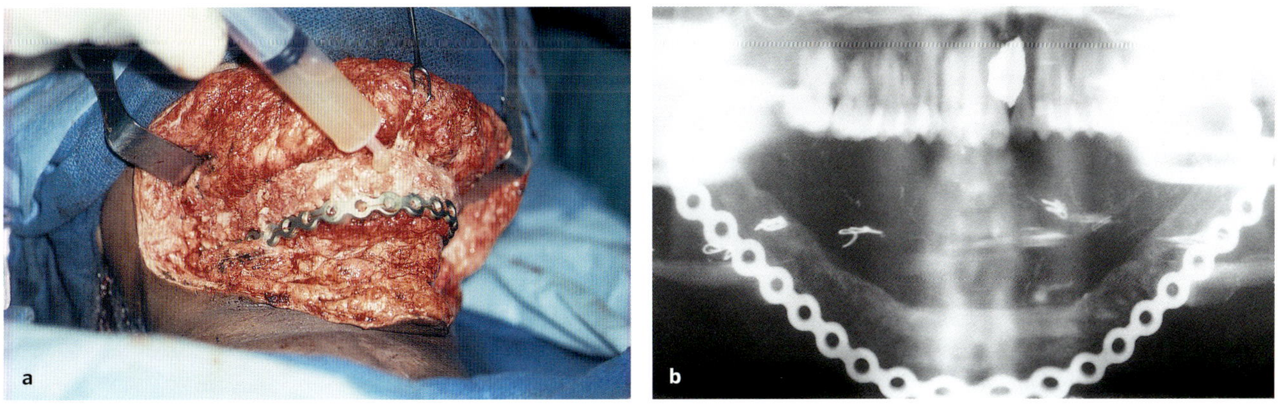

FIG 7-10 Total mandibular reconstruction with autogenous cancellous marrow contained by a rigid reconstruction plate. *(a)* PRP liquid is applied to the surface, where it will clot over the next 15 minutes and release an abundance of growth factors. *(b)* Radiographic appearance of consolidated bone graft (with reconstruction plate in place) enhanced by PRP at 3 months. (Reprinted with permission from Marx and Garg.[5])

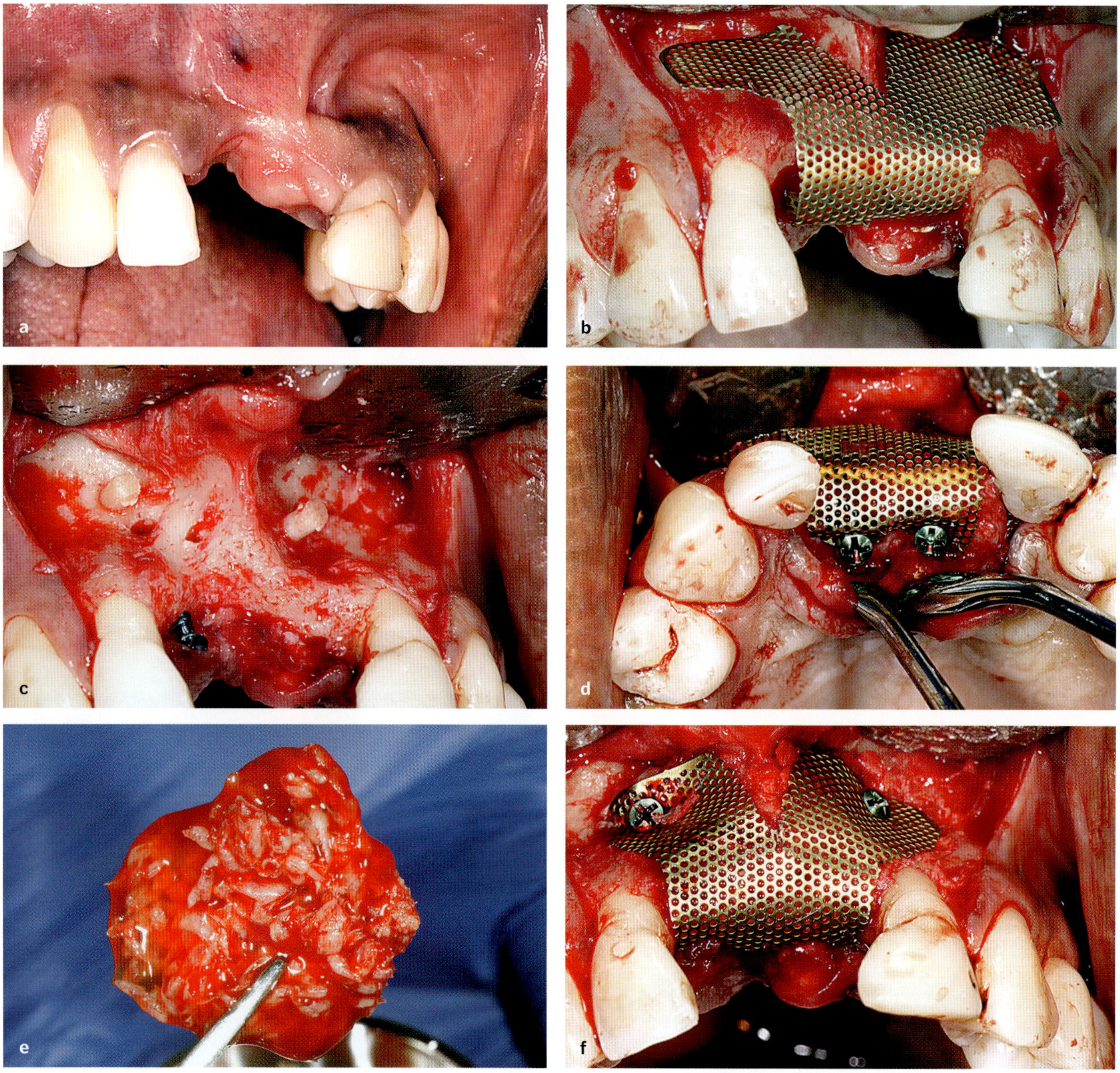

FIG 7-11 Use of PRP sticky bone and titanium mesh in a vertical and horizontal ridge augmentation procedure. *(a)* Vertical and horizontal bone loss in the area of the maxillary central incisors. *(b)* Titanium mesh trimmed and contoured over the areas of bone deficiency. *(c)* Cortical bone dowels are placed to provide facial support and a titanium screw to provide vertical support for the titanium mesh. *(d)* The titanium mesh is secured on the palatal aspect of the ridge with fixation screws. *(e)* Autogenous bone mixed with PRP to obtain sticky bone. *(f)* The autogenous/sticky bone graft material is placed into the prepared site and covered with the precut and preformed titanium mesh. To complete the procedure, a PRP membrane is placed over the site, the flap is advanced over it, and passive primary closure is obtained.

A titanium-reinforced membrane could be used for smaller defects (Fig 7-11). The graft material should be mixed with PRP, and layers of PRP should be used to completely cover the basket/membrane prior to ridge closure.[5]

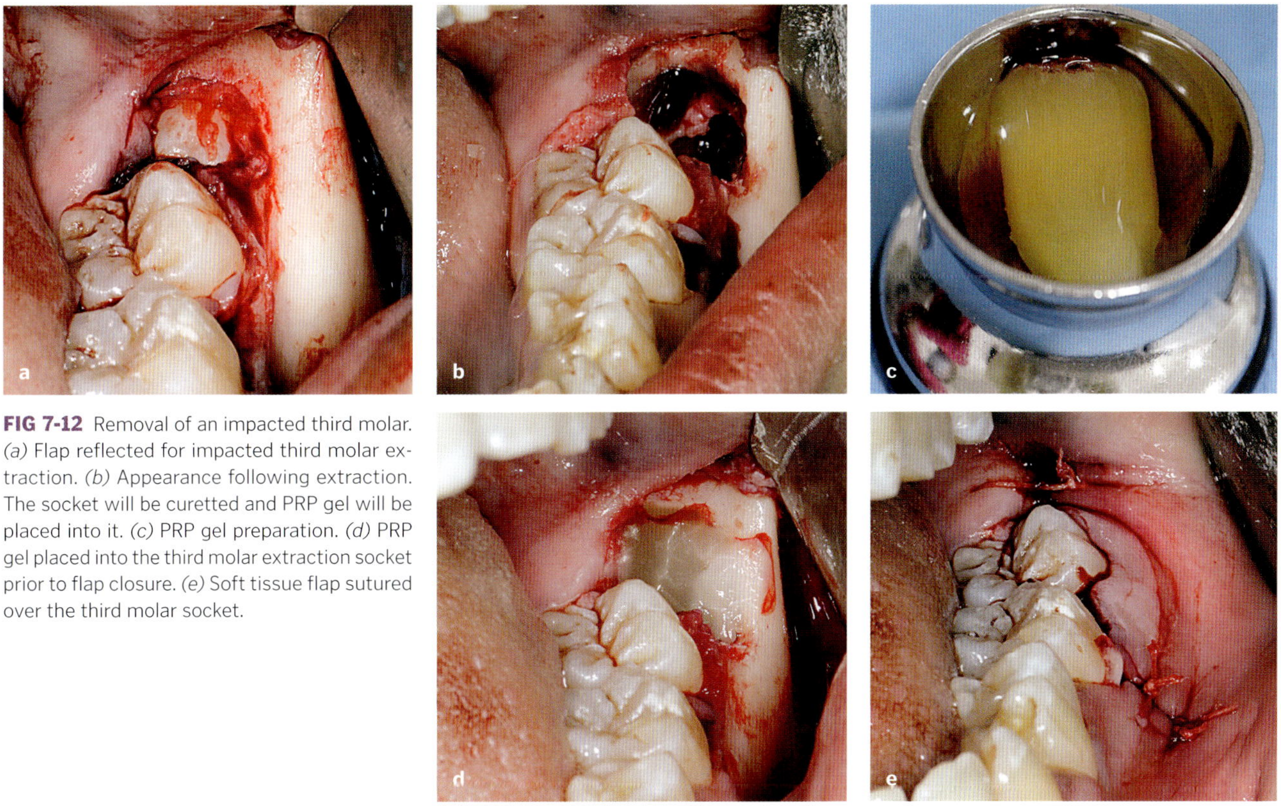

FIG 7-12 Removal of an impacted third molar. *(a)* Flap reflected for impacted third molar extraction. *(b)* Appearance following extraction. The socket will be curetted and PRP gel will be placed into it. *(c)* PRP gel preparation. *(d)* PRP gel placed into the third molar extraction socket prior to flap closure. *(e)* Soft tissue flap sutured over the third molar socket.

Third Molars

Alveolar osteitis (dry socket) and reduced bone regeneration (often adjacent to the second molar)—two complications that can occur following third molar socket surgery—may be addressed with PRP for susceptible patients (eg, smokers, patients older than 30 years, patients undergoing chemotherapy, and patients with diabetes). PRP can both lower the risk of dry socket and facilitate bone regeneration. Though traditional non-PRP treatment can help to avert alveolar osteitis,[24] bone height can still be reduced, and the distal root of the second molar may be affected by bone absence, which could lead to root caries and periodontitis. The destruction of the blood clot in the third molar extraction site by alveolar osteitis causes bone reduction due to the absence of platelet growth factors and cell adhesion molecules needed for osteoconduction and bone regeneration via the socket's bony walls. Administering PRP can help compensate for the loss of the blood clot by reducing the risk of dry socket and by increasing final bone height[25-27] (Fig 7-12). PRP use can also help to eliminate the periodontal defect adjacent to the second molar and to increase final bone height as well. PRP's acidic pH, concentrated leukocytes, and creation of an oxygen-rich environment through increased vascularization (macrophages and neutrophils) may account for obstruction of the bacteria that causes alveolar osteitis.[5]

Platelet-rich fibrin can also be used to mitigate the complications of impacted third molar surgery—namely, pain, trismus, swelling, and edema (Fig 7-13). A study performed in 2020 by Cardarelli and Garg concluded that the combination of using piezosurgery in place of a handpiece and applying PRP to the extraction socket reduced the amount of pain and swelling experienced by patients compared with patients in the control group, who underwent traditional surgery to remove their impacted third molars.[28]

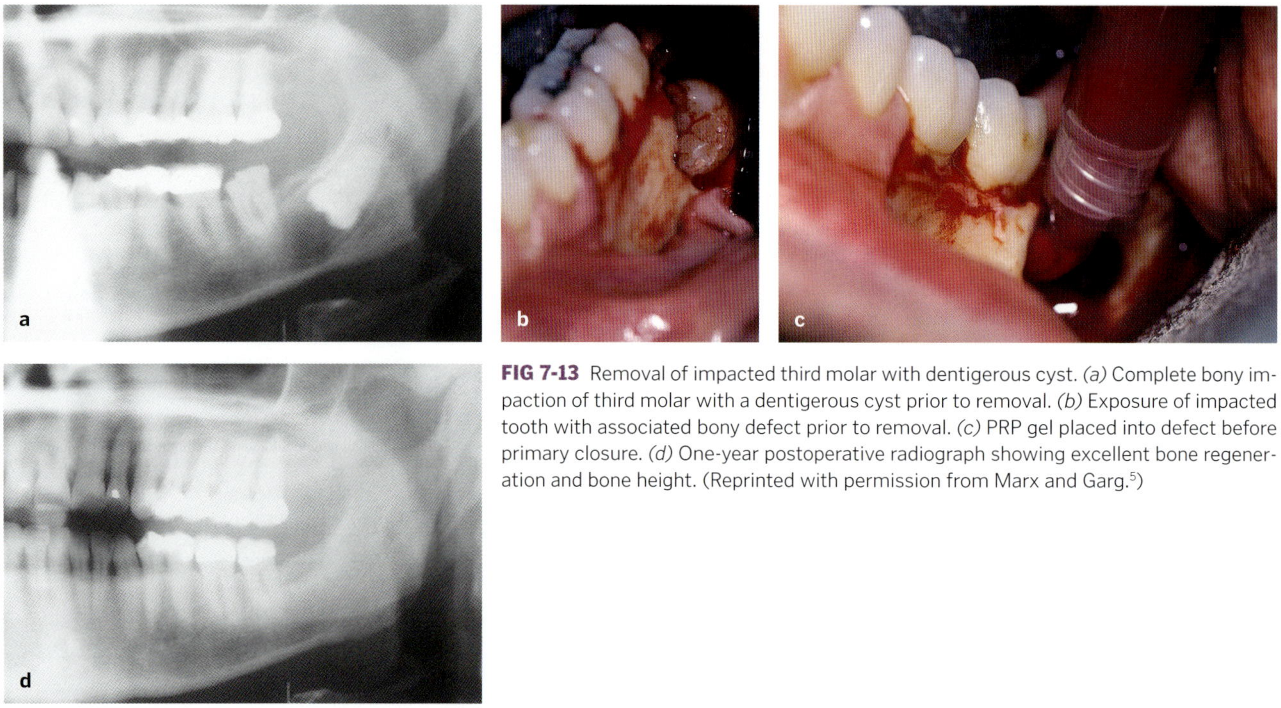

FIG 7-13 Removal of impacted third molar with dentigerous cyst. *(a)* Complete bony impaction of third molar with a dentigerous cyst prior to removal. *(b)* Exposure of impacted tooth with associated bony defect prior to removal. *(c)* PRP gel placed into defect before primary closure. *(d)* One-year postoperative radiograph showing excellent bone regeneration and bone height. (Reprinted with permission from Marx and Garg.[5])

Sinus Elevation Grafting

Sinus elevation bone grafting surgery is often necessary to increase maxillary bone height (between 8 and 18 mm) and width from the alveolar crest to the sinus floor for the support of one or more implants. For a variety of reasons, autogenous bone or freeze-dried bone allograft (FDBA) is the best grafting material for this procedure.[29,30] In situations that can lead to poorer clinical results, PRP should be an adjunct to the sinus elevation bone grafting procedure (Fig 7-14).

Contraindications for sinus elevation procedures include soft tissue density in the sinus, allergies, upper respiratory infections, sinusitis, and damage from previous sinus surgeries, all of which can lead to conditions that compromise success, such as sinus membrane friability and edema. Sinus inflammation is the most common reason for graft loss and failure of bone regeneration after a sinus elevation procedure.[5] Such inflammation may occur as a result of perforations in edematous and inflamed sinus membranes during the sinus elevation procedure. Obviously, such perforations must absolutely be avoided. Unless chronic sinus conditions or surgical mishaps like these can be avoided or treated successfully, alternatives to sinus elevation grafting, such as vertical and horizontal ridge augmentation, should be sought.[30,31]

The sinus elevation procedure employs local anesthesia and other sedation as required for the particular case.[32–34] The lateral sinus wall is exposed via a broad-based, full-thickness mucoperiosteal flap, extending from the maxillary tuberosity to a releasing incision in the canine fossa region. The flap should extend superiorly to just below the infraorbital foramen to provide access to the lateral wall of the sinus.

A large oval finishing bur or diamond bur is used to create an oval hole in the lateral cortical bone for access to the sinus membrane (Fig 7-15). To preserve its integrity, the sinus membrane is directly reflected in a series of stages, beginning at the sinus floor—superiorly, anteriorly, and posteriorly around the edges of the oval entry—until the sinus membrane is completely elevated posteriorly to the tuberosity and stretching over the sinus floor to

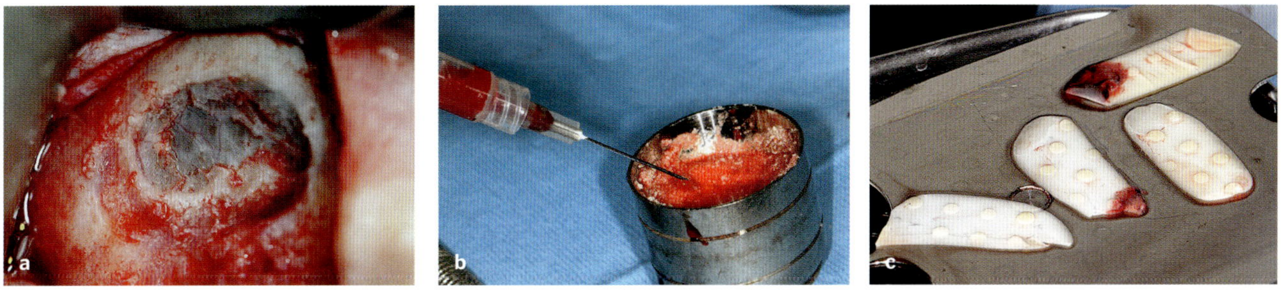
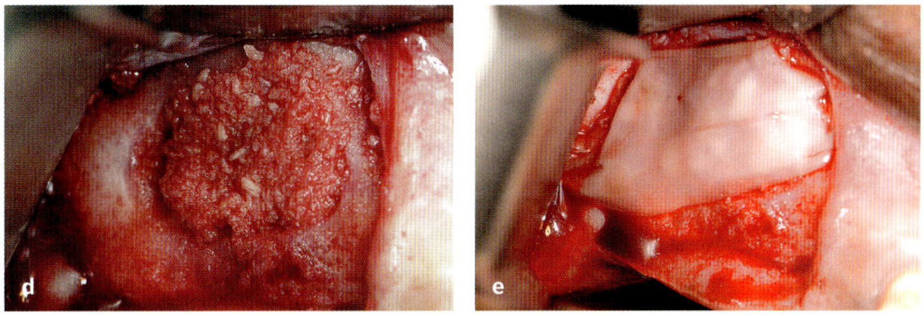

FIG 7-14 Use of PRP in sinus elevation procedure to increase the height of the maxillary bone. *(a)* Reflection of a flap and creation of a sinus window, exposing the sinus membrane. *(b)* Hydration of the FDBA with PRP liquid to create sticky bone. *(c)* PRP membranes formed in the membrane box. (d) Sinus membrane elevation followed by placement of the graft material. *(e)* The graft material is covered with PRP membranes prior to primary flap closure.

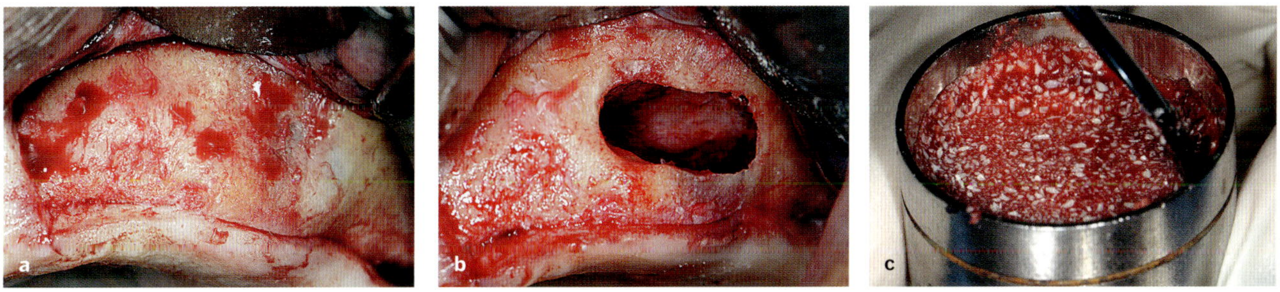
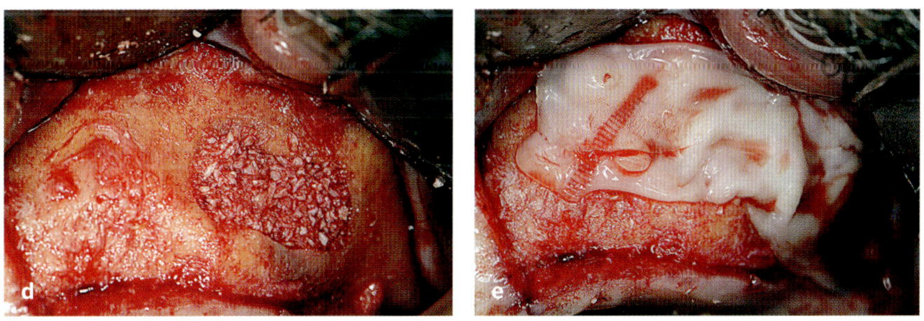

FIG 7-15 Use of PRP to increase the maxillary bone height via a lateral sinus approach. *(a)* Reflection of a tissue flap on the lateral ridge. *(b)* Preparation of the lateral sinus window and elevation of the sinus membrane. *(c)* Hydration of FDBA particles with PRP liquid to create sticky bone. *(d)* Placement of sticky bone graft into the prepared sinus elevation site. *(e)* Placement of PRP membranes over the graft material prior to flap advancement and primary closure.

the medial wall. A synthetic gauze material soaked in an anesthetic/epinephrine or cocaine solution is then placed in the sinus cavity to achieve complete hemostasis of the reflected area before it is further reflected and made visible, allowing for optimal placement of sufficient grafting material.[5]

The crestal sinus elevation procedure is an alternative approach that is less invasive than the lateral approach but usually results in smaller gains in maxillary bone height. In this technique, an osteotomy is created using a series of osteotomes, and the crestal bone and PRP gel are positioned into the maxillary sinus, effectively elevating the sinus floor and tenting the sinus membrane. Additional PRP gel is then introduced into the sinus cavity, and an implant is placed in the osteotomy (Fig 7-16).

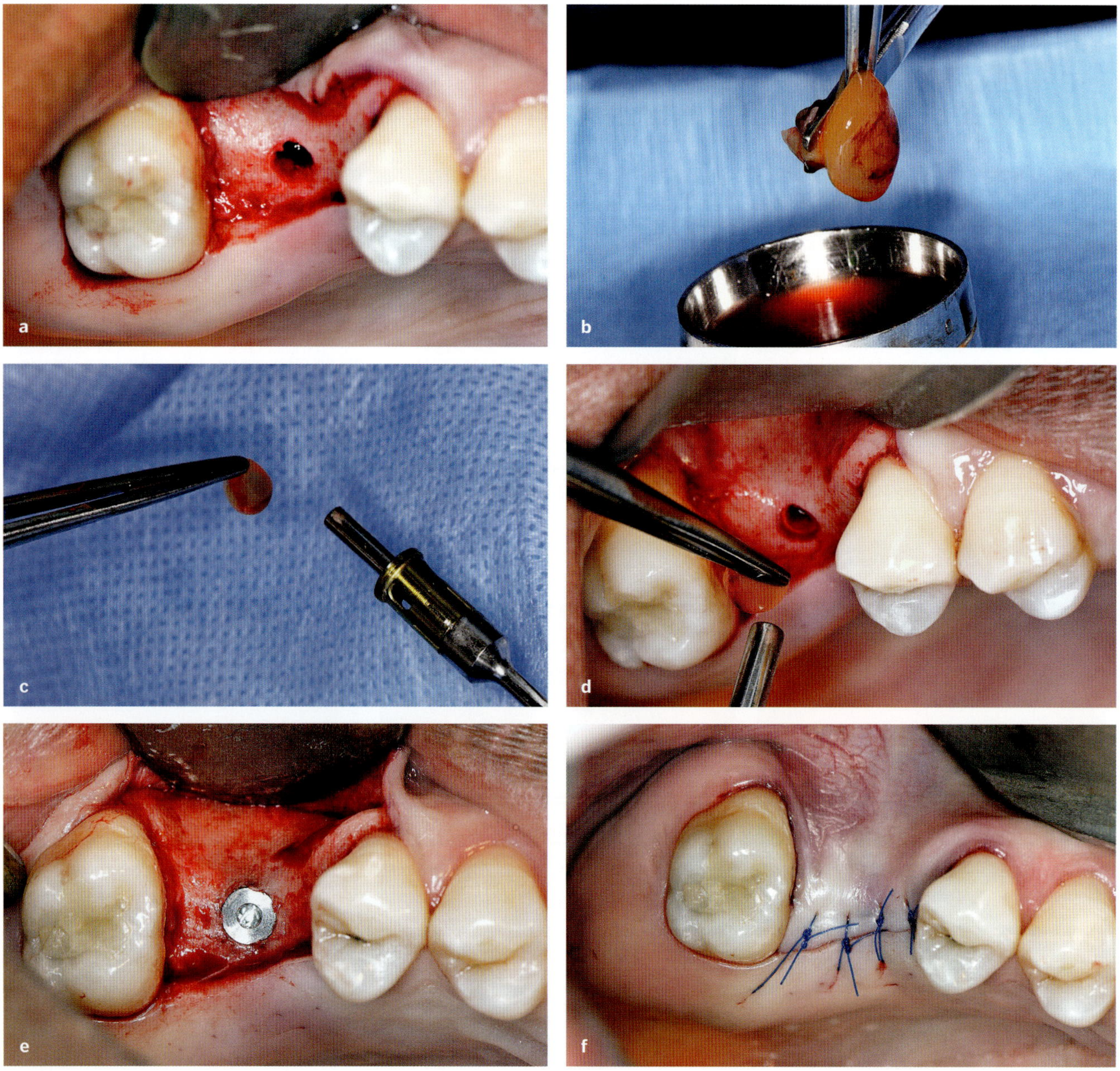

FIG 7-16 Application of PRP gel in a sinus elevation procedure using a crestal approach to increase the maxillary bone height in preparation for implant placement. *(a)* Flap reflection, exposure of the crestal bone, and implant osteotomy. *(b)* Preparation of the PRP gel, which is cut into small pieces of approximately 2 to 3 mm each. *(c)* Appearance of the size and consistency of the gel as it is being delivered to the graft site, where it will be condensed into the sinus. *(d)* Placement of the PRP gel into the osteotomy. *(e)* Immediate implant placement following grafting of the maxillary sinus with PRP gel. *(f)* Primary closure and suturing of the implant site.

If a healthy sinus membrane is inadvertently perforated,[35] a PRP membrane can be used to cover a perforation smaller than 3 mm and to serve as a fibrin scaffold that will release growth factors to facilitate membrane healing (Fig 7-17). Collagen membranes infused with PRP can be used to cover perforations greater than 3 mm while also preventing seepage of graft material and accelerating the healing/migration of the sinus membrane.

PRP can be a valuable adjunct to the grafting process for the sinus elevation procedure, whether autogenous bone, allogeneic bone, synthetic bone, or bone graft composites are being used.[36-38] For example, the mitogenic growth factors in PRP can complement the regenerative properties of endosteal osteoblasts and marrow stem cells in the autogenous graft.

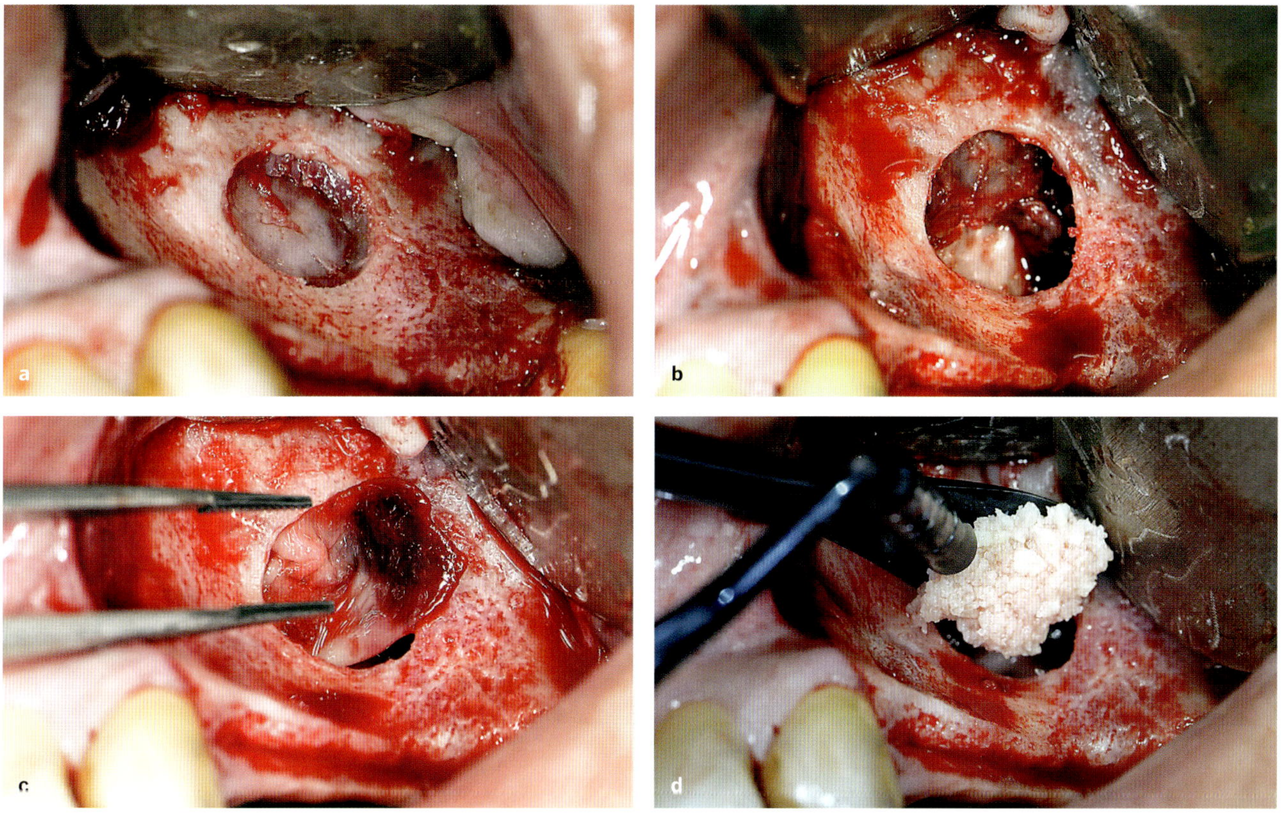

FIG 7-17 Maxillary sinus elevation procedure with membrane repair. *(a)* Flap reflected and lateral sinus window created in preparation for sinus elevation. *(b)* Elevation of the sinus membrane, resulting in inadvertent perforation of the sinus membrane. *(c)* Placement of a PRP membrane into the sinus to patch and repair the torn sinus membrane. *(d)* Placement of PRP sticky bone into the sinus cavity.

The clinician can mechanically and biochemically increase the numbers of endosteal osteoblasts and marrow stem cells by:

- Densely packing the autogenous cancellous marrow grafts in the syringe used to deliver the material to the sinus cavity grafting site.
- Packing the material again tightly into the graft site.
- Applying a final PRP "cover" to the grafting material. The growth factors in the autogenous bone and in the PRP bind the graft cell membranes and graft particles via cell adhesion proteins in fibrin, fibronectin, and vitronectin. PRP administered in this fashion also aids soft tissue healing in the surrounding mucosa.[5]

However, if only allogeneic bone (which is very low in bone morphogenetic protein) or a bone substitute is used in a sinus elevation graft, their osteoconductive properties must be complemented by the osteoinductive properties supplied by activated PRP. PRP is first used to incubate the graft material before it is placed in the sinus cavity. In both cases, PRP's growth factors can enable the healing action of the endosteal osteoblasts and marrow stem cells that have migrated from the bony sinus walls, while the allogeneic bone particles are bound by fibrin, fibronectin, and vitronectin, facilitating cell migration and bone regeneration.

Alternatively, a graft composite of autogenous bone and bone substitutes or allogeneic bone mixed with the mature PRP clot will combine the bone regeneration/healing properties of the separate graft materials with adjunctive PRP. A PRP membrane can be used to close the oval window of the lateral sinus wall once the graft is in place.[5]

Conclusion

PRP is useful in a variety of procedures that are performed to preserve and/or regenerate bone to facilitate implant placement. The growth factors released by PRP are believed to facilitate osteogenesis, and PRP's cell adhesion molecules can enhance osteoconduction. The healing of bone after damage or injury or grafting requires a combination of processes, including osteogenesis, osteoconduction, and osteoinduction. PRP can provide the biologic assistance to facilitate bone healing for dental implant procedures using autogenous bone, allogeneic bone, synthetic bone, and bone graft composites to augment alveolar ridges.

References

1. Cornell CN, Lane JM. Current understanding of osteoconduction in bone regeneration. Clin Orthop Relat Res 1998;(355 suppl):S267–S273.
2. Albrektsson T, Johansson C. Osteoinduction, osteoconduction and osseointegration. Eur Spine J 2001;10(suppl 2):S96–S101.
3. Hayrapetyan A, Jansen JA, van den Beucken JJ. Signaling pathways involved in osteogenesis and their application for bone regenerative medicine. Tissue Eng Part B Rev 2015;21:75–87.
4. García-Gareta E, Coathup MJ, Blunn GW. Osteoinduction of bone grafting materials for bone repair and regeneration. Bone 2015;81:112–121.
5. Marx RE, Garg A. Dental and Craniofacial Applications of Platelet-Rich Plasma. Chicago: Quintessence, 2005.
6. Yang LC, Hu SW, Yan M, Yang JJ, Tsou SH, Lin YY. Antimicrobial activity of platelet-rich plasma and other plasma preparations against periodontal pathogens. J Periodontol 2015;86:310–318.
7. Geurs N, Ntounis A, Vassilopoulos P, Van der Velden U, Loos BG, Reddy M. Using growth factors in human extraction sockets: A histologic and histomorphometric evaluation of short-term healing. Int J Oral Maxillofac Implants 2014;29:485–496.
8. Hatakeyama I, Marukawa E, Takahashi Y, Omura K. Effects of platelet-poor plasma, platelet-rich plasma, and platelet-rich fibrin on healing of extraction sockets with buccal dehiscence in dogs. Tissue Eng Part A 2014;20(3–4):874–882.
9. Döri F, Arweiler N, Húszár T, Gera I, Miron RJ, Sculean A. Five-year results evaluating the effects of platelet-rich plasma on the healing of intrabony defects treated with enamel matrix derivative and natural bone mineral. J Periodontol 2013;84:1546–1555.
10. Menezes LM, Rao J. Long-term clinical evaluation of platelet-rich plasma in the treatment of human periodontal intraosseous defects: A comparative clinical trial. Quintessence Int 2012;43:571–582.
11. Park OJ, Yi H, Jeon JH, et al. Pyrosequencing analysis of subgingival microbiota in distinct periodontal conditions. J Dent Res 2015;94:921–927.
12. Mlachkova AM, Popova CL. Efficiency of nonsurgical periodontal therapy in moderate chronic periodontitis. Folia Med (Plovdiv) 2014;56:109–115.
13. Albandar JM. Aggressive and acute periodontal diseases. Periodontol 2000 2014;65:7–12.
14. Panda S, Doraiswamy J, Malaiappan S, Varghese SS, Del Fabbro M. Additive effect of autologous platelet concentrates in treatment of intrabony defects: A systematic review and meta-analysis. J Investig Clin Dent 2016;7:13–26.
15. Kurikchy MQ, Al-Rawi NH, Ayoub RS, Mohammed SS. Histological evaluation of bone healing using organic bovine bone in combination with platelet-rich plasma (an experimental study on rabbits). Clin Oral Investig 2013;17:897–904.
16. Yilmaz S, Cakar G, Kuru B, Dirikan S, Yildirim B. Platelet-rich plasma in combination with bovine derived xenograft in the treatment of deep intrabony periodontal defects: A report of 20 consecutively treated patients. Platelets 2009;20:432–440.
17. Tavarez RR, Calixto AM, Maia Filho EM, et al. Atraumatic extraction, implant placement and immediate provisionalization. J Contemp Dent Pract 2014;15:513–517.
18. Leblebicioglu B, Hegde R, Yildiz VO, Tatakis DN. Immediate effects of tooth extraction on ridge integrity and dimensions. Clin Oral Investig 2015;19:1777–1784.
19. Iyer SS, Haribabu PK. Minimizing alveolar bone loss during and after extractions (Part I)—Review of techniques: Atraumatic extraction, root retention. Alpha Omegan 2013;106(3–4):67–72.
20. Talreja PS, Suvarna CR, Talreja PP. Alveolar ridge splitting for implant placement: A review of the procedure and report of 3 cases. Gen Dent 2015;63(1):e20–e23.
21. Milinkovic I, Cordaro L. Are there specific indications for the different alveolar bone augmentation procedures for implant placement? A systematic review. Int J Oral Maxillofac Surg 2014;43:606–625.
22. Aloy-Prósper A, Peñarrocha-Oltra D, Peñarrocha-Diago M, Peñarrocha-Diago M. The outcome of intraoral onlay block bone grafts on alveolar ridge augmentations: A systematic review. Med Oral Patol Oral Cir Bucal 2015;20:e251–e258.
23. Aghaloo TL, Moy PK. Which hard tissue augmentation techniques are the most successful in furnishing bony support for implant placement? Int J Oral Maxillofac Implants 2007;22(suppl):49–70 [erratum 2008;23:56].
24. Kolokythas A, Olech E, Miloro M. Alveolar osteitis: A comprehensive review of concepts and controversies. Int J Dent 2010;2010:249073.
25. Barona-Dorado C, González-Regueiro I, Martín-Ares M, Arias-Irimia O, Martínez-González JM. Efficacy of platelet-rich plasma applied to post-extraction retained lower third molar alveoli. A systematic review. Med Oral Patol Oral Cir Bucal 2014;19:e142–e148.
26. Alissa R, Esposito M, Horner K, Oliver R. The influence of platelet-rich plasma on the healing of extraction sockets: An explorative randomised clinical trial. Eur J Oral Implantol 2010;3:121–134.
27. Rutkowski JL, Fennell JW, Kern JC, Madison DE, Johnson DA. Inhibition of alveolar osteitis in mandibular tooth extraction sites using platelet-rich plasma. J Oral Implantol 2007;33:116–121.
28. Cardarelli A, Garg AK. Effects of platelet-rich plasma and piezosurgery on impacted mandibular third molar surgery outcomes: Preliminary study. Adv J Otolaryngol 2020;1:1–7.
29. Fretwurst T, Gad LM, Nelson K, Schmelzeisen R. Dentoalveolar reconstruction: Modern approaches. Curr Opin Otolaryngol Head Neck Surg 2015;23:316–322.
30. Sittitavornwong S, Gutta R. Bone graft harvesting from regional sites. Oral Maxillofac Surg Clin North Am 2010;22:317–330.
31. Dragan E, Rusa O, Nemtoi A, Melian G, Mihai C, Haba D. Maxillary sinus anatomic and pathologic CT findings in edentulous patients scheduled for sinus augmentation. Rev Med Chir Soc Med Nat Iasi 2014;118:1114–1121.

32. Beaumont C, Zafiropoulos GG, Rohmann K, Tatakis DN. Prevalence of maxillary sinus disease and abnormalities in patients scheduled for sinus lift procedures. J Periodontol 2005;76:461–467.
33. Kao SY, Lui MT, Cheng DH, Chen TW. Lateral trap-door window approach with maxillary sinus membrane lifting for dental implant placement in atrophied edentulous alveolar ridge. J Chin Med Assoc 2015;78:85–88.
34. Chen TW, Chang HS, Leung KW, Lai YL, Kao SY. Implant placement immediately after the lateral approach of the trap door window procedure to create a maxillary sinus lift without bone grafting: A 2-year retrospective evaluation of 47 implants in 33 patients. J Oral Maxillofac Surg 2007;65:2324–2328.
35. Al-Dajani M. Recent trends in sinus lift surgery and their clinical implications. Clin Implant Dent Relat Res 2016;18:204–212.
36. Schwarz L, Schiebel V, Hof M, Ulm C, Watzek G, Pommer B. Risk factors of membrane perforation and postoperative complications in sinus floor elevation surgery: Review of 407 augmentation procedures. J Oral Maxillofac Surg 2015;73:1275–1282.
37. Taschieri S, Testori T, Corbella S, et al. Platelet-rich plasma and deproteinized bovine bone matrix in maxillary sinus lift surgery: A split-mouth histomorphometric evaluation. Implant Dent 2015;24:592–597.
38. Kumar NK, Shaik M, Nadella KR, Chintapalli BM. Comparative study of alveolar bone height and implant survival rate between autogenous bone mixed with platelet rich plasma versus venous blood for maxillary sinus lift augmentation procedure. J Maxillofac Oral Surg 2015;14:417–422.

8

Facial Cosmetics and Autologous Blood Concentrates

Because of the simplicity and low cost of PRP production technology, more plastic surgeons are using autologous platelet derivatives in facial cosmetic surgery procedures to reduce swelling and ecchymosis as well as to facilitate healing in rhytidectomies, blepharoplasties, and laser skin resurfacing.[1,2] PRP also facilitates healing after dermatologic surgery while reducing scarring from the removal of skin lesions.[3] Finally, cosmetic dermatologists are increasingly using PRP in place of hyaluronic acid and other expensive dermal fillers to stimulate neocollagenesis and increase elasticity in patients seeking facial rejuvenation.

A 2015 critical review of the literature on topical and injectable platelet preparations for use in facial rejuvenation and wound healing concluded that a clear majority of studies report substantive, quantifiable enhancement not only of cellular changes but of wound healing and facial esthetic results.[4] In all cases, the clinician must realize that patient expectations for "perfect" outcomes generally exceed reality. Nevertheless, the popular demand for facial cosmetic treatments continues to grow, and along with it the use of platelet-derived products, not only as adjuncts in traditional plastic surgery procedures but also increasingly as standalone topical and injectable forms, or as complements to dermal fillers.

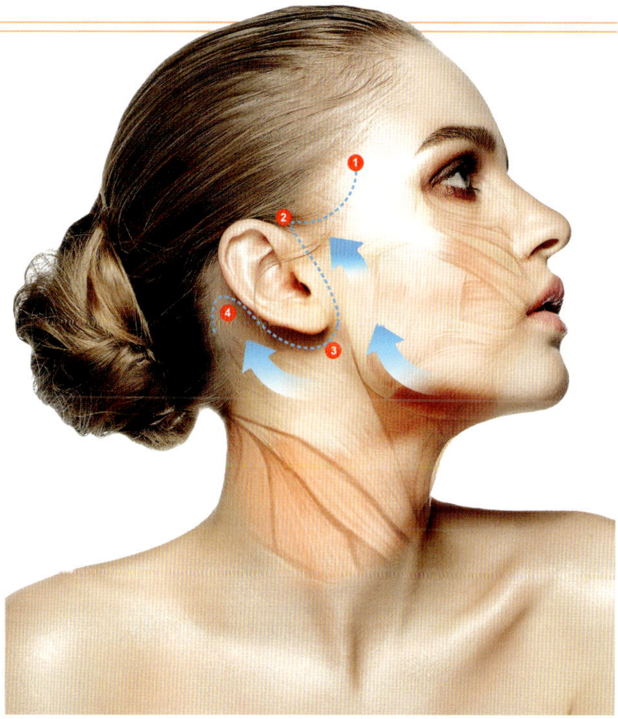

FIG 8-1 Facelift surgery begins with an incision in the temporal area (1) that extends vertically into the preauricular skinfold (2) before curving below the ear lobule (3) into the postauricular region (4). A preauricular flap lifts the cheek, whereas the postauricular flap lifts the neck and constricts the submental area *(blue arrows)*.

Traditional Cosmetic Surgery

Rhytidectomy, blepharoplasty, and dermal fat grafts represent popular traditional cosmetic surgery procedures that can be substantively enhanced by the use of PRP as an adjunctive agent.

Rhytidectomy

A facelift, or rhytidectomy, is the most common type of facial cosmetic surgery. In this surgery, an incision is made in the temporal area and extended vertically into the preauricular skinfold and then curved below the ear lobule into the postauricular region (Fig 8-1). A preauricular flap is created to lift the cheek and to remove

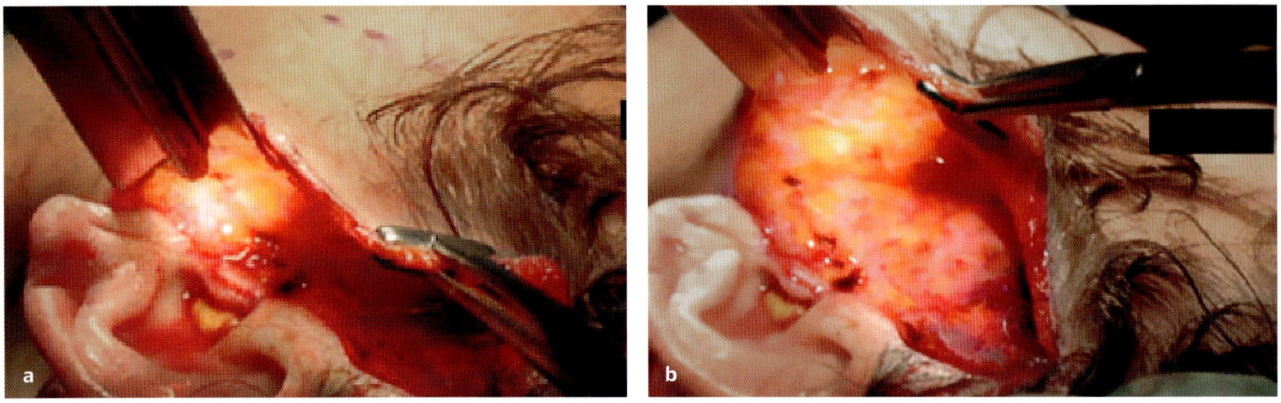

FIG 8-2 Surgical design of traditional facelift procedure. *(a)* Traditional facelift incision beginning in the hairline at the temples and continuing around the ear and ending in the lower scalp. *(b)* The tissues are undermined to allow for advancement, redraping, and trimming away of excess skin prior to suturing. PRP is applied to the undermined skin flap and the flap's edges to help them adhere to the deep fascia and to facilitate hemostasis and wound healing. PRP is also placed along the incision line to reduce scarring and the loss of hair follicles.

the jowls, while the postauricular flap lifts the neck and constricts the submental area. The clinician undermines the skin well beneath the superficial musculoaponeurotic system (SMAS), the fascia inferiorly continuous with the platysma muscle and superiorly with the superficial temporal fascia. The SMAS shrouds the face superficial to the parotid fascia and is connected to the fascial musculature nasolabially, periorally, and periorbitally. A beveling incision should be used in the temporal areas to reduce the chances for hair loss due to interrupted blood supply, and to avoid injury to the facial nerve branches, the plane of dissection should not be as deep as the temporal fascia.[5-8]

Once the skin flaps are pulled to determine the required lift, the surplus skin is excised, and the bleeding vessels are cauterized. Before closing, 4 to 5 mL of PRP is applied both to the undermined skin flap and to the flap's edges (Fig 8-2). Facial tightening via traction sutures is undertaken to suspend the SMAS to the temporal and mastoid areas. Subcuticular and skin closure employ PRP along the incision line, if needed, to position the skin flap evenly. Pressure dressing protects the treatment for 48 to 72 hours, followed by ice for 48 hours, head elevation, and hydrogen peroxide/antibiotic ointment at the incision areas.[5,7,8]

In addition to helping the skin flaps closely adhere to the deep fascia, the fibrin-fibronectin-vitronectin cell action of PRP facilitates hemostasis and wound healing generally, helping to prevent the development of ecchymosis and hematomas. Sclafani and Saman have used a platelet-rich fibrin (PRF) matrix for facial plastic surgery applications, not only for volume augmentation and fat transfer supplementation but also as an adjunct to open surgical procedures.[9,10] Scarring and the loss of hair follicles along the flap incision lines are reduced, as PRP also helps to quicken the growth of capillary, collagen, and nerve tissue from the underlying fascia. An additional, indirect effect of PRP cell action is the reduction of bacterial and viral infections in the treatment area. The rate and degree of normal skin sensation are accelerated by PRP's growth factors, which facilitate the ingrowth of capillaries and the synthesis of collagen (Box 8-1).

Box 8-1 Effects of PRP application in facelift surgery

- Helps skin flaps closely adhere to the deep fascia
- Reduces scarring and the loss of hair follicles along the incision line
- Facilitates hemostasis and wound healing
- Promotes the growth of capillary, collagen, and nerve tissue from the underlying fascia
- Reduces bacterial and viral infections in the treatment area
- Accelerates the return of natural skin sensation

While PRP can facilitate healing and help to prevent some complications associated with facelift surgery, there are a number of complications where PRP has no useful effect—a reminder that expert surgical technique is essential in this and other treatments where PRP could have adjunctive facility. For example, facial nerve injury, earlobe deformity, and certain causes of skin slough cannot be prevented by PRP use. However, PRP can be

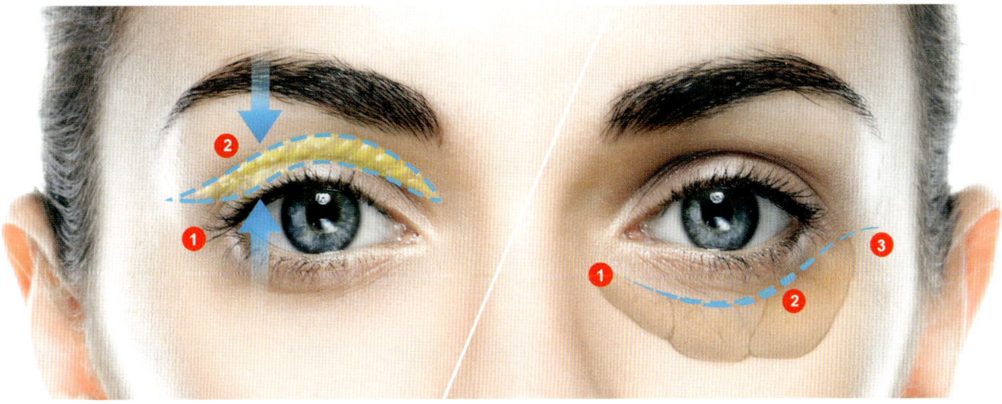

FIG 8-3 Incision design for blepharoplasty of upper and lower eyelids. *(left)* For the upper eyelid, an incision is made 7 to 10 mm above the ciliary margin following the curvature of the extra skin fold (1). A second incision is made superior and parallel to the first one but connecting medially and laterally (2). PRP is applied to the exposed orbital fat, muscle, and skin edges in conjunction with surface closure of the incisions. *(right)* For the lower eyelid, an incision is made parallel to and about 2 to 3 mm inferior to the ciliary margin (1). The incision gradually diverges at the lateral third of the lid an additional 1 mm below the ciliary margin (2), then merges into a natural lateral crease (3). PRP is applied to the wound site just prior to it being sutured closed.

useful in a number of other complications associated with rhytidectomy, as already noted. Ecchymosis (bruising) is the most common complication of a rhytidectomy but generally resolves itself naturally; however, it can develop into a hematoma or infection and lead to skin slough. Ecchymosis and hematoma can be prevented through the application of PRP, which promotes complete hemostasis and the elimination of space between the bottom of the skin flap and the deeper fascia. To prevent other unfortunate complications of the facelift procedure—namely, flap dehiscence and excessive scarring—the clinician should use PRP in conjunction with flap closure under mild tension.[11]

Blepharoplasty

The second most common facial cosmetic surgery, blepharoplasty, removes extra skin and muscle from beneath both the upper and lower eyelids. Surgical treatment planning, execution, and healing can lead to excellent results. Potential complications include excessive tearing, dry eyes, excessive scleral show, conjunctivitis, eversion/inversion of the eyelid, and corneal injury. PRP can facilitate healing and reduce complications in both types of blepharoplasty[5,12,13] (Fig 8-3).

Upper eyelid

With the patient gazing forward in a relaxed, seated position, the curvature of the extra skin fold is marked, 7 to 10 mm above the ciliary margin.[5,14] The line ends medially, slightly superior to the punctum, and laterally above the tarsal plate, in or parallel to the lateral crease, about a centimeter lateral to the lateral canthus. A second line is drawn superior and parallel to the first one, blending into it medially and laterally (Fig 8-3, left). Between the lines lies the extra tissue to be removed; the clinician grasps it with a tissue forceps until a lagophthalmos begins but with laxity of tension to prevent inferior movement of the eyebrow.

A precise incision is made along the planned lines, and the extra skin is removed to expose the fibers of the orbicularis oculi. Next, a smaller amount of muscle tissue, superior to the tarsal plate, is removed while the patient repeatedly opens and closes the eyes. Light pressure is used to help the exposed orbital fat protrude, which allows it to be sharply excised with little or no retraction. A 1.5-mL volume of PRP is applied to the exposed orbital fat, muscle, and skin edges to remove dead space in conjunction with surface closure of the incision, using a 6-0 nonresorbable suture. PRP-induced hemostasis reduces ecchymosis, excessive scarring, and lid retraction, while growth factors facilitate healing[5,14] (Box 8-2).

Box 8-2 Effects of PRP application in blepharoplasty procedure

- Facilitates healing
- Reduces complications
- Removes dead space during surface closure
- Reduces ecchymosis, excessive scarring, and lid retraction
- Preserves the excised fat for adjacent facial grafting, if required

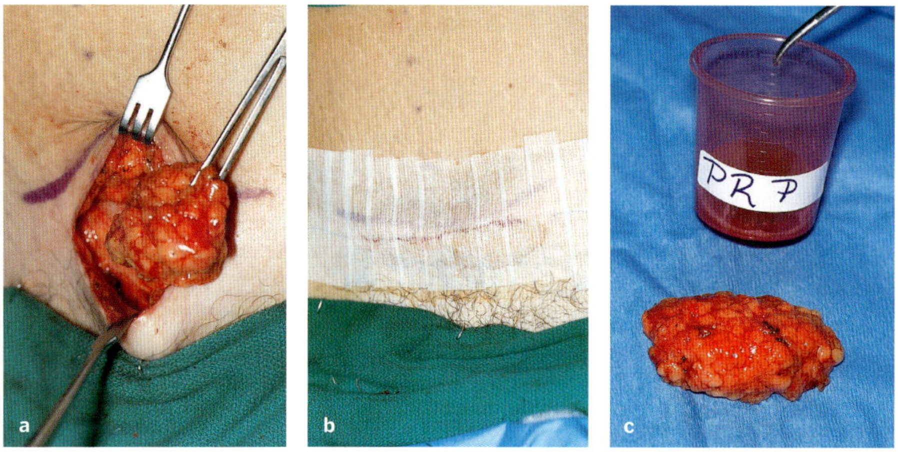

FIG 8-4 Harvest of dermal fat graft for facial augmentation. *(a)* A natural fold in the suprapubic area is a convenient and cosmetic site for harvesting dermal fat. *(b)* A linear closure of a periumbilical or suprapubic incision can be accomplished with dermal sutures supported by Steri-Strips (3M). The incision will eventually resemble a natural skin crease. *(c)* The dermal fat graft should not be particulated. Ideally, it should be taken directly from the donor site and placed into the recipient site. If it cannot be placed immediately into the recipient site, incubation in PRP liquid or even saline will maintain lipoblast and lipocyte survival. (Reprinted with permission from Marx and Garg.[5])

Lower eyelid

Lower eyelid blepharoplasty poses a higher risk of complications than upper eyelid procedures, including eyelid laxity, excessive scleral show, and ectropion.[15,16] Just below and lateral to the punctum of the inferior canaliculus and extending laterally beyond the lateral canthus into a natural crease, a line is drawn parallel to and approximately 2 to 3 mm below the ciliary margin. It gradually diverges at the lateral third of the lid an additional 1 mm below the ciliary margin, then merges into a natural lateral crease (Fig 8-3, right). Superficial to the orbicularis oculi, an inferior-based skin flap is developed via dissection. Extra muscle to excise is determined by the patient's opening and closing the eyes; however, a rim of pretarsal muscle must be left intact to preserve function. Protruding fat is removed as well. PRP can be used as an incubator to preserve the fat for adjacent facial grafting, if required. Next, the clinician approximates the skin flap to the subciliary incision line, removes excess skin, and drapes the skin flap passively over the wound. As with the upper lid blepharoplasty, a 1.5-mL volume of activated PRP is applied to the wound healing site, and a 6-0 nonresorbable suture is used to close the skin flap.

Autologous fat transfer

Nonparticulated dermal graft

Because of its ability to facilitate rapid capillary ingrowth, PRP has enabled clinicians to avoid many complications traditionally linked to dermal fat grafting. As always, the adjunctive use of PRP with free nonvascularized dermal fat grafts for facial cosmetic surgery does not obviate the need for precise surgery techniques to be used for harvesting, handling, and placing the grafts. Harvesting requires absolute sterility and optimal harvest sites such as the gluteal crease or an abdominal skin fold (Fig 8-4).[5,17-19] Fat should be harvested immediately prior to site grafting and then placed in saline or nonactivated PRP to prevent dehydration and to preserve lipoblasts and lipocytes. At the recipient site, the epidermis and full thickness of the dermis should be maintained as a minimum thickness and may include as much of the subcutaneous layer as possible (Fig 8-5).[5]

PRP liquid should be applied to the graft before it is sutured to the base of the grafting site. The edges of the graft should be tapered, and external bolsters should be used to avoid creating a line of demarcation between the graft and the adjacent native tissue. Once the final contours of the graft and the recipient site have been attained, about 5 to 10 mL of PRP should be added to the surface of the graft and over the entire recipient site to enhance stability (ie, reduce contraction of the graft and avoid loss of contour) via lipoblast and lipocyte survival. PRP also reduces infection, skin dehiscence, swelling, and bruising.[20-22] A 2014 study concluded that adding PRP to facial lipofilling reduces recovery time and improves the overall esthetic outcome in a minimal-access cranial suspension lift and fat grafting procedure (Fig 8-6).[5,23]

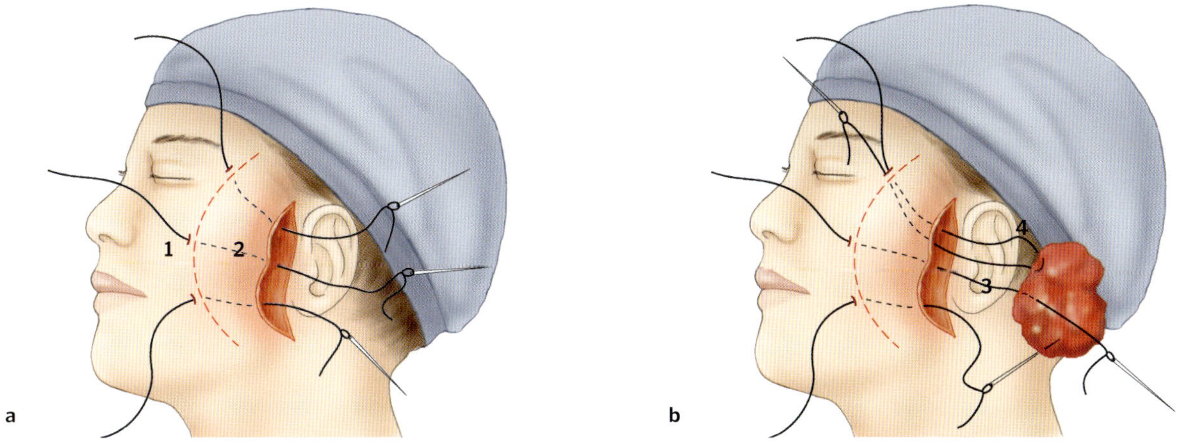

FIG 8-5 Technique for securing dermal fat graft. *(a)* Dermal fat grafting for facial augmentation begins with the needle entering the skin 1 to 2 cm from the edge of the wound space (1). From there, it enters the wound space at the very edge, where the native tissue joins the wound space (2). *(b)* Next, the needle passes in at one edge of the fat graft and then out, engaging a 5- to 8-mm section in a fashion similar to a horizontal mattress suture (3). The needle then passes through the wound space at the point where it adjoins native tissue and out the skin 5 to 8 mm from its entrance point (again similar to a mattress suture) (4). (Reprinted with permission from Marx and Garg.[5])

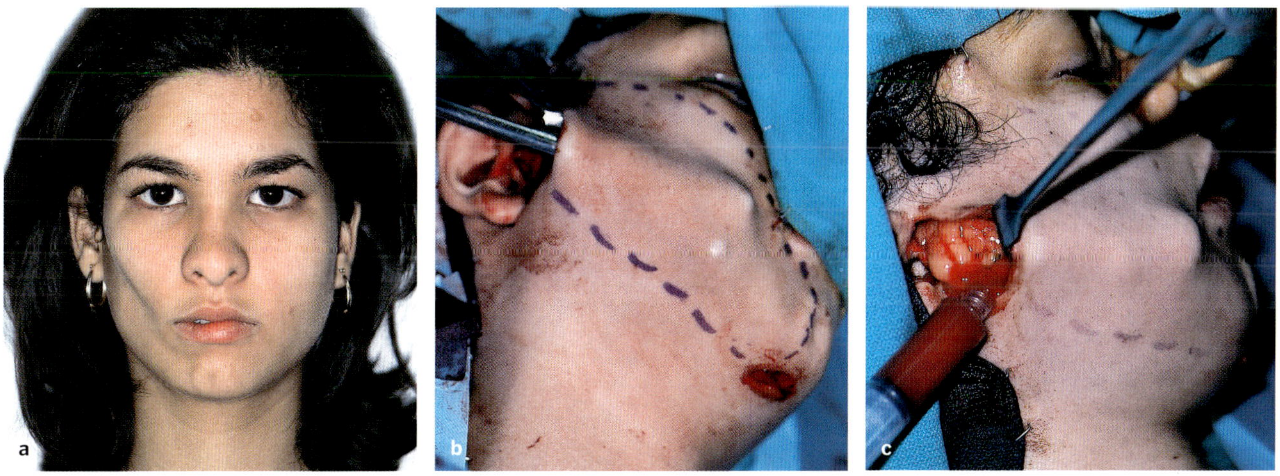

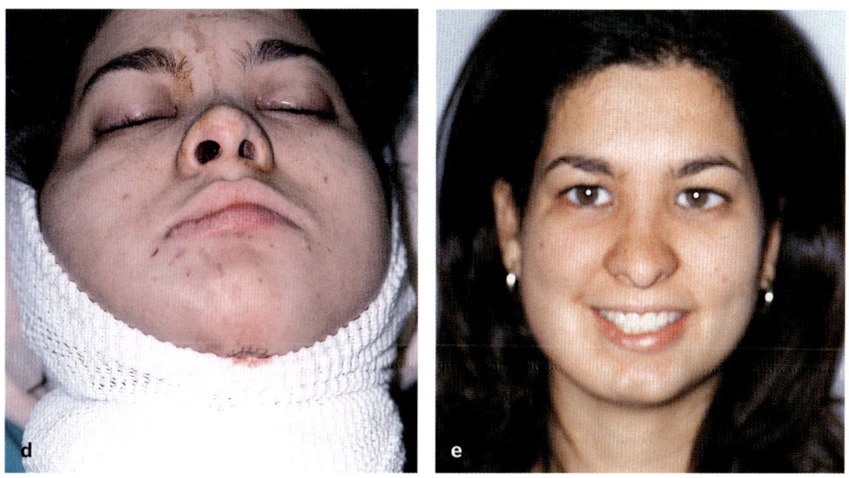

FIG 8-6 Dermal fat grafting with PRP for correction of a congenital defect. *(a)* Facial atrophy of the right side due to Parry-Romberg syndrome. *(b)* A rhytidectomy type of dissection is used to develop recipient tissue for a dermal fat graft. *(c)* Dermal fat graft harvested from the gluteal crease is placed with PRP liquid over and around it. *(d)* When a dermal fat graft is placed, pressure dressings are required to reduce dead space and to stabilize the graft. *(e)* Predictable correction of facial atrophy to normal contours is achieved using dermal fat grafting with PRP enhancement. (Reprinted with permission from Marx and Garg.[5])

FIG 8-7 Lipofilling procedure using dermal fat combined with PRP. *(a)* Preoperative view of a middle-aged patient with common signs of aging, including wrinkles, deep folds, dermal atrophy, and photoaging. *(b)* Markup of the areas targeted for lipofilling, including the malar regions, nasolabial folds, marionette lines, and forehead. *(c)* Syringes of micronized autologous fat that has been centrifuged and combined with PRP. *(d)* Appearance 1 year posttreatment demonstrates the lasting benefits of lipofilling therapy. (Courtesy of Keith Jeffords, MD, DDS.)

Facial lipofilling

Lipofilling is a form of autologous fat grafting that differs from nonparticulated fat grafting in several ways. First, the fat can be harvested from the patient via vacuum aspiration (liposuction), syringe aspiration, or surgical excision. Methods that are less traumatic result in superior adipocyte viability and graft survival. Second, the harvested fat must be prepared for transplantation because it contains not only adipocytes but also collagen fibers, blood, and other substances. The most common methods of fat preparation are sedimentation, filtering, washing, and centrifugation. Since being introduced by Coleman in 1987,[24] centrifugation has remained the gold standard. Aspirated fat in syringes is centrifuged at 3,000 rpm for 3 minutes, separating it into three distinct layers of lipids, fatty tissue, and blood. The middle layer is used for grafting, and the other two layers are discarded. Third, the processed fat is transferred to the recipient site via injection of small aliquots through a small-gauge cannula delivered in a fanning pattern.

The primary esthetic indications for lipofilling for facial rejuvenation are the correction of dark circles; replacement of lost facial volume, especially in the periorbital regions as a result of aging; and correction of deep wrinkles and folds (Fig 8-7). Moreover, because it contains stem cells, fat grafting is not only a more affordable and longer-lasting alternative to dermal fillers, but it also possesses potent regenerative capacity that significantly increases its value as an anti-aging therapy.

Historically, fat grafting is associated with a number of clinical shortcomings that can only be overcome by the addition of PRP. Dermal fat has a well-documented tendency to resorb that has often led to poor outcomes. This problem is partly a result of delayed capillary ingrowth,

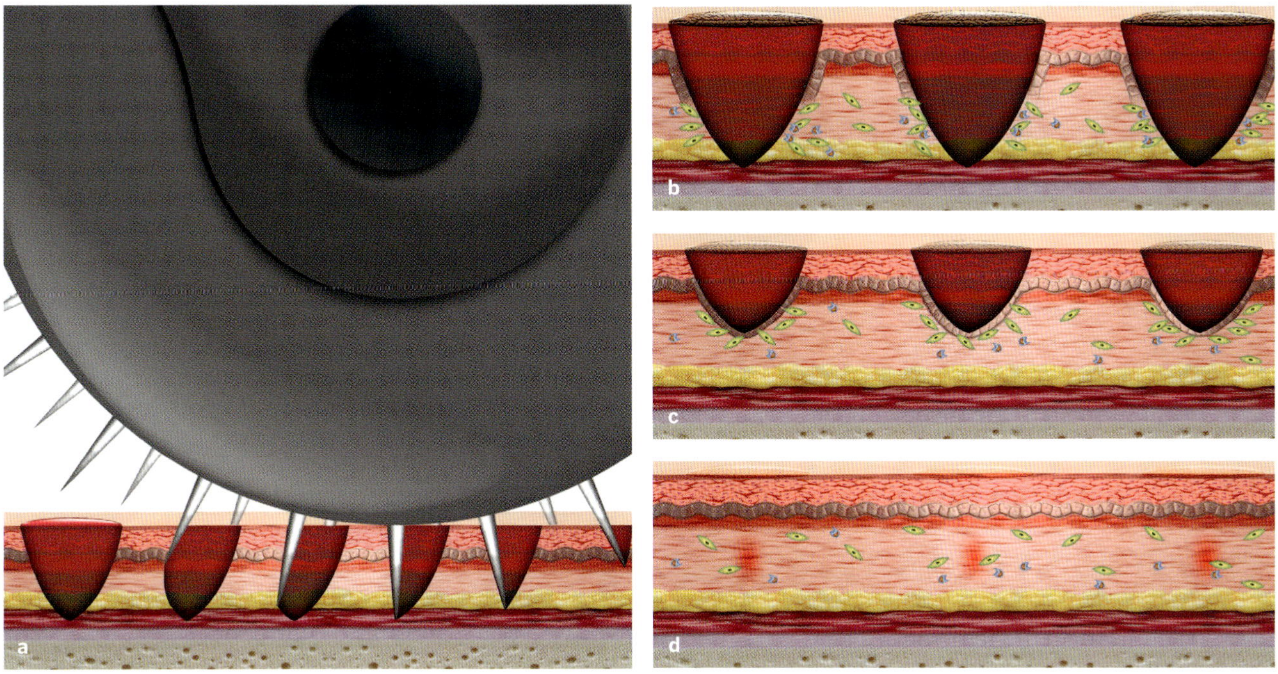

FIG 8-8 The four-stage wound healing cycle. *(a)* Bleeding. Microneedles penetrate the epidermis, rupturing blood vessels and triggering a release of platelets. Blood clots are formed, and the bleeding stops. *(b)* Inflammation. White blood cells enter the tiny wounds to facilitate inflammation during vasodilation. Neutrophils and fibroblasts migrate into the area to clear bacteria and other debris. *(c)* Proliferation. Healing begins as granulation tissue—consisting of fibroblasts, inflammatory cells, new blood vessels, and collagen—grows across the surface of each wound, closing and protecting it. *(d)* Maturation. As tissue remodeling continues, additional collagen production increases the skin's strength and elasticity. However, because the wounds are superficial, no scars are formed. (Reprinted with permission from Garg and Rossi.[27])

Box 8-3 Benefits of PRP in fat grafting procedures

- Promotes rapid capillary ingrowth
- Enhances lipoblast and lipocyte survival
- Improves dimensional graft stability
- Lowers the incidence of infections and skin dehiscence
- Reduces swelling and skin bruising

which results in the death of lipoblasts and lipocytes contained in the graft that are needed for its survival. Necrosis of these fat cells provokes an intense inflammatory response that produces swelling, pain, possible skin dehiscence, and additional fat necrosis. In consequence, the graft shrinks, develops internal abscesses, may calcify, and might even result in unsightly scarring. PRP's capacity to promote a rapid capillary ingrowth addresses all of these shortcomings, making lipofilling for esthetic facial augmentation a straightforward and successful treatment option[5,25,26] (Box 8-3).

Facial Rejuvenation Procedures

The first part of this chapter focused on the unique role of PRP liquid as an adjunct to traditional cosmetic plastic surgery procedures. In this section we examine how different configurations of PRP can be used as a standalone agent to stimulate neocollagenesis and combat the natural effects of aging in patients seeking facial rejuvenation.

Microneedling

Microneedling, also known as *collagen induction therapy*, uses a simple handheld device to deliver hundreds of tiny, invisible punctures to the top layer of facial skin. These micro-injuries stimulate the body's natural wound healing processes, leading to cell turnover and the production of new collagen and elastin (Fig 8-8).[27]

Although it is a relatively new treatment option in cosmetic therapy, microneedling offers a less-invasive alternative to more conventional procedures such as laser

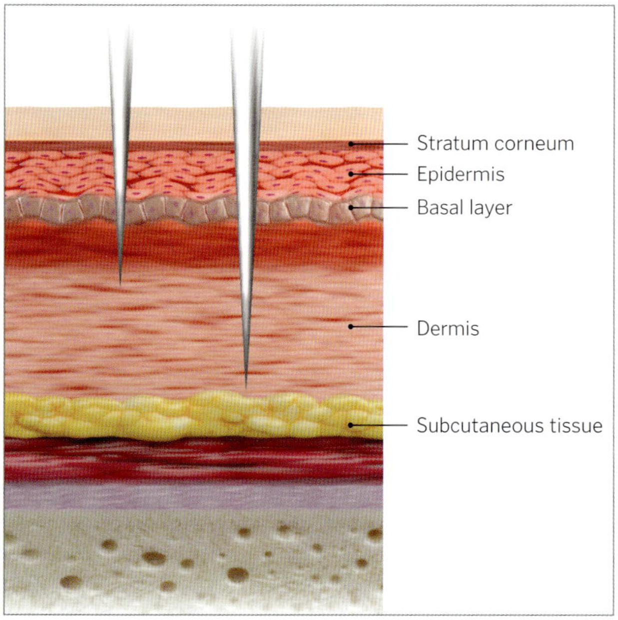

FIG 8-9 The needles of a microroller should reach within the dermal layer and no further. (Reprinted with permission from Garg and Rossi.[27])

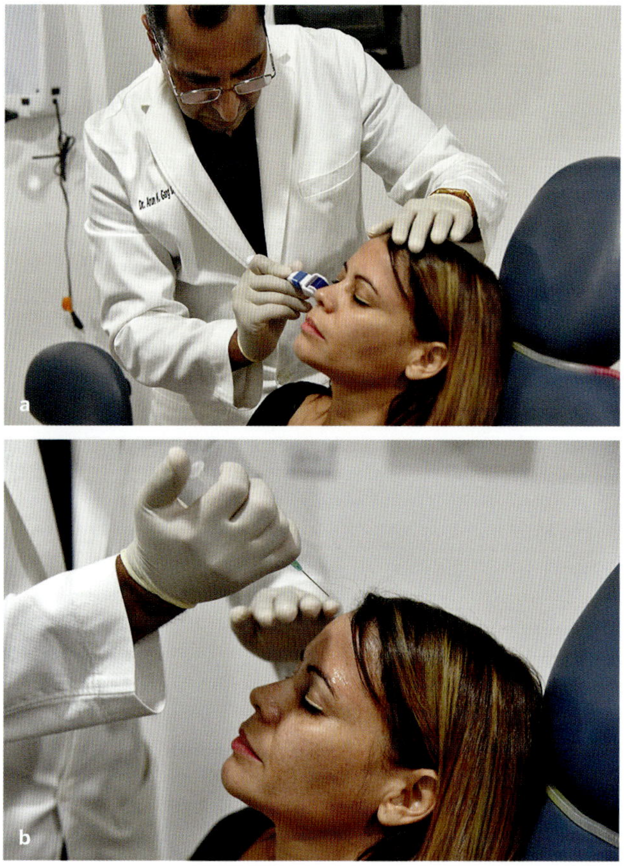

FIG 8-10 ➤ Dermal microneedling followed by topical PRP application. *(a)* Dermal microneedling treatment using a microroller device, which creates approximately 200 punctures per square centimeter of facial skin. *(b)* Topical application of PRP liquid immediately after the microneedling treatment, while the microchannels remain open.

skin resurfacing and deep chemical peeling for patients seeking skin rejuvenation or treatment for many other common dermatologic conditions.

Facial rejuvenation

Numerous studies have demonstrated the effectiveness of reducing and softening facial rhytids through microneedling treatment. Clinically, the microneedles penetrate 1.5 mm to 2 mm into the dermis (Fig 8-9)[27]; at this depth, the needles cut through the epidermis but do not destroy it. Although the epidermis is only about 0.2 mm thick, it serves as the body's sole protection from the environment. Therefore, the epidermis, and particularly the stratum corneum, must remain intact.

When applied topically following microneedling treatment, PRP has been shown to improve skin texture, color homogeneity, elasticity, and firmness[11] (Fig 8-10). PRP works by stimulating the proliferation of mesenchymal stem cells and fibroblasts, which secrete hyaluronic acid and type 1 collagen. The adjunctive use of PRP with microneedling is low-cost, minimally invasive, and low-risk. For best results, the procedure should be repeated two to three times at intervals of 8 to 10 weeks.

The goal of microneedling is to create as many microchannels as possible. Normal saline is then used to wash away the flow of blood in the needling area. Because the needling channels remain open for no more than 15 minutes, PRP should be topically applied to the treated areas immediately after microneedling is completed (see Fig 8-10b). Although serum may ooze from the skin for up to 20 minutes after the procedure, bleeding should cease almost immediately.

Microneedling combined with topical PRP has many advantages over other traditional cosmetic procedures, such as laser resurfacing or skin peeling. It is noninvasive, does not damage the skin, and can be performed using topical anesthesia, making it extremely attractive to patients. The technique is easy to master and requires only a few tools to perform, which minimizes the cost of the procedure. When combined with PRP, the procedure leaves the skin thicker, with more collagen and elastin.

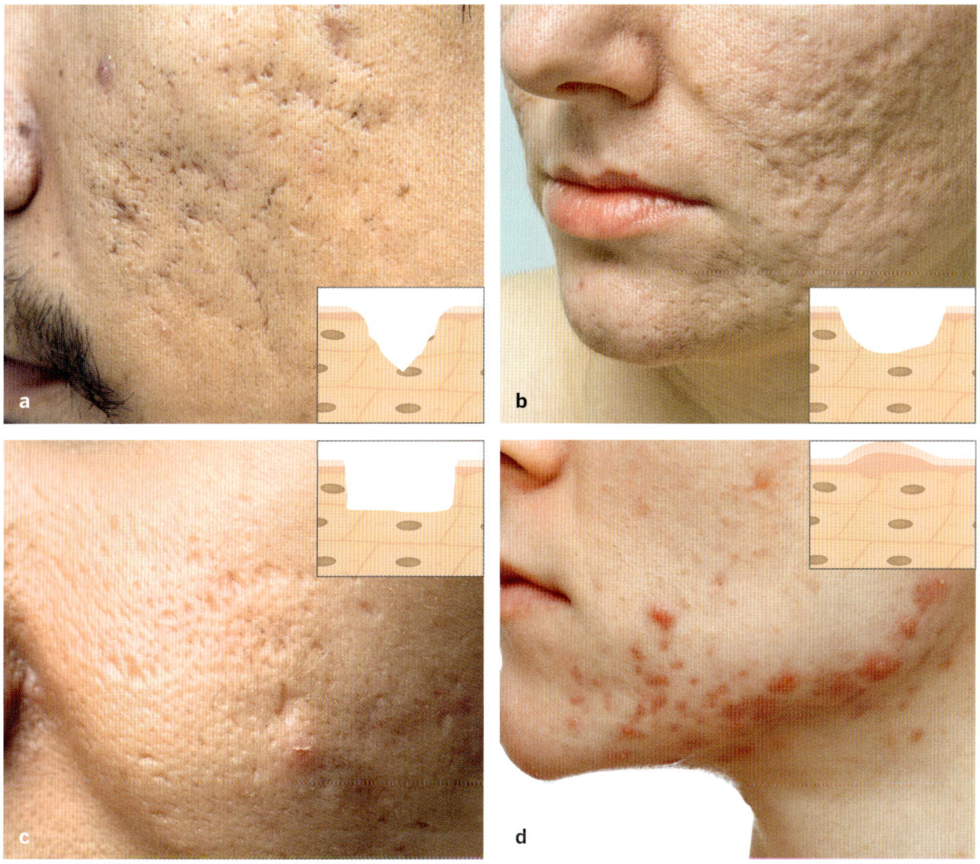

FIG 8-11 Microneedling prior to application of PRP for improvement of acne scars. Microneedling is more effective in improving boxcar and rolling patterns than ice pick scars. *(a)* Ice pick atrophic scar. *(b)* Rolling atrophic scar. *(c)* Boxcar atrophic scar. *(d)* Hypertrophic scar. (Reprinted with permission from Garg and Rossi.[27])

A 2018 randomized controlled trial (RCT) by El-Domyati et al compared the effectiveness of microneedling alone vs microneedling combined with PRP vs microneedling combined with trichloroacetic acid (TCA) peeling for facial rejuvenation. In this split-face comparative study, 24 volunteers with photoaging skin received one session every 2 weeks for six sessions of treatment. Photography and punch biopsies were performed before and after 3 months of treatment for clinical, histometric, and histologic evaluation. Based on the clinical evaluation of wrinkle appearance, skin texture, and overall satisfaction, the authors concluded that there was a significant increase in the improvement of the PRP-treated side when compared with both microneedling alone and microneedling with TCA peeling after 3 months. Assessment of skin biopsies demonstrated significantly increased thickness in the epidermis after all three treatments, although the increased thickness was slightly greater in the group that received dermaroller combined with TCA peeling treatment. Overall, the authors concluded that microneedling combined with PRP is more beneficial for facial rejuvenation than microneedling combined with TCA peeling or microneedling alone.[28]

Acne

Microneedling improves the appearance of scars not only by stimulating the regeneration of new collagen but also by disrupting the parallel pattern in which the collagen fibers are laid down.

More studies have been conducted on the effectiveness of microneedling for acne scarring than for any other condition. Clinical improvements ranging from 50% to 70% have been reported following a typical protocol of three to five treatments at 2- to 4-week intervals.[29,30] These clinical results have also been substantiated by histologic skin changes.[31] However, combining microneedling with topical application and/or injection of PRP has been shown to achieve better cosmetic outcomes for the treatment of acne scars than either treatment alone.

Scarring is the product of overexpression of type 1 collagen during the last stage of the wound healing process. Scars develop different physical characteristics according to the amounts of collagen overexpressed (Fig 8-11). An atrophic scar appears as a light or white sunken recess in the skin. This type of scar is most commonly associated with acne or chickenpox. Atrophic scars are further

Box 8-4 Benefits of PRP in the treatment of acne scars

- Stimulates regeneration of new collagen
- Disrupts the parallel pattern in which the collagen fibers are laid down
- Promotes tissue remodeling

subclassified as ice pick, rolling, or boxcar in shape, based on their physical appearance. The results of clinical studies have demonstrated that microneedling therapy improves the appearance of rolling and boxcar scars more effectively than ice pick scars.[32] A hypertrophic scar is red and appears to be raised above the surrounding skin. This type of scar will often develop as a result of a burn or traumatic injury. In multiple studies, microneedling produced improvements in the appearance of hypertrophic scars that were comparable to other mainstream treatment modalities, such as fractional laser, but with the advantage of a lower risk of side effects.[33]

The results of several recent studies suggest that the upregulation of growth factors associated with PRP enhances the effects of microneedling to promote tissue remodeling (Box 8-4). A 2017 RCT assigned 90 patients with acne scars to one of three treatment groups: *(1)* microneedling, *(2)* intradermal PRP, or *(3)* alternating microneedling and intradermal PRP treatments.[34] The PRP was injected only within the scars. Although all three groups showed improved scar appearance, combination treatment earned the highest mean improvement rating, followed by microneedling, then PRP. Patient satisfaction was also significantly higher for the combined PRP plus microneedling group.

A 2016 split-face study of 50 patients also evaluated microneedling with and without PRP for acne scars.[35] In this study, microneedling of the entire face was followed by intradermal PRP injection into acne scars and topical PRP application to one side of the face; distilled water was applied to the other side. While both sides showed clinical improvement, the PRP-treated side showed a superior response according to Goodman's qualitative and quantitative scoring systems as well as independent physician assessment. Patient satisfaction was similarly greater. At study completion, almost all patients stated that the addition of PRP reduced the visibility of acne scars and roughness of the skin. The synergistic effects of microneedling and PRP therapies offer a unique and effective treatment approach.

Hair loss

Androgenetic alopecia (AGA), also known as *male pattern baldness*, is a hereditary disease that affects half of the population. (Contrary to conventional wisdom, AGA affects women as well as men, although the effect is slightly different.) In these individuals, the scalp hair follicles are sensitive to dihydrotestosterone, causing them to shrink and gradually disappear and thereby resulting in hair loss. Once the follicle is gone, hair cannot be regenerated by any known treatment, including PRP. Hair loss has both physical and psychologic effects, and there is a high prevalence of depression in patients with AGA. The standard of care in the treatment of AGA is oral finasteride and topical minoxidil, but the negative side effects and the need for long-term treatment leads to considerable patient noncompliance.

Microneedling followed by topical PRP application has been found to induce hair regrowth in adults with AGA through stimulation of stem cells and activation of growth factors, which in turn stimulates the dermal papillae (Fig 8-12 and Box 8-5). However, a significant number of patients cannot receive microneedling treatment of the scalp (Fig 8-13) because of the length of their hair, which gets tangled in the microroller device. Similar to microneedling but more effective, subcutaneous injection of PRP into the scalp of AGA patients has shown great promise with no adverse effects (Fig 8-14). A recent survey of the literature revealed a total of 132 articles on PRP treatment of AGA, most of them published in the past 3 years, a finding that attests to the growing amount of interest in this novel therapy (Fig 8-15).

To evaluate its efficacy, in 2019 Mao et al performed a comprehensive meta-analysis of studies in which the results of PRP injections for AGA were compared against a control. Based on the data extracted from 11 articles comprising 262 AGA patients, the investigators found a statistically significantly increased density of hairs per square centimeter after PRP injections in the treatment group versus the control group (mean difference of 39 hairs). Four of the studies examined the effect of PRP on hair cross section, and all of them considered the treatment to be effective, although these results were not statistically significant. Patient satisfaction surveys showed that 70% of patients in the PRP group experienced an improvement in hair quality and thickness, and 55% reported an increase in hair density. Finally, all five of the histologic or immunohistochemical examinations concluded that the injection of

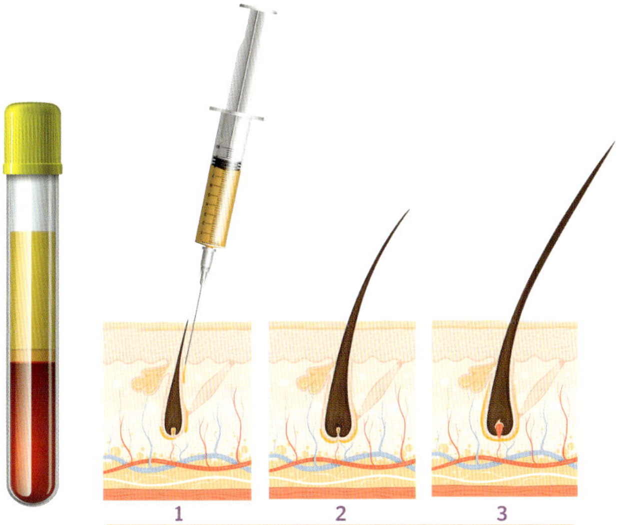

Box 8-5 Effects of PRP on hair loss resulting from AGA

- Stimulates stem cells
- Activates growth factors
- Stimulates the dermal papilla

FIG 8-12 When PRP is injected into the hair follicle, it induces the proliferation of dermal papilla cells (1) and increases phosphorylated extracellular signal-regulated kinases (pERKs) and Akt expression in the dermal papilla cells (2), which in turn induces cell growth, promotes cell survival, and prevents apoptosis (3).[36]

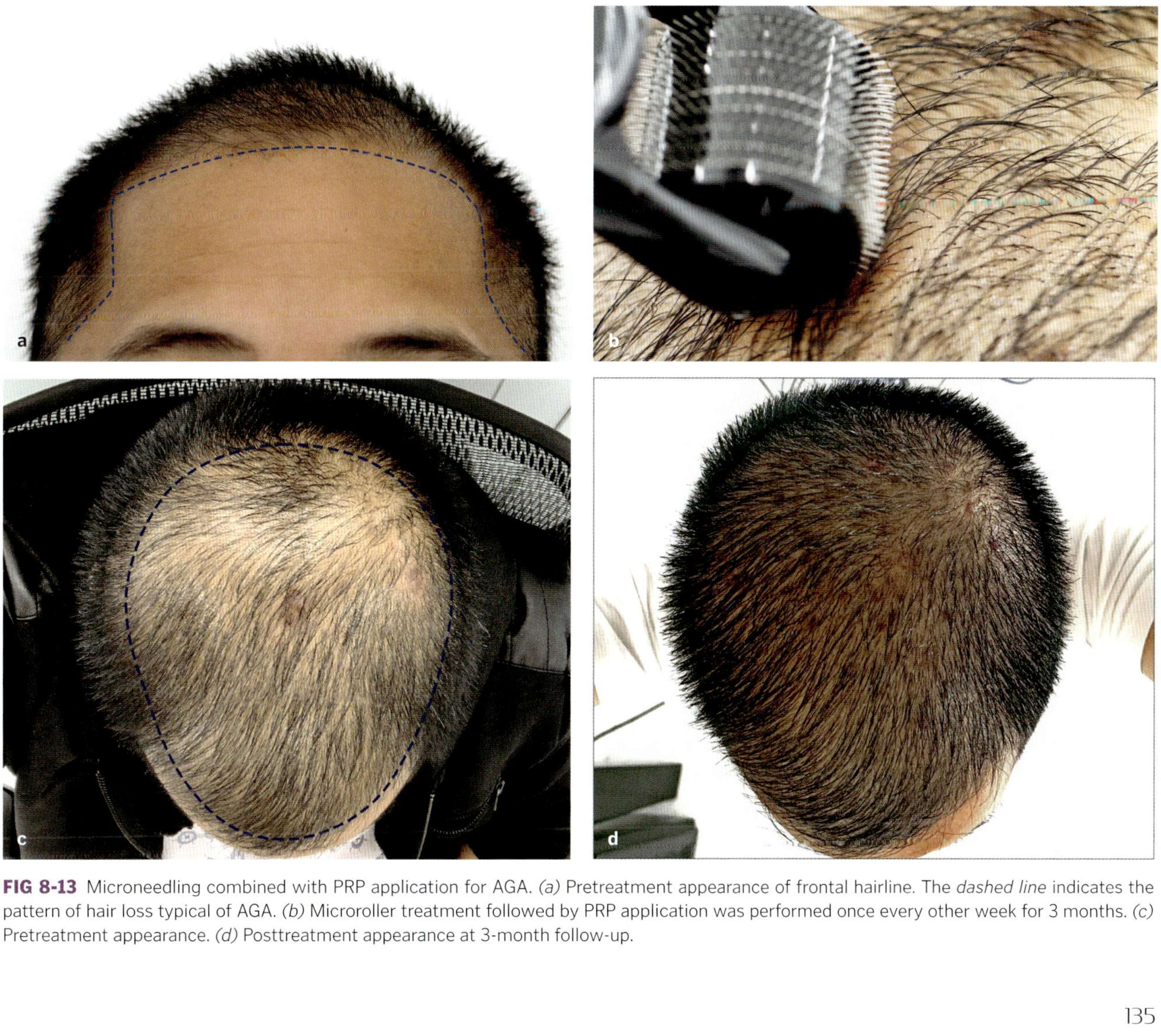

FIG 8-13 Microneedling combined with PRP application for AGA. *(a)* Pretreatment appearance of frontal hairline. The *dashed line* indicates the pattern of hair loss typical of AGA. *(b)* Microroller treatment followed by PRP application was performed once every other week for 3 months. *(c)* Pretreatment appearance. *(d)* Posttreatment appearance at 3-month follow-up.

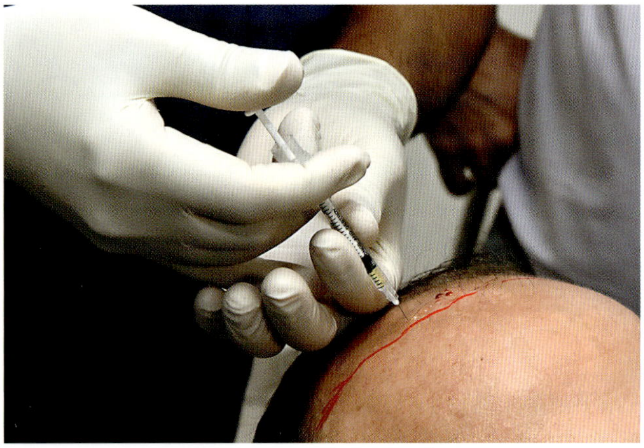

FIG 8-14 Subcutaneous injection of PRP into the scalp of a patient with AGA to induce hair regrowth.

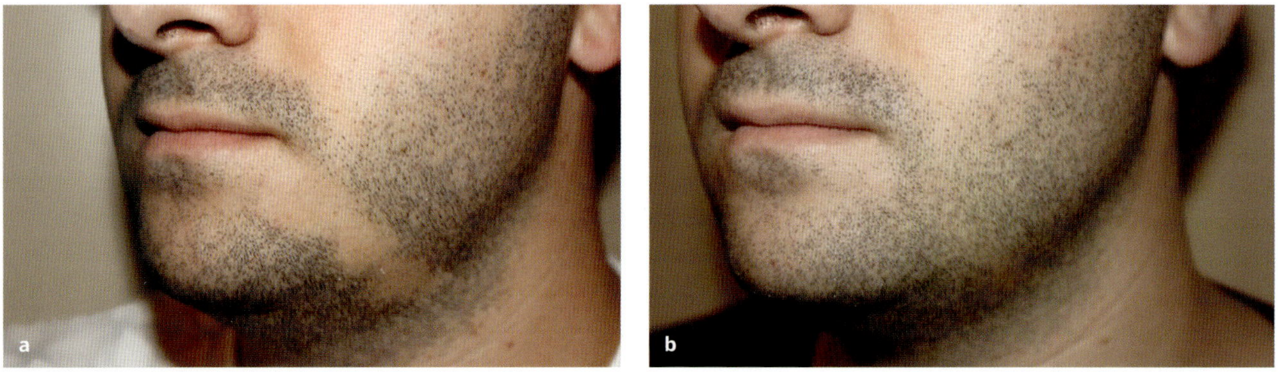

FIG 8-15 PRP application has been shown to promote facial hair growth as effectively as it does scalp hair. *(a)* Pretreatment photograph showing an area of inadequate facial hair growth on the face and chin area. *(b)* The effects of PRP injection are evident after just one treatment. (Courtesy of Keith Jeffords, MD, DDS.)

PRP resulted in a thicker epidermis, proliferation of collagen fibers and fibroblasts, and an increased number of blood vessels around these fibers.[37]

Dermal fillers

Demand for facial rejuvenation procedures that use dermal fillers has risen each year since 2006 and is forecast to grow by at least 8% in the next 5 years. Commercial products containing hyaluronic acid, such as Juvéderm (Allergan) and Restylane (Galderma), claim the biggest share of the dermal fillers market, but additional products with other active ingredients continue to earn US Food and Drug Administration (FDA) approval as the industry expands.

As a dermal filler, PRP performs as well as the best commercial products and better than most—at a fraction of the cost. The average cost of a syringe of Juvéderm is $620, and at least two syringes are required for most procedures, not including the clinician's fee. In contrast, the average cost of a syringe of PRP is about $0.50. Furthermore, no commercial filler product boasts the safety profile of PRP, as it is the only one derived from the patient's own blood.

All of the esthetic procedures described so far in this chapter are performed using PRP liquid. In contrast, PRP configured for dermal filler applications takes advantage of the plasma layer at the top of the test tube (Fig 8-16)—which is usually discarded—to create a biologic filler, or

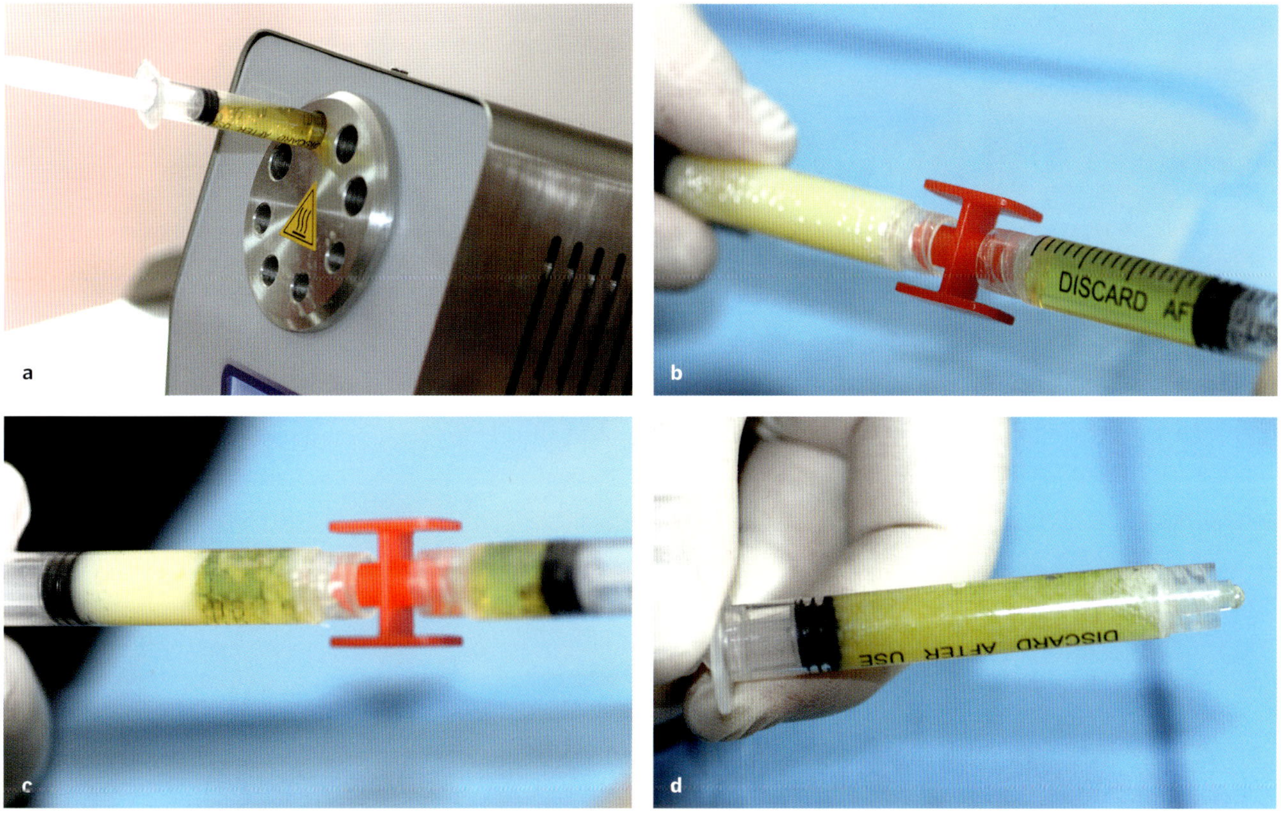

FIG 8-16 Combining PRP liquid with the heated plasma layer to create PRP biofiller. *(a)* The serum layer, which is usually discarded, is placed into a specially designed heater. *(b)* The plasma, which solidifies when subjected to heat, is combined with liquid PRP using a Luer-lock device. *(c)* The contents are pushed back and forth 10 times to obtain a smooth and creamy consistency. *(d)* Appearance of the heated serum combined with the liquid PRP, which is used as a dermal biofiller that achieves results lasting 8 months.

biofiller, that achieves two aims: *(1)* It provides immediate volume to smooth fine wrinkles and deep folds and restore the loss of facial contour associated with aging; and *(2)* it stimulates the production of new collagen and elastin to reverse the loss of hydration and elasticity in the skin (Fig 8-17 and Box 8-6). PRP biofiller can be used not only to smooth dynamic facial folds (Fig 8-18), lines (Fig 8-19), and wrinkles but also to eliminate dark circles under the eyes (Figs 8-20 and Fig 8-21); disrupt fibrotic adhesion in atrophic scars (Fig 8-22); change the contour of the chin (Fig 8-23); and add plumpness and definition to the lips (Fig 8-24), the malar regions (Fig 8-25), and other areas of the face (Fig 8-26).

Evidence regarding the safety and effectiveness of PRP in reducing the visible signs of aging was evaluated in a 2020 systematic review. A total of 24 studies (including 8 RCTs) representing 480 patients were included. Of the 24 studies, 10 provided treatment with PRP as monotherapy, and the other 14 combined PRP with laser resurfacing or lipofilling. Of the 180 patients who received PRP as monotherapy, 90% were satisfied: 67% achieved up to 50% improvement and 33% achieved greater than 50% improvement. These patients typically received three treatments by injection into the face and neck and reported improvements in skin elasticity, texture, hydration, wrinkles, and micropigmentation within 1 to 3 months. Their results lasted approximately 3 to 6 months before gradually returning to baseline.

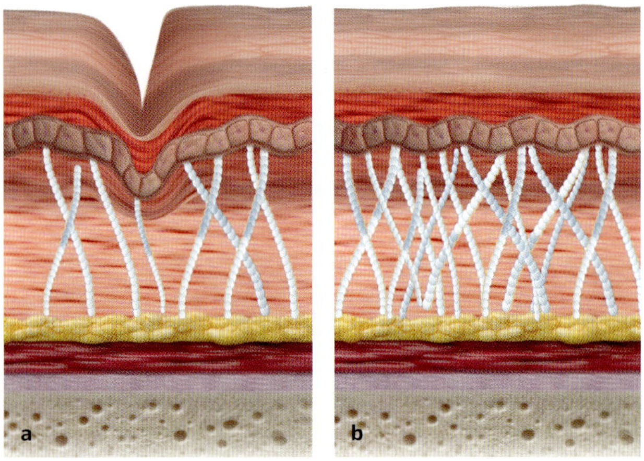

FIG 8-17 PRP stimulates the production of collagen and elastin to diminish the appearance of lines and wrinkles. *(a)* Appearance of skin section before application of PRP. *(b)* Appearance after application of PRP biofiller. (Reprinted with permission from Garg and Rossi.[27])

Box 8-6 Benefits of PRP in facial rejuvenation therapy

- Provides immediate volume to smooth fine wrinkles and deep folds
- Restores the loss of facial contour
- Stimulates the production of new collagen and elastin
- Reverses the loss of hydration and elasticity in the skin

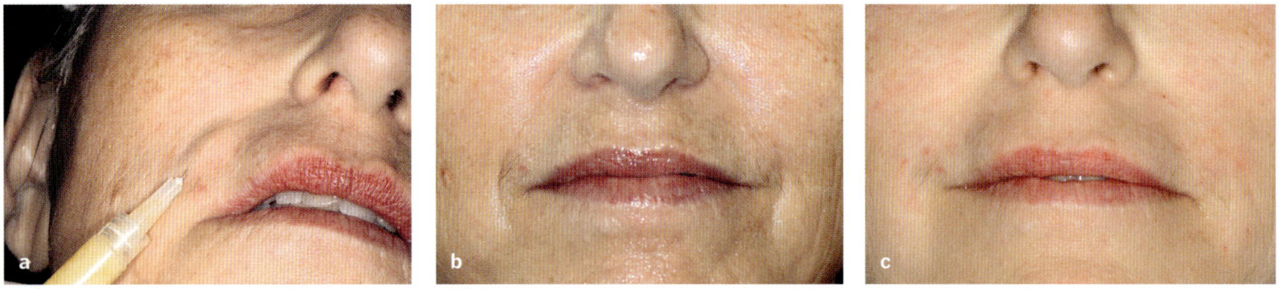

FIG 8-18 Treatment of moderate nasolabial folds. *(a)* PRP biofiller is injected into the nasolabial fold to make it appear less pronounced. *(b)* Preoperative view. *(c)* Immediate postoperative view. (Reprinted with permission from Garg and Rossi.[27])

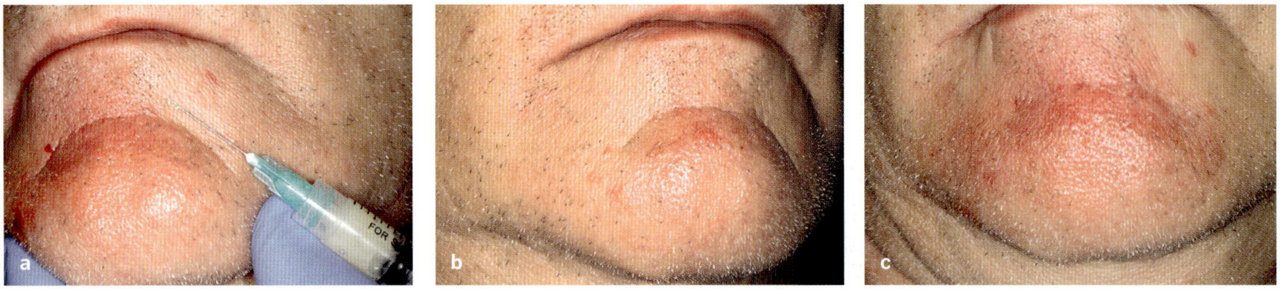

FIG 8-19 Treatment of mental crease. *(a)* The filler is injected into the mental crease. *(b)* Pretreatment view. *(c)* Posttreatment view. (Reprinted with permission from Garg and Rossi.[27])

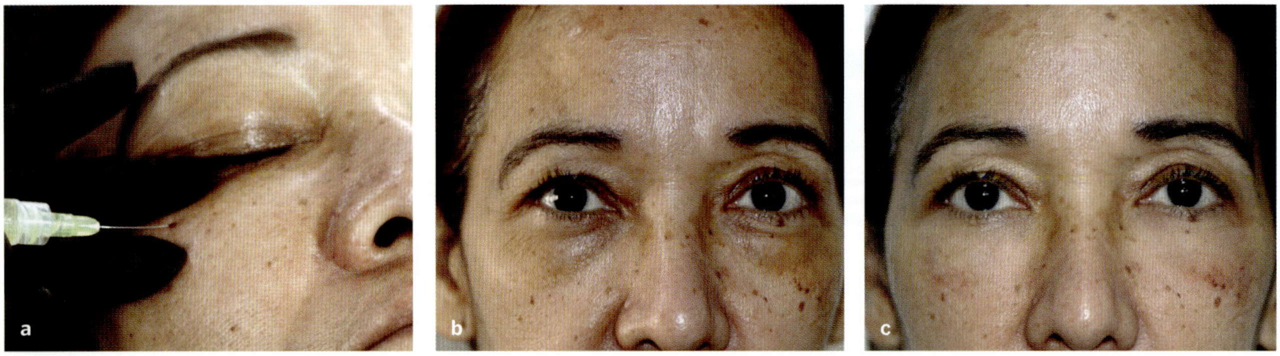

FIG 8-20 Treatment of dark circles beneath the eyes (tear troughs). *(a)* Injecting for the dark circles under the eyes. *(b)* Preoperative appearance. *(c)* Immediate postoperative appearance.

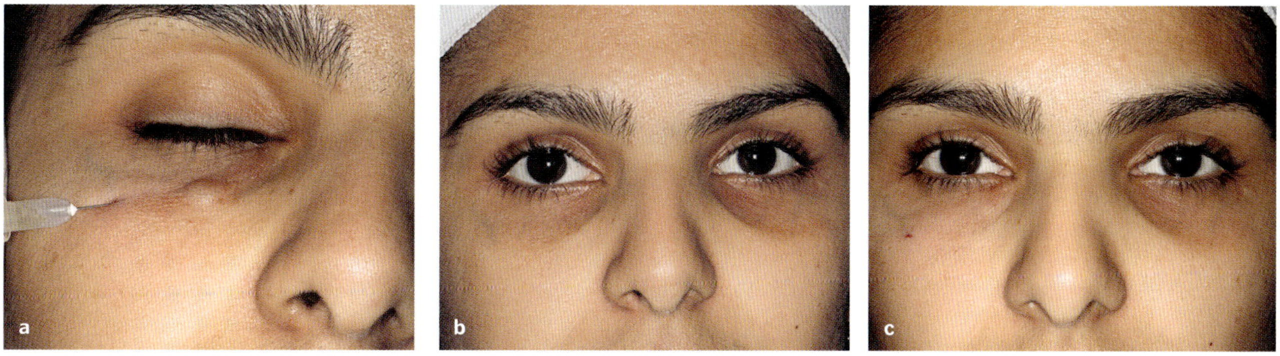

FIG 8-21 Treatment of tear trough with dermal biofiller. *(a)* The filler can be observed as it is deposited in the lower portion of the eyelid. *(b)* Preoperative appearance. *(c)* Immediate postoperative appearance following treatment of right eye. The contrast between the treated side and untreated side is striking. (Reprinted with permission from Garg and Rossi.[27])

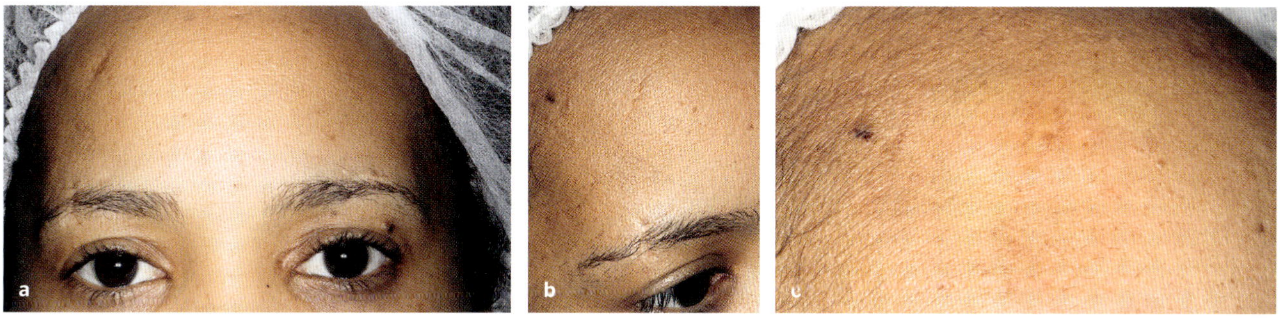

FIG 8-22 Treatment of a small scar. *(a)* Pretreatment appearance. *(b)* In this case, the scar is superficial and dermal filler is placed beneath it. *(c)* Postoperative view. (Reprinted with permission from Garg and Rossi.[27])

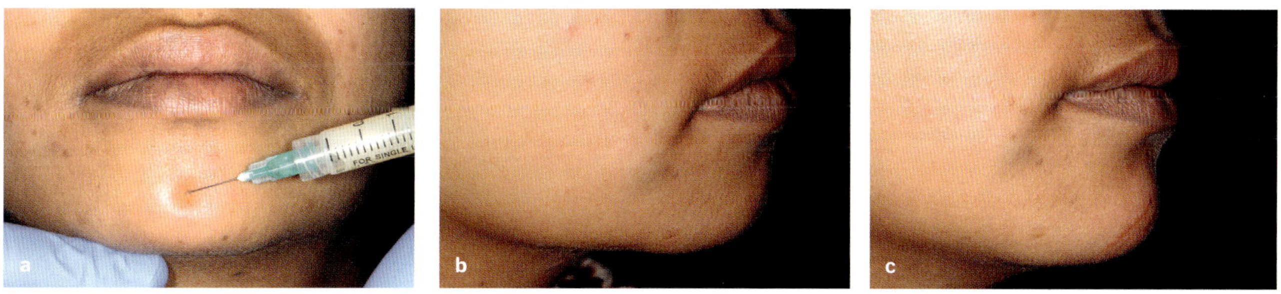

FIG 8-23 Chin augmentation with dermal filler. *(a)* The injection is made in the symphysis, and the needle is inserted until it touches the periosteum. *(b)* Pretreatment view. *(c)* Posttreatment view. (Reprinted with permission from Garg and Rossi.[27])

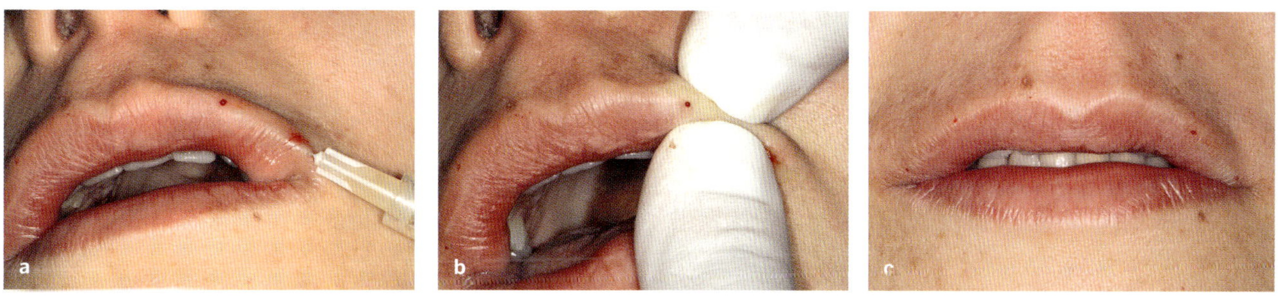

FIG 8-24 Lip border augmentation. *(a)* The amount of filler injected must be consistent throughout the contour. *(b)* Special attention is paid to the sculpting of the angle. *(c)* Immediate postoperative appearance. Observe the harmony of the new contour of the upper lip. (Reprinted with permission from Garg and Rossi.[27])

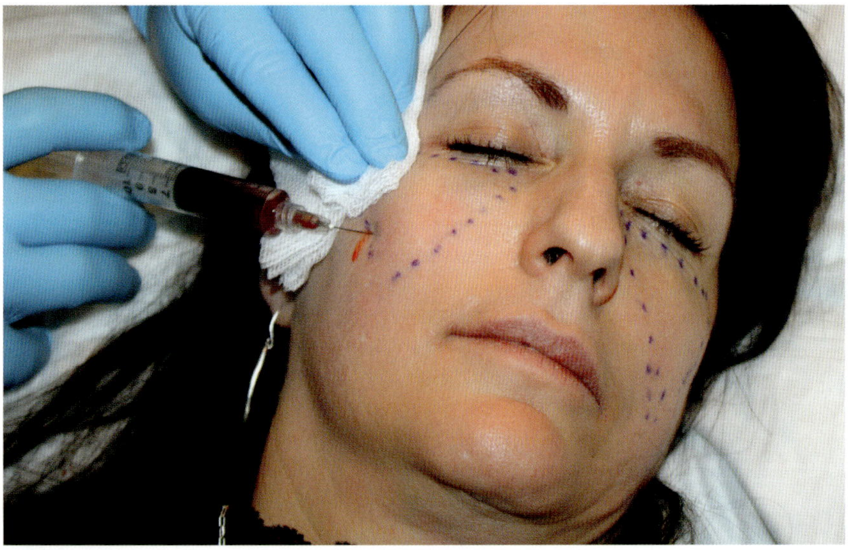

FIG 8-25 Treatment of tear troughs and malar regions. (Courtesy of Keith Jeffords, MD, DDS.)

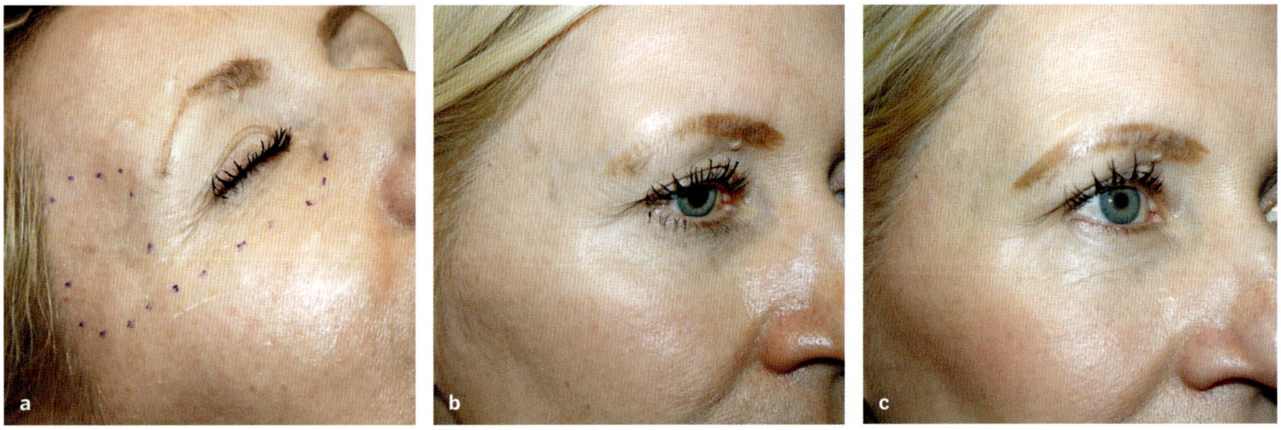

FIG 8-26 Treatment of infratemporal areas. *(a)* Area outlined to be filled in with micronized fat and liquid PRP. *(b)* Preoperative view showing the infratemporal area lacking in volume. *(c)* Postoperative view at 4 months. (Courtesy of Keith Jeffords, MD, DDS.)

The pooled results of the RCTs in which serial PRP injections were compared to saline injections in 127 patients demonstrated that PRP injections produced significantly greater improvements in skin texture, tone, wrinkles, and dermal collagen compared to saline. In the patients treated with PRP as an adjuvant to laser resurfacing, the PRP significantly decreased the intensity and duration of laser-associated side effects, including erythema, edema, dyspigmentation, and crusting. Some studies reported incremental clinical improvement in skin elasticity with PRP injections compared with laser alone.[38]

In one of the included studies, PRP was added to facial lipofilling, which significantly reduced postoperative recovery time (15 days vs 9 days) but did not improve skin elasticity or increase volume.[39] The authors concluded that injection of PRP induced improvements in facial skin appearance, texture, and lines with no significant adverse effects.

Recently, a group of dermatologists set out to assess the clinical effects of PRP on the physical appearance of photoaged facial skin. In this parallel, split-face RCT, 27 participants received intradermal injections of PRP to one cheek and an equivalent volume of sterile normal saline to the contralateral cheek. The exterior of the syringes were covered during the injections, which were made from the zygomatic area to the mandibular area and from the nasolabial folds to the preauricular area, for a total of 3 mL per cheek.

Primary outcomes were photoaging scores (fine lines, mottled pigmentation, roughness, and sallowness) as rated by two masked dermatologists. Secondary outcomes included participant self-assessment scores of improvement on a five-point scale (worsening, no change, mild improvement, moderate improvement, or significant improvement), participant overall satisfaction scores on a four-point scale (not satisfied, slightly satisfied, moderately satisfied, or very satisfied), and participant-reported or investigator-observed adverse events. Masked participants noted that both fine and coarse texture improved significantly more with a single treatment of PRP than with normal saline. Both participants and raters found PRP to be nominally but not significantly superior to normal saline.

Overall, participant self-assessment scores of improvement for pigmentation, texture, wrinkles, and telangiectasias were higher for PRP compared with normal saline at 6 months after injections, when participants rated the PRP-treated side as significantly more improved compared with the normal saline side for texture and wrinkles. The PRP treatments for pigmentation and telangiectasias were both nominally but not significantly improved compared with normal saline at 6 months.

Compared with prior work reported for PRP and facial rejuvenation, this study has reduced bias by design. Earlier work had predominantly looked only at the skin of those treated with PRP, with no comparator or control. Also, measurements have often not been obtained by masked participants or investigators. Moreover, there has been a reliance on intermediate outcome measures, such as biochemical markers, histologic features, skin elasticity, and other clinical or physiologic features detected by devices rather than the visual appearance of skin features as rated by humans. The present study was designed specifically to correct for these deficiencies. The authors join other researchers in arguing that a criterion standard for esthetic procedures is visible change noticeable by the patient.[40]

Conclusion

A renaissance in the use of PRP in facial cosmetic treatments has been engendered by the simplicity and lower cost of using platelet-derived products. Recent reviews of the literature in facial cosmetics confirm the efficacy of PRP for wound healing, survival of fat grafts, and regeneration of bone grafts. Injectable platelet preparations for facial rejuvenation and wound healing are providing particularly remarkable results and patient satisfaction. The use of PRP in facial cosmetic treatments will no doubt continue to grow, both as an adjunct in traditional plastic surgery procedures and as a standalone or complementary agent in topical and injectable applications.

References

1. Farrior E, Ladner K. A retrospective review of the use of autologous platelet gels for rhytidectomy. Arch Facial Plast Surg 2012;14:83–84.
2. Sclafani AP. Applications of platelet-rich fibrin matrix in facial plastic surgery. Facial Plast Surg 2009;25:270–276.
3. Kim DH, Je YJ, Kim CD, et al. Can platelet-rich plasma be used for skin rejuvenation? Evaluation of effects of platelet-rich plasma on human dermal fibroblast. Ann Dermatol 2011;23:424–431.
4. Sclafani AP, Azzi J. Platelet preparations for use in facial rejuvenation and wound healing: A critical review of current literature. Aesthetic Plast Surg 2015;39:495–505.
5. Marx RE, Garg A. Dental and Craniofacial Applications of Platelet-Rich Plasma. Chicago: Quintessence, 2005.
6. Hassouneh B, Brenner MJ. Systematic review and meta-analysis in facial plastic surgery. Facial Plast Surg Clin North Am 2015;23:273–283.
7. Mittelman H, Hershcovitch M. Management of the midface during rhytidectomy. Facial Plast Surg Clin North Am 2015;23:195–200.
8. Gordon NA, Adam SI 3rd. Deep plane face lifting for midface rejuvenation. Clin Plast Surg 2015;42:129–142.
9. Sclafani AP. Safety, efficacy, and utility of platelet-rich fibrin matrix in facial plastic surgery. Arch Facial Plast Surg 2011;13:247–251.
10. Sclafani AP, Saman M. Platelet-rich fibrin matrix for facial plastic surgery. Facial Plast Surg Clin North Am 2012;20:177–186.
11. Farrior E, Ladner K. Platelet gels and hemostasis in facial plastic surgery. Facial Plast Surg 2011;27:308–314.
12. Vick VL, Holds JB, Hartstein ME, Rich RM, Davidson BR. Use of autologous platelet concentrate in blepharoplasty surgery. Ophthal Plast Reconstr Surg 2006;22:102–104.
13. Drolet BC, Sullivan PK. Evidence-based medicine: Blepharoplasty. Plast Reconstr Surg 2014;133:1195–1205.
14. Branham G, Holds JB. Brow/upper lid anatomy, aging and aesthetic analysis. Facial Plast Surg Clin North Am 2015;23:117–127.
15. Baker S, LaFerriere K, Larrabee WF Jr. Lower lid blepharoplasty: Panel discussion, controversies, and techniques. Facial Plast Surg Clin North Am 2014;22:97–118.
16. Schwarcz RM, Kotlus B. Complications of lower blepharoplasty and midface lifting. Clin Plast Surg 2015;42:63–71.

17. Trojahn Kølle SF, Oliveri RS, Glovinski PV, Elberg JJ, Fischer-Nielsen A, Drzewiecki KT. Importance of mesenchymal stem cells in autologous fat grafting: A systematic review of existing studies. J Plast Surg Hand Surg 2012;46:59–68.
18. Ozkaya O, Egemen O, Barutça SA, Akan M. Long-term clinical outcomes of fat grafting by low-pressure aspiration and slow centrifugation (Lopasce technique) for different indications. J Plast Surg Hand Surg 2013;47:394–398.
19. Glasgold M, Glasgold R, Lam S. Autologous fat grafting for midface rejuvenation. Clin Plast Surg 2015;42:115–121.
20. Jin R, Zhang L, Zhang YG. Does platelet-rich plasma enhance the survival of grafted fat? An update review. Int J Clin Exp Med 2013;6:252–258.
21. Modarressi A. Platelet-rich plasma (PRP) improves fat grafting outcomes. World J Plast Surg 2013;2:6–13.
22. Liao HT, Marra KG, Rubin JP. Application of platelet-rich plasma and platelet-rich fibrin in fat grafting: Basic science and literature review. Tissue Eng Part B Rev 2014;20:267–276.
23. Willemsen JC, van der Lei B, Vermeulen KM, Stevens HP. The effects of platelet-rich plasma on recovery time and aesthetic outcome in facial rejuvenation: Preliminary retrospective observations. Aesthet Plast Surg 2014;38:1057–1063.
24. Coleman W 3rd. Complicated surgical techniques. II. Liposuction. Clin Dermatol 1987;5:110–119.
25. Fontes T, Brandão I, Negrão R, Martins MJ, Monteiro R. Autologous fat grafting: Harvesting techniques. Ann Med Surg 2018;36:212–218.
26. Sigmonacci F, Bertozzi N, Grieco MP, Grignaffini E, Raposio E. Procedure, applications, and outcomes of autologous fat grafting. Ann Med Surg 2017;20:49–60.
27. Garg AK, Rossi R Jr. Dermal Fillers for Dental Professionals. Chicago: Quintessence, 2021.
28. El-Domyati M, Abdel-Wahab H, Sossam A. Combining microneedling with other minimally invasive procedures for facial rejuvenation: A split-face comparative study. Int J Dermatol 2018;57:1324–1334.
29. El-Domyati M, Barakat M, Awad S, Medhat W, El-Fakahany H, Farag H. Multiple microneedling sessions for minimally invasive facial rejuvenation: An objective assessment. Int J Dermatol 2015;54:1361–1369.
30. Alam M, Han S, Pongrutthipan M, et al. Efficacy of a needling device for the treatment of acne scars: A randomized clinical trial. JAMA Dermatol 2014;150:844–849.
31. Harris AG, Naidoo C, Murrell DF. Skin needling as a treatment for acne scarring: An up-to-date review of the literature. Int J Women Dermatol 2015;1:77–81.
32. Bhargava S, Kumar U, Varma K. Subcision and microneedling as an inexpensive and safe combination to treat atrophic acne scars in dark skin: A prospective study of 45 patients at a tertiary care center. J Clin Aesthet Dermatol 2019;12:18–22.
33. Iosifidis C, Goutos I. Percutaneous collagen induction (microneedling) for the management of non-atrophic scars: Literature review. Scars Burn Heal 2019;5:2059513119880301.
34. Ibrahim ZA, El-Ashmawy AA, Shora OA. Therapuetic effect of microneedling and autologous platelet-rich plasma in the treatment of atrophic scars: A randomized study. J Cosmet Dermatol 2017;16:388-399.
35. Asif M, Kanodia S, Singh K. Combined autologous platelet-rich plasma with microneedling versus microneedling with distilled water in the treatment of atrophic acne scars: A concurrent split-face study. J Cosmet Dermatol 2016;15:434–443.
36. Li ZJ, Choi H-I, Choi D-K, et al. Autologous platelet-rich plasma: A potential therapeutic tool for promoting hair growth. Derm Surg 2012;38:1040–1046.
37. Mao G, Guohui Z, Weixin. Platelet-rich plasma for treating androgenic alopecia: A systematic review. Aesthet Plast Surg 2019;43: 1326–1336.
38. Maisel-Campbell AL, Ismail A, Reynolds KA, et al. A systematic review of the safety and effectiveness of platelet-rich plasma (PRP) for skin aging. Arch Dermatol Res 2020;312:301–315.
39. Willemsen JCN, Van Dongen J, Spiekman M, et al. The addition of platelet-rich plasma to facial lipofilling: A double-blind, placebo-controlled, randomized trial. Plast Reconstr Surg 2018;141:331–343.
40. Alam M, Hughart R, Champlain A, et al. Effect of platelet-rich plasma injection for rejuvenation of photoaged facial skin: A randomized clinical trial. JAMA Dermatol 2018;154:1447–1452.

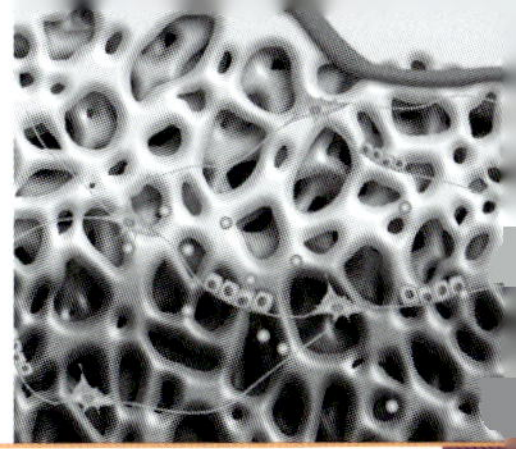

9

The Future of Autologous Blood Concentrates

Growth factor proteins released by PRP can no doubt exert a substantive positive influence on blood coagulation/clotting, immunities, angiogenesis, and wound healing.[1] The wide diversity of such medical uses for PRP reported in the literature for the past 30 years attests to its current clinical efficacy and future potential in general surgery,[2-10] plastic surgery and dermatology,[11-21] cardiothoracic surgery,[22,23] neurosurgery,[24-36] bone surgery,[37-41] ophthalmology,[42-47] otolaryngology,[48-52] orthopedics and sports medicine,[53-73] urology,[74-76] podiatry,[77-82] chronic wound healing,[83-95] and oral, periodontal, and maxillofacial surgeries.[96,97]

Because standard terminologies and protocols have not been clearly established for biologic/regenerative therapies, many studies view skeptically the clinical applications of PRP and several related autologous products.[39] The mid-1990s saw the steady growth of regenerative medical therapies based on platelet-released growth factors, and many of those therapies are still evolving.[98-109] Well known are the causative features of growth factors and the functional matrix that PRP provides to facilitate wound healing, but a great deal of work is still needed to replicate and standardize how platelet-derived products are created and applied. Nevertheless, because evidence suggests that lack of standardization and replication reveals no significant difference in the osteogenesis of growth factors,[110-114] perhaps the calls for reducing the variety of preparation techniques, products, and nomenclatures are not only premature but also unwarranted. Perhaps continued expansive exploration of the medical science of PRP and platelet-derived therapies is what is needed, not narrowed visions, however scientifically well intentioned. For example, nonstandardized nomenclature for autologous platelet-derived products could be seen as positive in a search for products appropriate for individual patient needs, determining whether the product is called PRP, platelet-rich fibrin (PRF), compressor-prepared PRF membrane (C-PRF), autologous PRP (aPRP), platelet gel (PG), autologous platelet concentrates (APCs), preparation rich in growth factors (PRGF), platelet-derived growth factor (PDGF), platelet-leukocyte gel (PLG), and other names yet to be determined (Fig 9-1).

FIG 9-1 The term *platelet-rich plasma* was coined by the author, in collaboration with Dr Robert Marx, more than 20 years ago. Since then, a plethora of terms have been used by researchers to describe similar (and in some cases identical) platelet-derived concentrates, attesting to the broad-ranging potential of the original concept.

Interestingly, and perhaps ironically, studies spanning nearly 20 years in innovative oral, periodontal, and maxillofacial surgeries often offer the best hope for standardized platelet-derived concentrates, especially for bone regeneration, though PRP experimentation in these fields of dentistry continues.[115–118] Additionally, platelet-derived concentration therapies across the medical spectrum often receive high marks from reviewers despite the steady calls from others for high-level studies and standardization. Praise for the therapies echoes the belief that as more and more clinicians become aware of the current, wide-ranging experiments with clinical applications of PRP and other platelet derivatives, these therapies will provide even small medical practices with the multiple advantages of biologics, especially for many of their otherwise medically compromised patients, including further experimentation with simpler, cost-effective production and application methods.[119] Some of the areas for continued experimentation and evolution of PRP preparation and application include platelet activation and enrichment, leukocyte inclusion/exclusion, and patient and product selection.[120] And even reviewers who unequivocally call for caution in the use of platelet derivatives due to the lack of definitive controlled studies simultaneously acknowledge the sound scientific theories for applying PRPs, noting in particular the lack of adverse effects from its use.[121,122] However, instead of seeing the number of variables in PRP preparation and use as positive signs of its potential clinical use, they would rather concentrate on how reports of PRP's efficacy are often misleading and its use unpredictable.[39]

Nevertheless, a 2015 study reminds us that the regulatory role and stance of the US Food and Drug Administration (FDA) can be instructive for clinicians regarding the increased clinical use of PRP.[123] Because PRP is a blood product, the FDA's Center for Biologics Evaluation and Research holds sway over its clinical use, as this division regulates products derived from human cells and tissues. Under the regulations governing the use of PRP, however, we find that blood products like PRP are not handled in a traditional way, particularly regarding the process that requires clinical and animal studies. Regarding preparatory systems specifically, the FDA so far has allowed PRP-prep systems that are essentially similar in their ability to produce PRP that can be used as an adjunct for bone grafting procedures. The authors of the report note that the FDA distinguishes this use of PRP from other uses described as "off label," such as an injection of PRP. This kind of nonrestricted use can be pursued so long as certain safety criteria are met. These criteria include the clinician's sufficient knowledge of the product, his or her ability to follow established scientific and medical reasoning when employing the product, and the clinician's intent to record the results of the product's use.

Many clinicians would be interested in knowing also that the FDA has not provided any guidelines or restrictions concerning whether or not activated PRP is used in such procedures other than the general criteria listed above concerning knowledge and sound scientific and medical practices. If anything, then, the FDA's decisions appear to encourage those who caution against too-hasty moves toward standardization of biologic therapies.[123]

PRP's regulation in Europe, by contrast, is handled by Directive 2002/98/EC of the European Parliament and Council of January 27, 2003.[124] This organization regulates the quality and safety criteria for the use of blood and blood components, including how it is collected, processed, and distributed. The usage laws of specific European countries concerning biomaterials such as PRP generally adhere to those put forth by this directive, but there is a great deal of nonconformity as well. So one of the tasks of the European medical community is to move toward more uniformity, at least regarding the safety and quality of PRP-related products.[124]

Thus, the question remains: Might a lack of uniformity actually expand the possibilities for real scientific/medical breakthroughs concerning PRP and other biomaterials, or should caution prevail via government regulations? Four promising areas of research are offered for consideration in this chapter.

3D Printing

3D printing has fundamentally changed how products are manufactured and services are delivered today compared to a decade ago, and dentistry is no exception. The use of 3D printing (and chairside milling) to fabricate dental appliances and prosthetics has radically simplified the traditional workflow in dental practices across the country. There are a variety of 3D printing technologies to choose from depending on how the product is to be used. In dentistry, stereolithography—which operates by means of an ultraviolet laser—is the most popular because it allows for high accuracy and provides a smooth surface finish.

Recently, a form of 3D printing known as *robocasting* was employed to fabricate a bioactive scaffold composed

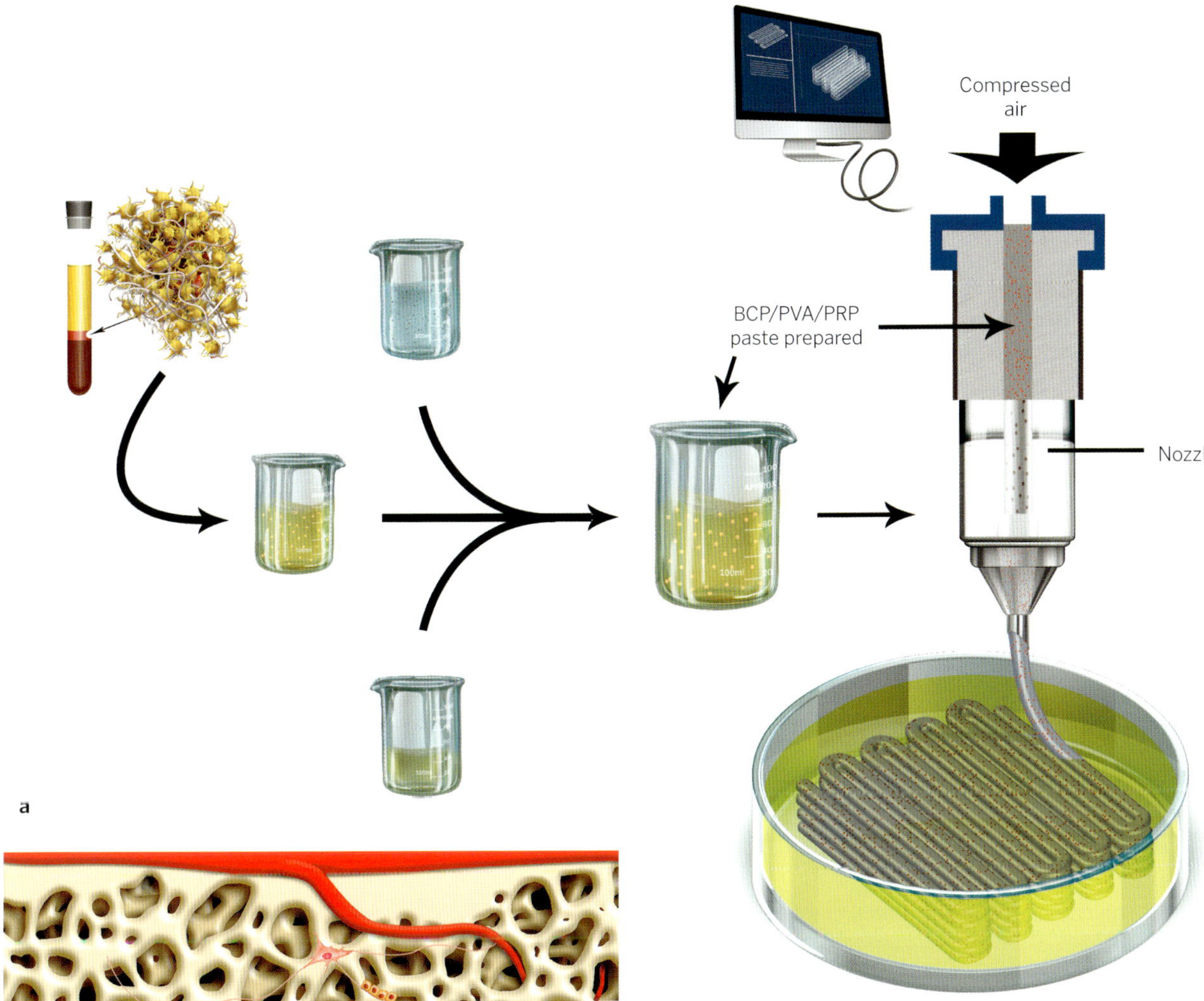

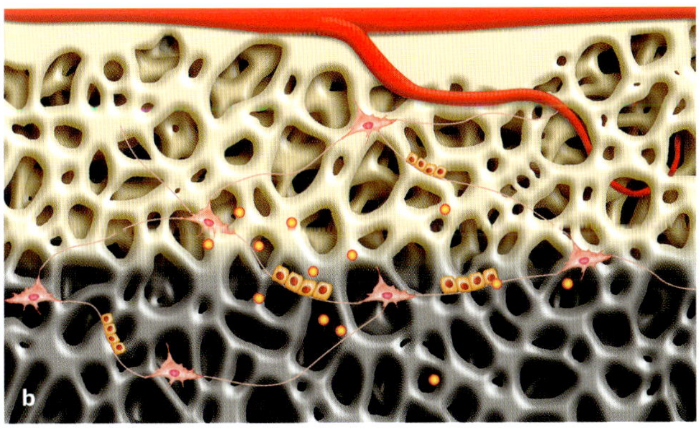

FIG 9-2 *(a)* Robocasting is a 3D printing technique that was recently used to print scaffolds for the repair of critical-sized bone defects using "ink" made of biphasic calcium phosphate (BCP), polyvinyl alcohol (PVA), and PRF. *(b)* The biologic activity of PRF was retained during the 3D printing process, and the presence of PRF in the biocompatible microenvironment of the scaffold promoted the adhesion, proliferation, and differentiation of bone marrow mesenchymal stem cells.

of nanobiphasic calcium phosphate, polyvinyl alcohol, and PRF for the treatment of critical-sized bone defects (Fig 9-2). Robocasting, also known as *direct ink writing* (DIW), is a 3D printing technique in which a thin filament of biomaterial (usually ceramic or metal) is dispensed from a nozzle to form an object one layer at a time. Because it prints at low temperatures, this technology allows for the incorporation of bioactive molecules that retain maximum levels of biologic activity.

In a proof of concept study, researchers demonstrated that the printed scaffolds exhibited good biologic activity and biocompatibility. The biologic activity of PRF was retained during the 3D printing process, and the presence of PRF in the biocompatible microenvironment of the scaffold promoted the adhesion, proliferation, and differentiation of bone marrow mesenchymal stem cells. When the cell/scaffold constructs were transplanted into critical-sized bone defects in the rabbit radius, the presence of PRF appeared to provide osteoconductive and osteoinductive stimuli that promoted bone regeneration and repair.[125]

Laser Therapy

A second promising avenue of research involves combining PRP with laser therapy to amplify the effects of each. As detailed in chapter 4, numerous studies have demonstrated the efficacy of PRP in promoting healing in injured or damaged tendons, muscles, and joints. The results of these studies show that PRP injection results in better mechanical, biochemical, and/or histologic properties of the repaired tissue. While the exact mechanism of PRP is not known, elevated levels of growth factors such as PDGF, transforming growth factor β (TGF-β), and insulin-like growth factor 1 (IGF-1) play a positive role in healing. Moreover, PRP injection has been shown to reduce the degree of inflammation at the site of injury or damage.

Low-level laser therapy (LLLT), another emerging treatment for injured or damaged tissue, has also been shown to accelerate the process of healing through growth factor proliferation, collagen synthesis, and reduced inflammation. Laser irradiation may increase cell growth via changes in mitochondrial physiology as well as RNA synthesis. It can stimulate angiogenesis, restore blood flow to the injury site, and limit ischemic necrosis, thereby accelerating tissue repair.

Based on the favorable results of several controlled animal studies, in which the combination of PRP and LLLT led to decreased healing times compared to either treatment alone,[126,127] a clinical study was conducted in 2019 in which PRP plus intra-articular LLLT was offered to patients with arthrosis of the knee, hip, glenohumeral, or first carpometacarpal joints who had received no benefit from PRP injection alone. The authors hypothesized that intra-articular laser treatment might be an effective augmentation to PRP injection and might increase the efficacy of a subsequent PRP injection (Fig 9-3).

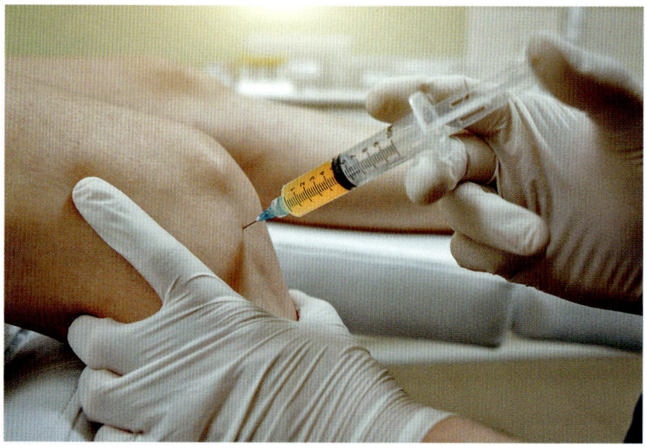

FIG 9-3 PRP injection combined with intra-articular LLLT was recently shown to be more effective than either treatment alone in reducing joint pain in patients with osteoarthrosis.

A total of 28 patients (30 joints) were enrolled, and 26 were followed for at least 2 years. (Two patients were lost to follow-up.) At 6 months posttreatment, nearly half of the joints (46%) had a "good result"—that is, more than 40% pain improvement, normal function with activities of daily living, and avoidance of joint replacement surgery. Thirty-two percent of the joints had more than 40% improvement at 1 year posttreatment and continued to improve at 2 years posttreatment. Thirteen joints had arthroplasty over a period of 10 months after the experimental treatment. The authors concluded that PRP plus intra-articular LLLT has significant efficacy in reducing arthrosis symptoms for up to 2 years in patients who had failed prior PRP injection, especially those with moderate rather than bone-on-bone arthrosis.

While the mechanism of action of LLLT, also known as *biophotomodulation*, is an area of continued research, the authors speculate that the cytochrome-C oxidase molecule acts as a chromophore for red and near-infrared light, resulting in increased energy and nitric oxide production and possibly an increase in TGF-β. As for the synergy between LLLT and PRP, the additive effects may exceed those of either one alone.[128]

Peripheral Nerve Regeneration

Research is currently underway to evaluate the use of PRP in peripheral nerve regeneration, which could have significant therapeutic potential for a number of health conditions, including spinal cord injury. In dentistry, this application is being investigated as a possible means of regenerating the nerve fibers and reconstructing the neural feedback pathways—originally disrupted by tooth extraction—around osseointegrated implants.

Several of the growth factors in PRP are known to influence neural cell survival and might have the potential to regenerate injured peripheral nerves. A preliminary clinical study involving a modified inferior alveolar nerve lateralization technique found that applying PRP over the

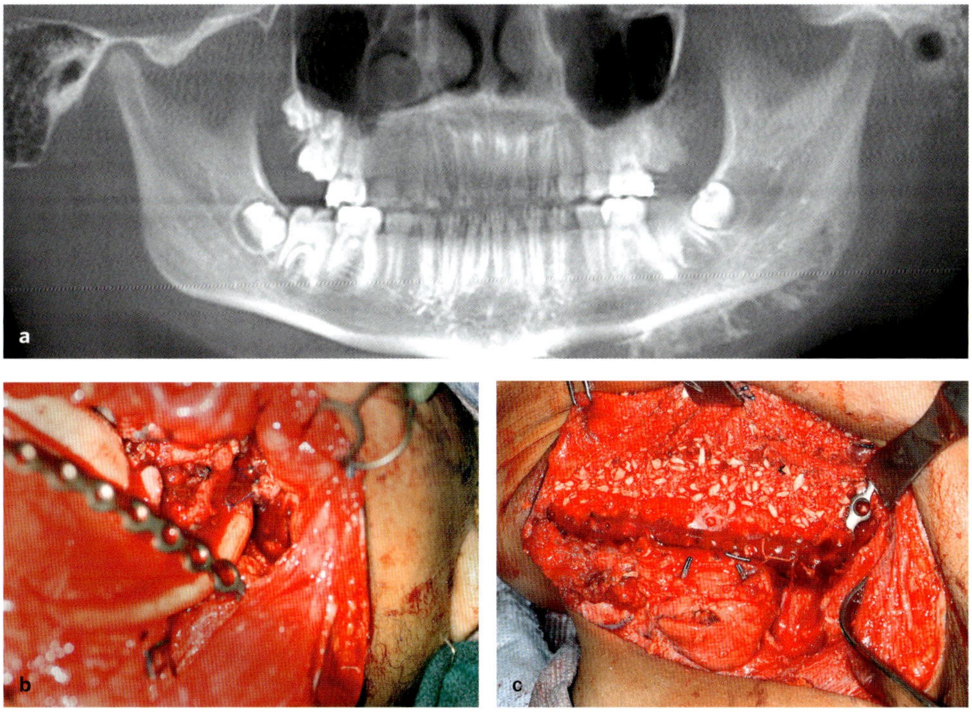

FIG 9-4 *(a)* Aggressive benign tumor of the left mandible in an 11-year-old girl. *(b)* Following tumor resection, the mandible was reconstructed with a bone graft and bone plate combined with PRP. The inferior alveolar nerve was also reconstructed and covered with PRP. *(c)* Functional neurosensory recovery was achieved. (Courtesy of Ramzey Tursun, DDS, FACS.)

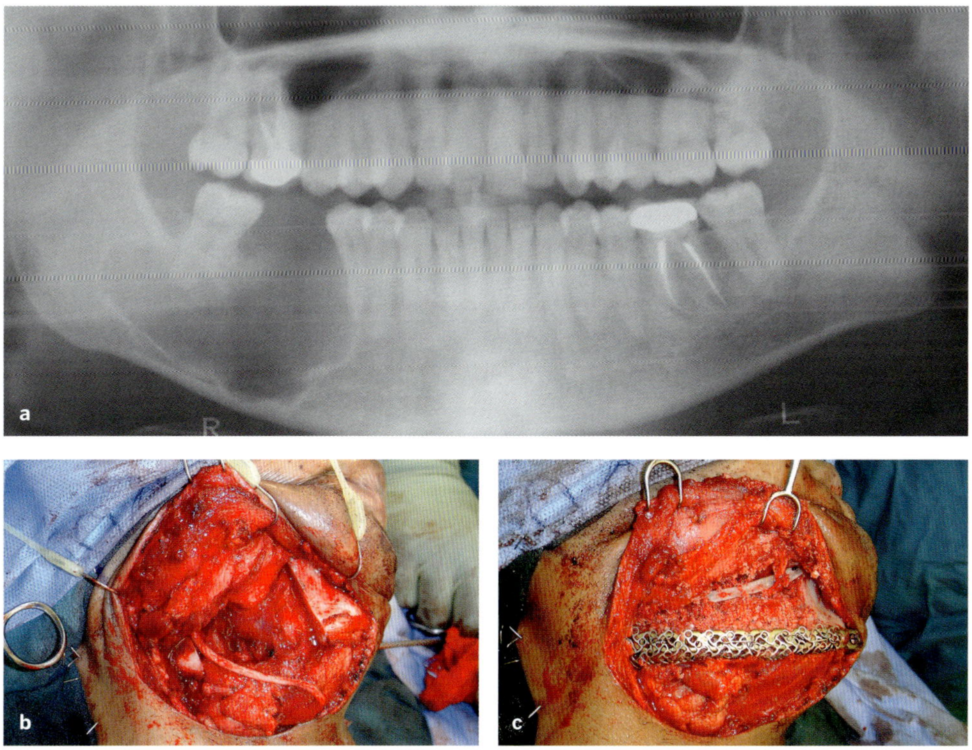

FIG 9-5 *(a)* Benign tumor of the right mandible. *(b)* A nerve pull-through technique was performed, and the tumor was resected. *(c)* PRP was applied to the bone graft as well as the inferior alveolar nerve. Full functional neurosensory recovery was achieved. (Courtesy of Ramzey Tursun, DDS, FACS.)

inferior alveolar nerve and covering it with a collagen membrane could accelerate the recovery of sensation[129] (Figs 9-4 and 9-5). Additionally, regrowth of both blood vessels and nerve axons in an extraction socket was reported by the authors of a recent split-mouth study on the use of PRF in preserving horizontal and vertical ridge dimension.[130]

PRP has already demonstrated its capacity to induce nerve regeneration when traumatic gaps of nerve structures are less than 3 cm in length. To regenerate axons across nerve gaps longer than 3 cm, grafting material that both supports and promotes axon regeneration is likely needed to bridge the gap. The fact that defects around dental implants are normally not as large as peripheral nerve defects might make the regeneration of peri-implant nerve fibers more feasible. PRP alone may be sufficient.[131]

Ultrasound Technology

Finally, a 2016 study describes the use of ultrasound waves (4.5 MHz) produced from piezoelectric ceramics to create PRP comparable to commercial centrifuges.[132] Once a syringe is filled with autologous blood, a standing wave created inside the syringe causes red blood cells to collect at certain pressure points. Platelet concentrates are greater than the centrifugal process, with similar growth factor production but with lower rates of red blood cell infiltration. The Mason-Weaver equation was used by these researchers to compare the centrifuge and ultrasound processes in hopes of further enhancement of this innovative method. The ultrasound method appears to prepare the PRP product more simply, in a much shorter time, and with lower contamination by red blood cells.[132]

Rethinking the Status Quo

Another recent study proposes an even more fundamental rethinking of the use of PRP. While noting the widespread use of PRP's concentrated growth factors as an appealing, unorthodox, and effective regenerative therapy, the authors also point to PRP's autologous-source drawbacks (for example, PRP's complex preparation, user/product/application variability, and the need for bovine thrombin as a coagulant). Improving preparation protocols and devices will help, they note, in the short run while autologous therapies are still used, but shifting from autologous to allogeneic cell therapy could eliminate a host of shortcomings associated with traditional PRP use.[133] This kind of inventive thinking could help advance the use of biologic therapies immensely, but it can only happen when clinicians are not "overtrained" in established protocols to the point that the status quo becomes mundane and restrictive.

It is appropriate to conclude by noting that PRP is only one of many biologic therapies in use today. While much progress has been made by using PRP independently via injections, for example, its original use as an adjunct to traditional therapies reminds us that much of its efficacy comes from its complementary use with other therapies, particularly other biologic therapies, including non-PRP growth factors and recombinant proteins, stem cells, cytokines, and biologic scaffolds.[122] Therefore, a preoccupation with standardizing the production and use of PRP could blind the medical community to ways that the current diverse PRP products and applications could be used with other biologic therapies to aid patients otherwise less well served.

Conclusion

Any rigid standardization of platelet sequestration, the nomenclature and classification of platelet-rich products, and the application of those products for wound healing could inhibit, not enhance, the treatment of patients requiring such therapies, especially those patients seeking alternatives to traditional surgery as well as those whose healing has been otherwise compromised. The dynamic nature of biologic research and application often yields contradictory results; such is the nature of PRP derivations and clinical applications across the medical spectrum. A mature scientific approach should acknowledge the dynamism and contradictions of such studies as a natural outgrowth of clinical vigor and the pursuit of greater scientific understanding of a field whose most productive days still lie ahead.

References

1. Lubkowska A, Dolegowska B, Banfi G. Growth factor content in PRP and their applicability in medicine. J Biol Regul Homeost Agents 2012;26(2 suppl 1):3S–22S.
2. Cervelli V, De Angelis B, Lucarini L, et al. Tissue regeneration in loss of substance on the lower limbs through use of platelet-rich plasma, stem cells from adipose tissue, and hyaluronic acid. Adv Skin Wound Care 2010;23:262–272.
3. Balbo R, Avonto I, Marenchino D, Maddalena L, Menardi G, Peano G. Platelet gel for the treatment of traumatic loss of finger substance. Blood Transfus 2010;8:255–259.
4. Iervolino V, Di Costanzo G, Azzaro R, et al. Platelet gel in cutaneous radiation dermatitis. Support Care Cancer 2013;21:287–293.
5. Kucharczyk A, Kołodziejczak M, Sudoł-Szopinska I, Bielecki K. Autologous growth factors used for the treatment of recurrent fistula-in-ano preliminary results. Tech Coloproctol 2014;18:317–318.

6. Cintron JR, Park JJ, Orsay CP, et al. Repair of fistulas-in-ano using fibrin adhesive: Long-term follow-up. Dis Colon Rectum 2000;43:944–950.
7. Ventakesh KS, Ramanujam P. Fibrin glue application in the treatment of recurrent anorectal fistulas. Dis Colon Rectum 1999;42:1136–1139.
8. Knighton DR, Ciresi KF, Fiegel VD, Austin LL, Butler EL. Classification and treatment of chronic nonhealing wounds. Successful treatment with autologous platelet-derived wound healing factors (PDWHF) Ann Surg 1986;204:322–330.
9. Sentovich SM. Fibrin glue for all anal fistulas. J Gastrointest Surg 2001;5:158–161.
10. van der Hagen SJ, Baeten CG, Soeters PB, van Gemert WG. Autologous platelet derived grow factors (platelet-rich plasma) as an adjunct to mucosal advancement flap in high crypto glandular peri-anal fistulae: A pilot study. Colorectal Dis 2011;13:215–218.
11. Cieslik-Bielecka A, Choukroun J, Odin G, Dohan Ehrenfest DM. L-PRP/L-PRF in esthetic plastic surgery, regenerative medicine of the skin and chronic wounds. Curr Pharm Biotechnol 2012;13:1266–1277.
12. Modarressi A. Platelet rich plasma (PRP) improves fat grafting outcomes. World J Plast Surg 2013;2:6–13.
13. Leo MS, Kumar AS, Kirit R, Konathan R, Sivamani RK. Systematic review of the use of platelet-rich plasma in aesthetic dermatology. J Cosmet Dermatol 2015;14:315–323.
14. Bhargava S, Kumar U, Varma K. Subcision and microneedling as an inexpensive and safe combination to treat atrophic acne scars in dark skin: A prospective study of 45 patients at a tertiary care center. J Clin Aesthet Dermatol 2019;12:18–22.
15. Iosifidis C, Goutos I. Percutaneous collagen induction (microneedling) for the management of non-atrophic scars: Literature review. Scars Burn Heal 2019;5:2059513119880301.
16. Ibrahim ZA, El-Ashmawy AA, Shora OA. Therepeutic effect of microneedling and autologous platelet-rich plasma in the treatment of atrophic scars: A randomized study. J Cosmet Dermatol 2017;16:388–399.
17. Asif M, Kanodia S, Singh K. Combined autologous platelet-rich plasma with microneedling versus microneedling with distilled water in the treatment of atrophic acne scars: A concurrent split-face study. J Cosmet Dermatol 2016;15:434–443.
18. Mao G, Zhang G, Fan W. Platelet-rich plasma for treating androgenic alopecia: A systematic review. Aesthet Plast Surg 2019;43:1326–1336.
19. Maisel-Campbell AL, Ismail A, Reynolds KA, et al. A systematic review of the safety and effectiveness of platelet-rich plasma (PRP) for skin aging. Arch Dermatol Res 2020;312:301–315.
20. Willemsen JCN, Van Dongen J, Spiekman M, et al. The addition of platelet-rich plasma to facial lipofilling: A double-blind, placebo-controlled, randomized trial. Plast Reconstr Surg 2018;141: 331–343.
21. Alam M, Hughart R, Champlain A, et al. Effect of platelet-rich plasma injection for rejuvenation of photoaged facial skin: A randomized clinical trial. JAMA Dermatol 2018;154:1447–1452.
22. Ferrari M, Zia S, Valbonesi M, et al. A new technique for hemodilution, preparation of autologous platelet-rich plasma and intraoperative blood salvage in cardiac surgery. Int J Artif Organs 1987;10:47–50.
23. Zhai Q, Wang Y, Yuan Z, Zhang R, Tian A. Effects of platelet-rich plasmapheresis during cardiovascular surgery: A meta-analysis of randomized controlled clinical trials. J Clin Anesth 2019;56:88–97.
24. Tsai CH, Hsu HC, Chen YJ, Lin MJ, Chen HT. Using the growth factors-enriched platelet glue in spinal fusion and its efficiency. J Spinal Disord Tech 2009;22:246–250.
25. Hartmann EK, Heintel T, Morrison RH, Weckbach A. Influence of platelet-rich plasma on the anterior fusion in spinal injuries: A qualitative and quantitative analysis using computer tomography. Arch Orthop Trauma Surg 2010;130:909–914.
26. Zurita M, Otero L, Aguayo C, et al. Cell therapy for spinal cord repair: Optimization of biologic scaffolds for survival and neural differentiation of human bone marrow stromal cells. Cytotherapy 2010;12:522–537.
27. Takeuchi M, Kamei N, Shinomiya R, et al. Human platelet-rich plasma promotes axon growth in brain-spinal cord coculture. Neuroreport 2012;23:712–716.
28. Li H, Hamza T, Tidwell JE, Clovis N, Li B. Unique antimicrobial effects of platelet-rich plasma and its efficacy as a prophylaxis to prevent implant-associated spinal infection. Adv Healthc Mater 2013,2.1277–1284.
29. Singh R, Rohilla RK, Dhayal RK, Sen R, Sehgal PK. Role of local application of autologous platelet-rich plasma in the management of pressure ulcers in spinal cord injury patients. Spinal Cord 2014;52:809–816.
30. Biglari B, Reitzel T, Swing T, et al. A pilot study on the effectiveness of platelet-rich plasma and debridement for the treatment of nonhealing fistulas in spinal cord-injured patients. Adv Skin Wound Care 2015;28:123–128.
31. Zhang Y, Ying G, Ren C, et al. Administration of human platelet-rich plasma reduces infarction volume and improves motor function in adult rats with focal ischemic stroke. Brain Res 2015;1594:267–273.
32. Kubota G, Kamoda H, Orita S, et al. Platelet-rich plasma enhances bone union in posterolateral lumbar fusion: A prospective randomized controlled trial. Spine J 2019;19:e34–e40.
33. Manini DR, Shega FD, Guo C-F, Wang Y-X. Role of platelet-rich plasma in spinal fusion surgery: Systematic review and meta-analysis. Adv Orthop 2020;2020:8361798.
34. Basso M, Cavagnaro L, Zanirato A, et al. What is the clinical evidence on regenerative medicine in intervertebral disc degeneration? Muscoloskeletal Surg 2017;101:93–104.
35. Tuakli-Wosornu YA, Terry A, Boachie-Adjei K, et al. Lumbar intradiskal platelet-rich plasma (PRP) injections: A prospective, double-blind, randomized controlled study. PM R 2016;8:1–10.
36. Levi D, Horn S, Tyszko S, Levin J, Hecht-Leavitt C, Walko E. Intradiscal platelet-rich plasma injection for chronic discogenic low back pain: Preliminary results from a prospective trial. Pain Med 2016;17:1010–1022.
37. Griffin XL, Wallace D, Parsons N, Costa ML. Platelet rich therapies for long bone healing in adults. Cochrane Database Syst Rev 2012;7:CD009496.
38. Lenza M, Ferraz Sde B, Viola DC, Santos OF, Cendoroglo Neto M, Ferretti M. Platelet-rich plasma for long bone healing. Einstein (Sao Paulo) 2013;11:122–127.
39. Malhotra A, Pelletier MH, Yu Y, Walsh WR. Can platelet-rich plasma (PRP) improve bone healing? A comparison between the theory and experimental outcomes. Arch Orthop Trauma Surg 2013;133:153–165.
40. Zhang N, Wu YP, Qian SJ, Teng C, Chen S, Li H. Research progress in the mechanism of effect of PRP in bone deficiency healing. Sci World J 2013;2013:134582.
41. Campana V, Milano G, Pagano E, et al. Bone substitutes in orthopaedic surgery: From basic science to clinical practice. J Mater Sci Mater Med 2014;25:2445–2461.
42. Caramella CM, Sandri G, Rossi S, et al. New therapeutic platforms for the treatment of epithelial and cutaneous lesions. Curr Drug Deliv 2013;10:18–31.
43. Sandri G, Bonferoni MC, Ferrari F, et al. An in situ gelling buccal spray containing platelet lysate for the treatment of oral mucositis. Curr Drug Discov Technol 2011;8:277–285.
44. Sandri G, Bonferoni MC, Rossi S, et al. Thermosensitive eyedrops containing platelet lysate for the treatment of corneal ulcers. Int J Pharm 2012;426:1–6.
45. Alio JL, Arnalich-Montiel F, Rodriguez AE. The role of "eye platelet rich plasma" (E-PRP) for wound healing in ophthalmology. Curr Pharm Biotechnol 2012;13:1257–1265.

46. Anitua E, Muruzabal F, De la Fuente M, Merayo-Lloves J, Orive G. Effects of heat-treatment on plasma rich in growth factors-derived autologous eye drop. Exp Eye Res 2014;119:27–34.
47. Alio J, Rodriguez AE, Ferreira-Oliveira R, Wrobel-Dudzinska D, Abdelghany AA. Treatment of dry eye disease with autologous platelet-rich plasma: A prospective, interventional, non-randomized study. Ophthalmol Ther 2017;6:285–293.
48. Hom DB, Maisel RH. Angiogenic growth factors: Their effects and potential in soft tissue wound healing. Ann Otol Rhinol Laryngol 1992;101:349–354.
49. Hirano S. Current treatment of vocal fold scarring. Curr Opin Otolaryngol Head Neck Surg 2005;13:143–147.
50. Kumai Y, Kobler JB, Herrera VL, Zeitels SM. Perspectives on adipose-derived stem/stromal cells as potential treatment for scarred vocal folds: Opportunity and challenges. Curr Stem Cell Res Ther 2010;5:175–181.
51. Ohno S, Hirano S, Kanemaru S, et al. Implantation of an atelocollagen sponge with autologous bone marrow-derived mesenchymal stromal cells for treatment of vocal fold scarring in a canine model. Ann Otol Rhinol Laryngol 2011;120:401–408.
52. Hom DB, Sun GH, Elluru RG. A contemporary review of wound healing in otolaryngology: Current state and future promise. Laryngoscope 2009;119:2099–2110.
53. Shen L, Yuan T, Chen S, Xie X, Zhang C. The temporal effect of platelet-rich plasma on pain and physical function in the treatment of knee osteoarthritis: Systematic review and meta-analysis of randomized controlled trials. J Orthop Surg Res 2017;12:16.
54. Ye Y, Zhou X, Mao S, Zhang J, Lin B. Platelet rich plasma versus hyaluronic acid in patients with hip osteoarthritis: A meta-analysis of randomized controlled trials. Int J Surg 2018;53:279–287.
55. Williams PN, Moran G, Bradley JP, ElAttrache NS, Dines JS. Platelet-rich plasma and other cellular strategies in orthopedic surgery. Curr Rev Musculoskelet Med 2015;8:32–39.
56. Nourissat G, Ornetti P, Berenbaum F, Sellam J, Richette P, Chevalier X. Does platelet-rich plasma deserve a role in the treatment of tendinopathy? Joint Bone Spine 2015;82:230–234.
57. Corti L. Nonpharmaceutical approaches to pain management. Top Companion Anim Med 2014;29:24–28.
58. Halper J. Advances in the use of growth factors for treatment of disorders of soft tissues. Adv Exp Med Biol 2014;802:59–76.
59. Scala M, Lenarduzzi S, Spagnolo F, et al. Regenerative medicine for the treatment of Teno-desmic injuries of the equine. A series of 150 horses treated with platelet-derived growth factors. In Vivo 2014;28:1119–1123.
60. Broeckx S, Zimmerman M, Crocetti S, et al. Regenerative therapies for equine degenerative joint disease: A preliminary study. PLoS One 2014;9:e85917.
61. Abrams GD, Frank RM, Fortier LA, Cole BJ. Platelet-rich plasma for articular cartilage repair. Sports Med Arthrosc 2013;21:213–219.
62. Wong S, Ning A, Lee C, Feeley BT. Return to sport after muscle injury. Curr Rev Musculoskelet Med 2015;8:168–175.
63. Chahal J, Van Thiel GS, Mall N, et al. The role of platelet-rich plasma in arthroscopic rotator cuff repair: A systematic review with quantitative synthesis. Arthroscopy 2012;28:1718–1727.
64. Charousset C, Zaoui A, Bellaïche L, Piterman M. Does autologous leukocyte-platelet-rich plasma improve tendon healing in arthroscopic repair of large or massive rotator cuff tears? Arthroscopy 2014;30:428–435.
65. Abtahi AM, Granger EK, Tashjian RZ. Factors affecting healing after arthroscopic rotator cuff repair. World J Orthop 2015;6:211–220.
66. Figueroa D, Figueroa F, Calvo R, Vaisman A, Ahumada X, Arellano S. Platelet-rich plasma use in anterior cruciate ligament surgery: Systematic review of the literature. Arthroscopy 2015;31:981–988.
67. Middleton KK, Barro V, Muller B, Terada S, Fu FH. Evaluation of the effects of platelet-rich plasma (PRP) therapy involved in the healing of sports-related soft tissue injuries. Iowa Orthop J 2012;32:150–163.
68. Mishra A, Harmon K, Woodall J, Vieira A. Sports medicine applications of platelet rich plasma. Curr Pharm Biotechnol 2012;13:1185–1195.
69. Wasterlain AS, Braun HJ, Harris AH, Kim HJ, Dragoo JL. The systemic effects of platelet-rich plasma injection. Am J Sports Med 2013;41:186–193.
70. McCarrel TM, Mall NA, Lee AS, Cole BJ, Butty DC, Fortier LA. Considerations for the use of platelet-rich plasma in orthopedics. Sports Med 2014;44:1025–1036.
71. Han C, Na Y, Zhu Y, et al. Is platelet-rich plasma an ideal biomaterial for arthroscopic rotator cuff repair? A systematic review and meta-analysis of randomized controlled trials. J Orthop Surg Res 2019;14:183.
72. Miller LE, Parrish WR, Roides B, Bhattacharyya S. Efficacy of platelet-rich plasma injections for symptomatic tendinopathy: Systematic review and meta-analysis of randomised injection-controlled trials. BMJ Open Sport Exerc Med 2017;3(1):e000237.
73. Andriolo L, Altamura SA, Reale D, Candrian C, Zaffagnini S, Filardo G. Nonsurgical treatments of patellar tendinopathy: Multiple injections of platelet-rich plasma are a suitable option: A systematic review and meta-analysis. Am J Sports Med 2019;47:1001–1018.
74. Wu CC, Wu YN, Ho HO, Chen KC, Sheu MT, Chiang HS. The neuroprotective effect of platelet-rich plasma on erectile function in bilateral cavernous nerve injury rat model. J Sex Med 2012;9:2838–2848.
75. Wu YN, Wu CC, Sheu MT, Chen KC, Ho HO, Chiang HS. Optimization of platelet-rich plasma and its effects on the recovery of erectile function after bilateral cavernous nerve injury in a rat model. J Tissue Eng Regen Med 2016;10:E294–E304.
76. Shirvan MK, Alamdari DH, Ghoreifi A. A novel method for iatrogenic vesicovaginal fistula treatment: Autologous platelet rich plasma injection and platelet rich fibrin glue interposition. J Urol 2013;189:2125–2129.
77. Chew KT, Leong D, Lin CY, Lim KK, Tan B. Comparison of autologous conditioned plasma injection, extracorporeal shockwave therapy, and conventional treatment for plantar fasciitis: a randomized trial. PM R 2013;5:1035–1043.
78. Owens RF Jr, Ginnetti J, Conti SF, Latona C. Clinical and magnetic resonance imaging outcomes following platelet rich plasma injection for chronic midsubstance Achilles tendinopathy. Foot Ankle Int 2011;32:1032–1039.
79. Murawski CD, Smyth NA, Newman H, Kennedy JG. A single platelet-rich plasma injection for chronic midsubstance achilles tendinopathy: A retrospective preliminary analysis. Foot Ankle Spec 2014;7:372–376.
80. Grambart ST. Sports medicine and platelet-rich plasma: Nonsurgical therapy. Clin Podiatr Med Surg 2015;32:99–107.
81. Vannini F, Di Matteo B, Filardo G, Kon E, Marcacci M, Giannini S. Platelet-rich plasma for foot and ankle pathologies: A systematic review. Foot Ankle Surg 2014;20:2–9.
82. Di Matteo B, Filardo G, Kon E, Marcacci M. Platelet-rich plasma: Evidence for the treatment of patellar and Achilles tendinopathy–A systematic review. Musculoskelet Surg 2015;99:1–9.
83. Lacci KM, Dardik A. Platelet-rich plasma: Support for its use in wound healing. Yale J Biol Med 2010;83:1–9.
84. Carter MJ, Fylling CP, Parnell LK. Use of platelet-rich plasma gel on wound healing: A systematic review and meta-analysis. Eplasty 2011;11:e38.
85. Martinez-Zapata MJ, Martí-Carvajal AJ, Solà I, et al. Autologous platelet-rich plasma for treating chronic wounds. Cochrane Database Syst Rev 2012;10:CD006899.
86. Skórkowska-Telichowska K, Czemplik M, Kulma A, Szopa J. The local treatment and available dressings designed for chronic wounds. J Am Acad Dermatol 2013;68:e117–e126.
87. Moraes VY, Lenza M, Tamaoki MJ, Faloppa F, Belloti JC. Platelet-rich therapies for musculoskeletal soft tissue injuries. Cochrane Database Syst Rev 2013;12:CD010071.

88. Mehrannia M, Vaezi M, Yousefshahi F, Rouhipour N. Platelet rich plasma for treatment of nonhealing diabetic foot ulcers: A case report. Can J Diabetes 2014;38:5–8.
89. Tolar J, Tolar M. A living band-aid for epidermolysis bullosa. Blood Transfus 2015;13:1–2.
90. Tadini G, Guez S, Pezzani L, et al. Preliminary evaluation of cord blood platelet gel for the treatment of skin lesions in children with dystrophic epidermolysis bullosa. Blood Transfus 2015;13:153–158.
91. Picard F, Hersant B, Bosc R, Meningaud J. The growing evidence for the use of platelet-rich plasma on diabetic chronic wounds: A review and a proposal for a new standard care. Wound Repair Regen 2015;23:638–643.
92. Goda AA, Metwally M, Ewada A, Ewees H. Platelet-rich plasma for the treatment of diabetic foot ulcer: A randomized, double-blind study. Egypt J Surg 2018;37:178–184.
93. Hersant B, SidAhmed-Mezi M, Bosc R, Meningaud JP. Autologous platelet-rich plasma/thrombin gel combined with split-thickness skin graft to manage postinfectious skin defects: A randomized controlled study. Adv Skin Wound Care 2017;30:502–508.
94. Ling Y, Wang S. Effects of platelet-rich plasma in the treatment of plantar fasciitis. A meta-analysis of randomized controlled trials. Medicine 2018;97:e12110.
95. Yang W-Y, Han Y-H, Cao X-W, et al. Platelet-rich plasma as a treatment for plantar fasciitis. A meta-analysis of randomized controlled trials. Medicine 2017;96:e8475.
96. Del Corso M, Vervelle A, Simonpieri A, et al. Current knowledge and perspectives for the use of platelet-rich plasma (PRP) and platelet-rich fibrin (PRF) in oral and maxillofacial surgery part 1: Periodontal and dentoalveolar surgery. Curr Pharm Biotechnol 2012;13:1207–1230.
97. Simonpieri A, Del Corso M, Vervelle A, et al. Current knowledge and perspectives for the use of platelet-rich plasma (PRP) and platelet-rich fibrin (PRF) in oral and maxillofacial surgery part 2: Bone graft, implant and reconstructive surgery. Curr Pharm Biotechnol 2012;13:1231–1256.
98. Ross R, Glomset J, Kariya B, Harker L. A platelet-dependent serum factor that stimulates the proliferation of arterial smooth muscle cells in vitro. Proc Natl Acad Sci U S A 1974;71:1207–1210.
99. Witte LD, Kaplan KL, Nossel HL, Lages BA, Weiss HJ, Goodman DS. Studies of the release from human platelets of the growth factor for cultured human arterial smooth muscle cells. Circ Res 1978;42:402–409.
100. Ross R, Nist C, Kariya B, Rivest MJ, Raines E, Callis J. Physiological quiescence in plasma-derived serum: influence of platelet-derived growth factor on cell growth in culture. J Cell Physiol 1978;97(3 pt 2 suppl 1):497–508.
101. Kaplan DR, Chao FC, Stiles CD, Antoniades HN, Scher CD. Platelet alpha granules contain a growth factor for fibroblasts. Blood 1979;53:1043–1052.
102. Assoian RK, Komoriya A, Meyers CA, Miller DM, Sporn MB. Transforming growth factor-beta in human platelets. Identification of a major storage site, purification, and characterization. J Biol Chem 1983;258:7155–7160.
103. Derynck R. Transforming growth factor-alpha: Structure and biological activities. J Cell Biochem 1986;32:293–304.
104. 104 Karey KP, Marquardt H, Sirbasku DA. Human platelet-derived mitogens. I. Identification of insulinlike growth factors I and II by purification and N alpha amino acid sequence analysis. Blood 1989;74:1084–1092.
105. Karey KP, Sirbasku DA. Human platelet-derived mitogens. II. Subcellular localization of insulinlike growth factor I to the alpha-granule and release in response to thrombin. Blood 1989;74:1093–1100.
106. Brunner G, Nguyen H, Gabrilove J, Rifkin DB, Wilson EL. Basic fibroblast growth factor expression in human bone marrow and peripheral blood cells. Blood 1993;81:631–638.
107. Dooley DC, Oppenlander BK, Spurgin P, et al. Basic fibroblast growth factor and epidermal growth factor downmodulate the growth of hematopoietic cells in long-term stromal cultures. J Cell Physiol 1995;165:386–397.
108. Banks RE, Forbes MA, Kinsey SE, et al. Release of the angiogenic cytokine vascular endothelial growth factor (VEGF) from platelets: Significance for VEGF measurements and cancer biology. Br J Cancer 1998;77:956–964.
109. Hormbrey E, Gillespie P, Turner K, et al. A critical review of vascular endothelial growth factor (VEGF) analysis in peripheral blood: Is the current literature meaningful? Clin Exp Metastasis 2002;19:651–663.
110. Sánchez M, Anitua E, Andia I. Poor standardization in platelet-rich therapies hampers advancement. Arthroscopy 2010;26:725–726.
111. Taylor DW, Petrera M, Hendry M, Theodoropoulos JS. A systematic review of the use of platelet-rich plasma in sports medicine as a new treatment for tendon and ligament injuries. Clin J Sport Med 2011;21:344–352.
112. Moraes VY, Lenza M, Tamaoki MJ, Faloppa F, Belloti JC. Platelet-rich therapies for musculoskeletal soft tissue injuries. Cochrane Database Syst Rev 2014;4:CD010071.
113. Kim TH, Kim SH, Sándor GK, Kim YD. Comparison of platelet-rich plasma (PRP), platelet-rich fibrin (PRF), and concentrated growth factor (CGF) in rabbit-skull defect healing. Arch Oral Biol 2014;59:550–558.
114. Sandrey MA. Autologous growth factor injections in chronic tendinopathy. J Athl Train 2014;49:428–430.
115. Marx RE, Carlson ER, Eichstaedt RM, Schimmele SR, Strauss JE, Georgeff KR. Platelet-rich plasma: Growth factor enhancement for bone grafts. Oral Surg Oral Med Oral Pathol Oral Radiol Endod 1998;85:638–646.
116. Shah P, Keppler L, Rutkowski J. A review of platelet derived growth factor playing pivotal role in bone regeneration. J Oral Implantol 2014;40:330–340.
117. Panda S, Doraiswamy J, Malaiappan S, Varghese SS, Del Fabbro M. Additive effect of autologous platelet concentrates in treatment of intrabony defects: A systematic review and meta-analysis. J Investig Clin Dent 2016;7:13–26.
118. Del Fabbro M, Corbella S, Taschieri S, Francetti L, Weinstein R. Autologous platelet concentrate for post-extraction socket healing: A systematic review. Eur J Oral Implantol 2014;7:333–344.
119. Davis VL, Abukabda AB, Radio NM, et al. Platelet-rich preparations to improve healing. Part I: workable options for every size practice. J Oral Implantol 2014;40:500–510.
120. Davis VL, Abukabda AB, Radio NM, et al. Platelet-rich preparations to improve healing. Part II: Platelet activation and enrichment, leukocyte inclusion, and other selection criteria. J Oral Implantol 2014;40:511–521.
121. Kuffler DP. Platelet-rich plasma promotes axon regeneration, wound healing, and pain reduction: Fact or fiction. Mol Neurobiol 2015;52:990–1014.
122. Salamanna F, Veronesi F, Maglio M, Della Bella E, Sartori M, Fini M. New and emerging strategies in platelet-rich plasma application in musculoskeletal regenerative procedures: General overview on still open questions and outlook. Biomed Res Int 2015;2015:846045.
123. Beitzel K, Allen D, Apostolakos J, et al. US definitions, current use, and FDA stance on use of platelet-rich plasma in sports medicine. J Knee Surg 2015;28:29–34.
124. Fiorentino S, Roffi A, Filardo G, Marcacci M, Kon E. European definitions, current use, and EMA stance of platelet-rich plasma in sports medicine. J Knee Surg 2015;28:51–54.
125. Song Y, Lin K, He S, et al. Nano-biphasic calcium phosphate/polyvinyl alcohol composites with enhanced bioactivity for bone repair via low-temperature three-dimensional printing and loading with platelet-rich fibrin. Int J Nanomed 2018;13:505–523.

126. Allahverdi A, Sharifi D, Abedi G, Hesaraki S, Fattahiyan H. Effect of platelet-rich plasma, low-level laser therapy (650 nm) or their combination on the healing of Achilles tendon in rabbits: A histopathological study. Eur J Experimental Biol 2014;4:201–208.
127. Garcia TA, Camargo RCT, Koike TE, et al. Histological analysis of the association of low level laser therapy and platelet-rich plasma in regeneration of muscle injury in rats. Braz J Phys Ther 2017;21:425–433.
128. Prodromos CC, Finkle S, Dawes A, Dizon A. Intra-articular laser treatment plus platelet-rich plasma (PRP) significantly reduces pain in many patients who had failed prior PRP treatment. Medicines 2019;6:75.
129. Khojasteh A, Hosseinpour S, Nazeman P, Dehghan MM. The effect of a platelet-rich fibrin conduit on neurosensory recovery following inferior alveolar nerve lateralization: A preliminary clinical study. Int J Oral Maxillofac Surg 2016;45:1303–1308.
130. Temmerman A, Vandessel J, Castro A, et al. The use of leucocyte and platelet-rich fibrin in socket management and ridge preservation: A split-mouth, randomized, controlled clinical trial. J Clin Periodontol 2016;43:990–999.
131. Huang Y, Bornstein MM, Lambrichts I, Yu HY, Politis C, Jacobs R. Platelet-rich plasma for regeneration of neural feedback pathways around dental implants: A concise review and outlook on future possibilities. Int J Oral Sci 2017;9:1–9.
132. Wu Y, Kanna M, Liu C, Zhou Y, Chan C. Generation of autologous platelet-rich plasma by the ultrasonic standing waves. IEEE Trans Biomed Eng 2016;63:1642–1652.
133. Kawase T. Platelet-rich plasma and its derivatives as promising bioactive materials for regenerative medicine: Basic principles and concepts underlying recent advances. Odontology 2015;103:126–135.

10

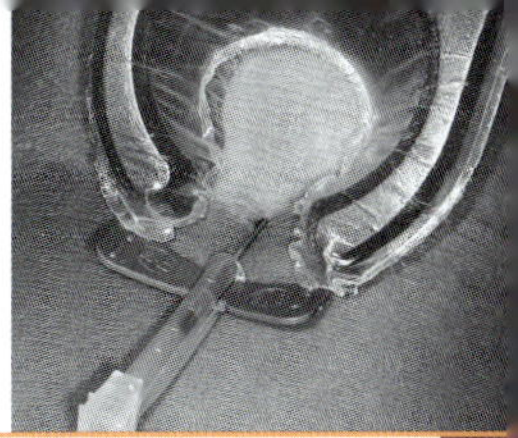

Principles and Practice of Phlebotomy

Some people may be tempted to skip this chapter. Others, however, will recognize that the ability to perform phlebotomy safely and successfully is an absolute prerequisite to incorporating PRP therapy into their practice and will take the time to read it carefully.

Anyone reading this book will have undergone venipuncture at least once, and it is useful to reflect on that experience. Regardless of whether the setting was a doctor's office, hospital laboratory, or blood donation center, your overriding concern upon meeting the phlebotomist was likely: how much experience does this person have? About 10% of the population suffers from needle phobia, but no one looks forward to having their skin punctured once, much less two or three times. The phlebotomist, on the other hand, must learn to recognize and be sensitive to the apprehensiveness of the patient without being affected by it. For most people, this requires a level of confidence that only comes with guidance and experience.

The point is that performing phlebotomy for the purpose of preparing PRP can induce a whole new level of anxiety in a surgeon or surgical assistant, regardless of their reputation or other clinical accomplishments. Because the scheduled operation depends on their ability to obtain a blood specimen, it often becomes the most stressful part of the whole procedure. There is no mathematic formula that can tell you which is the right vein, the right needle, the right angle, and the right force that will yield the perfect blood draw. Successful phlebotomy requires a combination of patience, empathy, and skill, plus a certain amount of finesse.

The best advice is to be as prepared as you can: Read this chapter carefully, familiarize yourself with each piece of blood collection equipment, follow standard operating procedures, practice quality control, and use effective communication. Practice your new skill on a phlebotomy simulation arm or a willing coworker and, in the beginning, examine your patients' arms at the initial consultation and choose those with the most prominent veins until you gain experience and confidence. Finally, always remember that the safety and comfort of your patient is your first priority. After two failed attempts, admit defeat; you will earn your patient's respect and gratitude.

Box 10-1 Routine venipuncture equipment

- Blood collection tray
- Evacuated tube system holders
- Winged blood collection sets
- Needles
- Needle disposal sharps containers
- Evacuated collection tubes
- Tourniquets
- Gloves
- 70% isopropyl alcohol, iodine swabs, chlorhexidine gluconate swabs
- 2 × 2-inch gauze pads
- Bandages
- Antimicrobial hand gel

Adapted from Strassinger and Di Lorenzo.[1]

Venipuncture Equipment

The clinician or surgical assistant who performs the venipuncture must have a thorough understanding of the tools used in the procedure (Box 10-1). There are different blood collection systems available, and the reader must consider

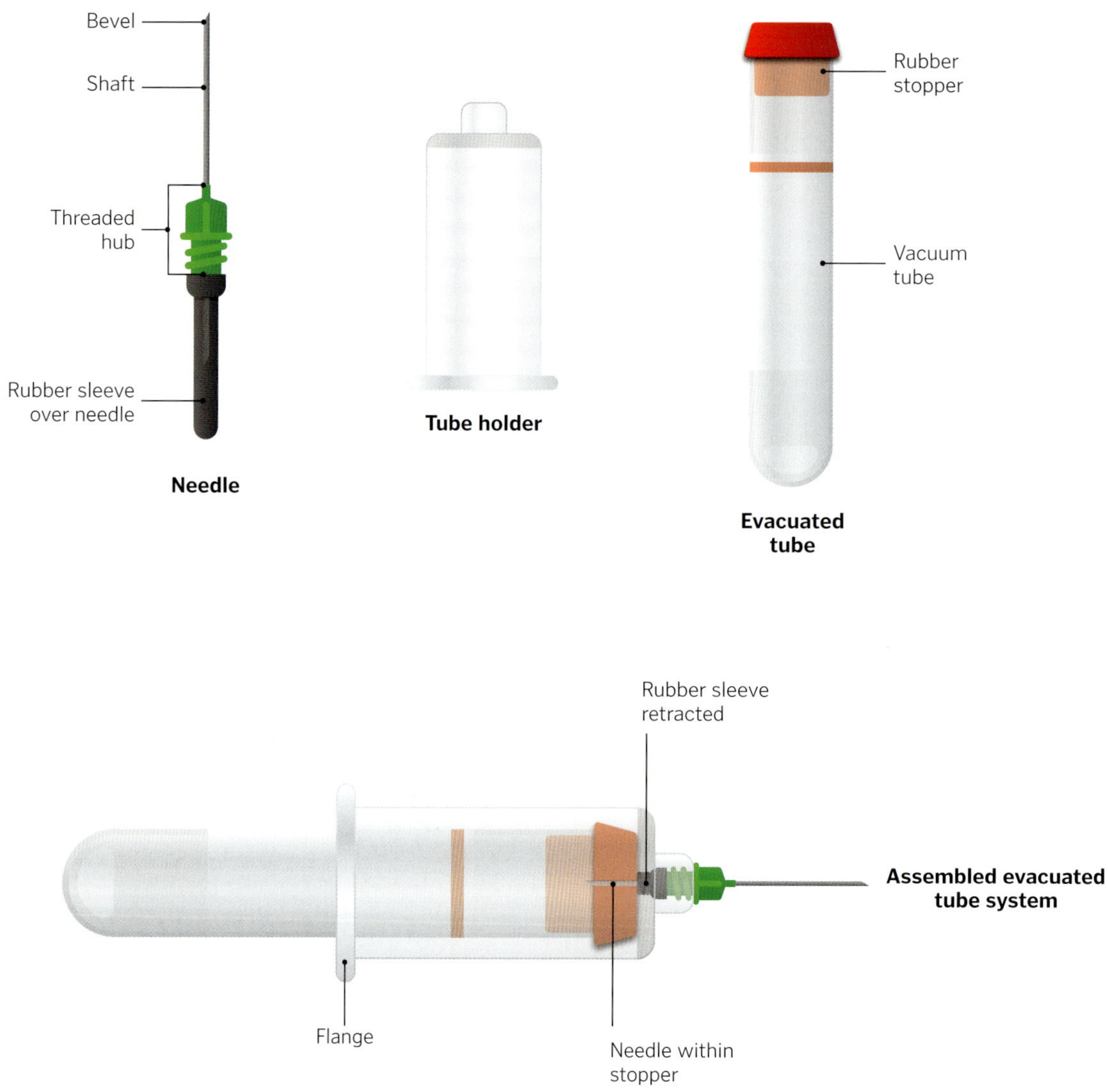

FIG 10-1 Evacuated tube system.

the advantages and disadvantages of each to determine the one best suited to their needs.

Blood collection

To collect blood for the purpose of preparing PRP, one of two types of needle systems should be used.

Evacuated tube system

Most adults who have had their blood drawn are probably familiar with the evacuated tube system (ETS), as it is the method used most frequently for venipuncture. The ETS system consists of a needle that has a beveled point at each end and is separated in the middle by a threaded hub. One end of the needle is for venipuncture and the other end penetrates the rubber stopper of the blood collection tube. The latter end is covered by a retractable rubber sheath that prevents leakage of blood when a collection tube is being removed from it (Fig 10-1). The blood collection tubes are color coded and vacuum-sealed. The primary advantage of the ETS is that the blood is collected directly into the color-coded tube that will

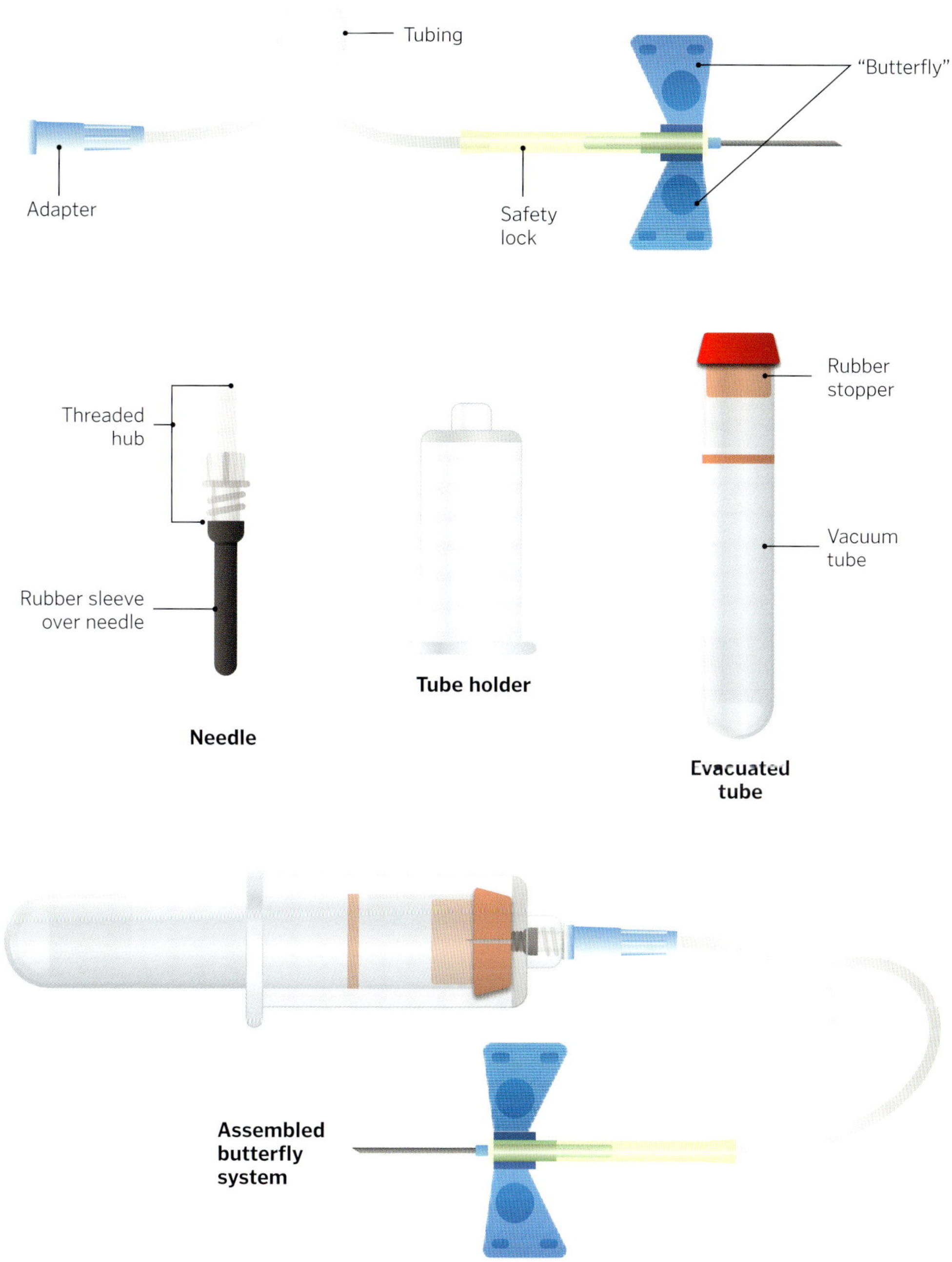

FIG 10-2 Winged blood collection system.

be placed into the centrifuge, eliminating the need for transferal and minimizing the risk of contamination or biohazard exposure.

Winged blood collection set

Beginners often find the winged system easier to use than the ETS. This system consists of a needle to which a plastic holder resembling butterfly wings has been attached (hence the nickname "butterfly needle"). The plastic wings are used to hold the needle during insertion into the vein. Attached to the back end of the needle is a short length of flexible plastic tubing. When the plastic tubing is connected to an evacuated tube holder, the blood can be collected directly into the color-coded tube that will be placed into the centrifuge (Fig 10-2). Again, this is important to avoid the risk of blood contamination or

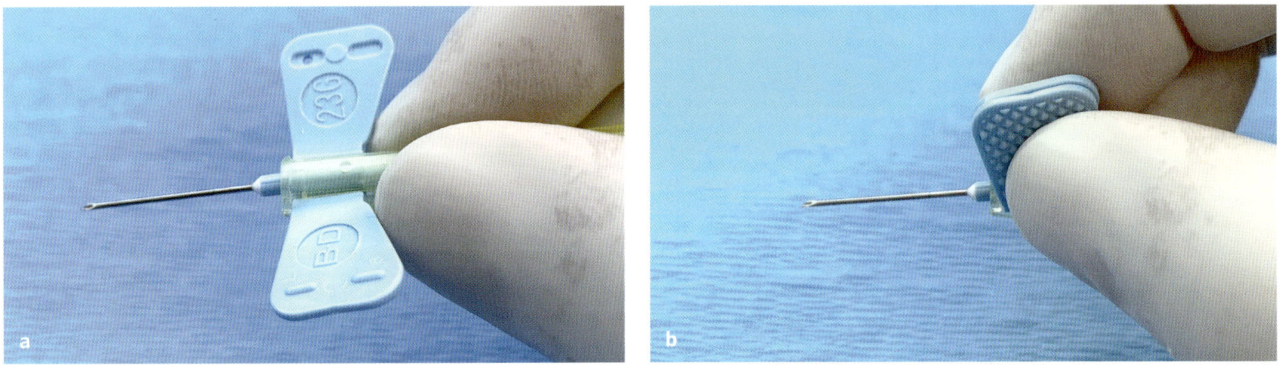

FIG 10-3 Winged needle with butterfly wings open *(a)* and with wings closed and used as a handle to allow a shallow angle of entry into the vein *(b)*.

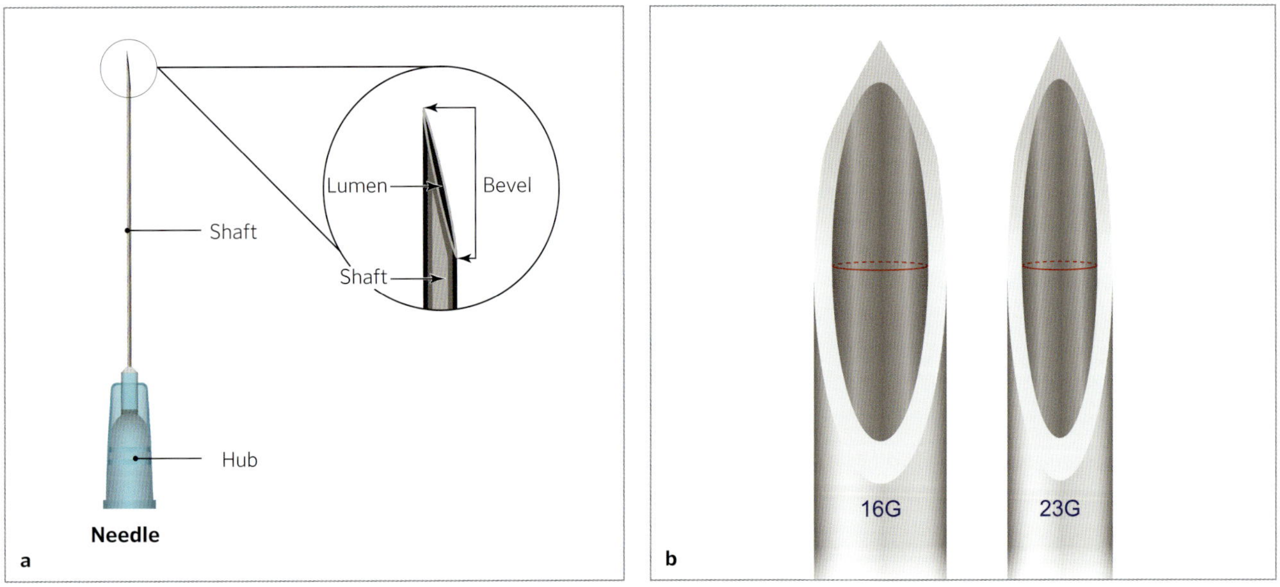

FIG 10-4 *(a)* Enlarged view of the venipuncture end of the needle. *(b)* Illustration of the diameter of a 16-gauge needle compared with that of a 23-gauge needle.

biohazard exposure. The advantage of this system, especially for beginners, is that the butterfly wings make it easier to insert the needle at a shallow angle, particularly when working with very small veins (Fig 10-3).

Needles

Hypodermic and butterfly needles designed for venipuncture are sterile and disposable and may only be used once. Hypodermic needles are available in a range of sizes, but a 1- or 1.5-inch length is appropriate for most patients. Butterfly needles, which range from 0.5 to 0.75 inches in length, are selected according to the depth of the vein.

Gauge refers to the diameter of the needle lumen (Fig 10-4); the smaller the gauge number, the larger the diameter of the needle. Large (eg, 16-gauge) needles are used to collect blood for transfusions, whereas small (eg, 25-gauge) needles are used for patients with small veins, children, and the elderly. When drawing blood to prepare PRP, a 21- or 23-gauge needle is most commonly used.

The gauge of the needle used to draw blood for PRP (or for any other purpose) can negatively affect the quality of the PRP and/or the health of the patient. For example, unless special circumstances require it, a 25-gauge needle should not be used because it can damage the platelets (hemolysis) as they move through the needle. The small-gauge needle

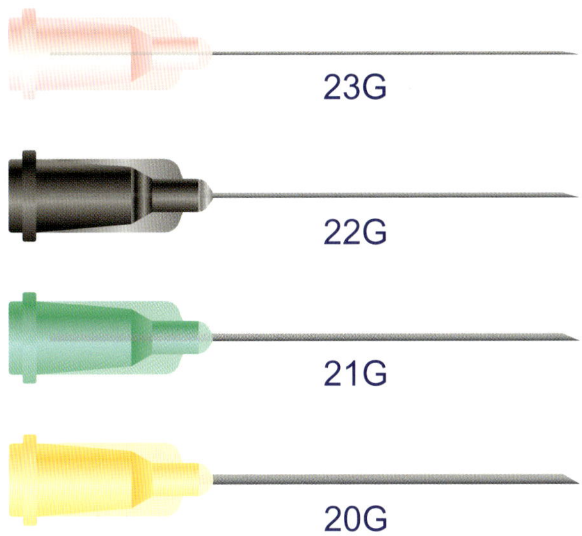

FIG 10-5 Color coding of needle hubs according to the gauge of the needle.

FIG 10-6 Containers approved for disposal of used needles are available in a variety of sizes.

FIG 10-7 A syringe with a plunger is not recommended for use when drawing blood to prepare PRP.

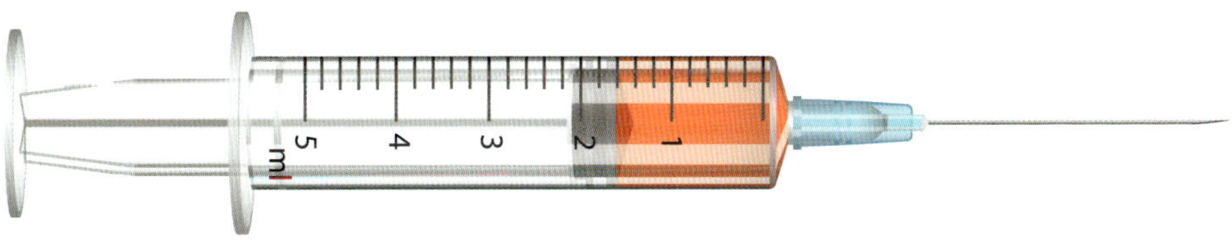

also must spend a longer time in the vein, prolonging the patient's discomfort. Conversely, a needle gauge that is larger than necessary (eg, 16- to 20-gauge) should be avoided as well because patients on blood thinners can have bleeding episodes or develop a hematoma after the venipuncture.

Hypodermic and butterfly needles are individually packaged in sealed sterile shields that are color-coded by gauge for quick identification. Generally, 20-gauge needles are yellow, 21-gauge needles are green, 22-gauge needles are black, and 23-gauge needles are pink (Fig 10-5), although this can vary by manufacturer.

Needle disposal

Dental professionals are well versed in the OSHA (Occupational Safety and Health Administration) mandate to dispose of all used or uncapped needles in a puncture-resistant, leakproof "sharps" container that can be sealed and locked when it is full (Fig 10-6). Hypodermic and butterfly needles used for venipuncture must be disposed of in the same type of container as is used for needles to deliver anesthesia.

Collection tubes

Both of the venipuncture needle systems described earlier use an evacuated tube to collect the blood. The ETS was introduced by Beckton and Dickinson in 1949 under the trade name Vacutainer.[2] Prior to that, phlebotomists used a simple syringe with a plunger that drew the blood into a barrel attached to the end of the needle (Fig 10-7). Although this system is still in use today, it is not recommended for preparing PRP because the blood must be transferred from the syringe barrel into a test tube or collection tube that fits inside a standard centrifuge. This added step introduces the risk of damaging or

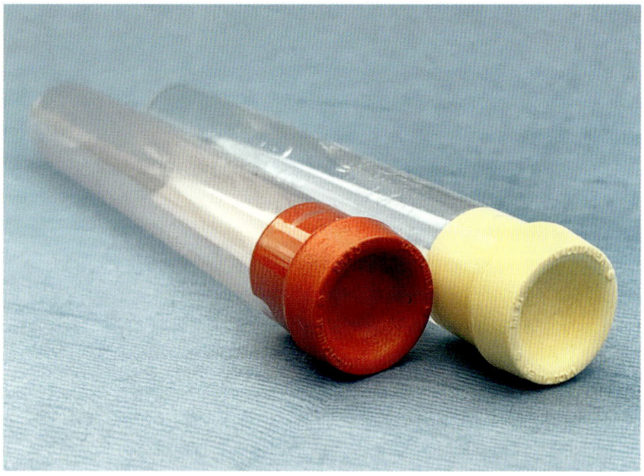

FIG 10-8 Evacuated tubes for blood collection. Tubes with a red stopper contain no anticoagulants or additives, whereas those with a yellow stopper contain ACD-A to prevent the blood from clotting.

FIG 10-9 Filled evacuated tubes should be placed in a metal blood collection rack both before and after being centrifuged.

contaminating the platelets. Additionally, pulling back on the plunger while the needle is in the vein can cause the needle to move, which is painful and potentially damaging, and the blood draw may not be as smooth compared with the even flow that is possible with the ETS.

In the ETS, there is no plunger, and the barrel is replaced by a glass or plastic tube that is vacuum-sealed with a rubber stopper (hence the term "evacuated" tube). This collection tube is securely attached to a needle holder as follows: The puncturing end (opposite the beveled insertion end) of the needle screws securely into the small opening at one end of the holder, while the evacuated tube is placed into the large opening at the opposite end of the holder.

Once the needle is in the patient's vein, the tube is advanced all the way into the holder so that the needle pierces the tube's rubber stopper, allowing the blood to flow into the tube. After the blood has been drawn, the tube is removed from the holder. In the process, the rubber stopper reseals itself, thus preventing the collected blood from leaking out. This system allows the filling of multiple tubes without interrupting the blood withdrawal. As noted earlier, the ETS ensures maximum safety because the collected blood never leaves the closed system.

Evacuated tubes are sterile, and most are coated with silicone to prevent cell adherence. The presence of any additive in the tube is indicated by the color of the tube's rubber stopper, based on a universal coding system. For the present purposes, only two colors are relevant: red and yellow. Red-stopper tubes are free of any additive or anticoagulant and are commonly referred to as plain tubes or simply "red tops." Any blood collected in red-stopper tubes will clot naturally within 30 to 60 minutes. Yellow-stopper tubes contain anticoagulant citrate dextrose-A (ACD-A) to prevent the collected blood from clotting by binding the calcium contained in it (Fig 10-8). The choice of evacuated container will depend on which type of PRP is being prepared (see chapter 4).

Blood collection rack

When the blood collection tubes are removed from the centrifuge, they must be kept in an upright position to maintain separation of the three layers (plasma; white blood cells, platelets, and fibrinogen; and red blood cells). A steel blood collection rack (Fig 10-9), which is autoclavable, will hold the evacuated tubes securely and can also be used to transport the tubes if the centrifuge is housed in a separate room from the operatory.

FIG 10-10 Single-use disposable tourniquet.

FIG 10-11 A plentiful supply of gloves, gauze pads, and adhesive tape should always be stocked in the blood collection cart or tray.

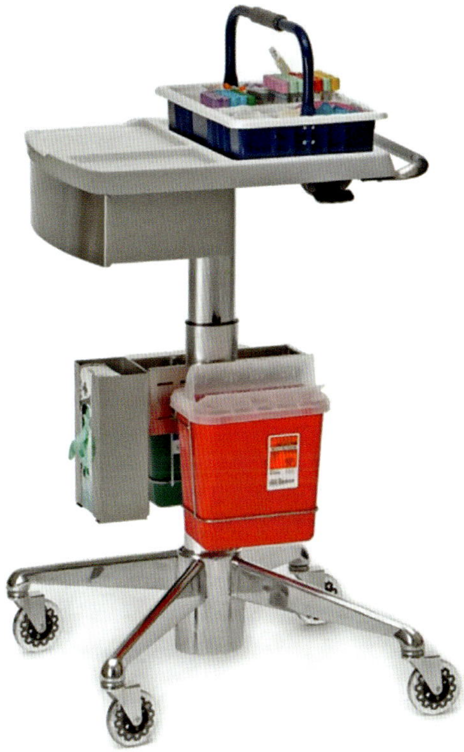

FIG 10-12 Mobile phlebotomy workstation.

Tourniquet

A tourniquet is a (nonlatex) strip of nitrile or vinyl that is tied to the patient's arm, about 3 to 4 inches above the venipuncture site, to make the vein easier to locate (Fig 10-10). Tourniquets work by impeding venous (but not arterial) blood flow. As a result, the vein becomes more visible or palpable.

Tourniquets are for single-use only and should be discarded immediately after venipuncture. In one study, analysis of the bacteria recovered from tourniquets used for routine venipuncture revealed that 25% were multiantibiotic resistant.

Gloves, gauze pads, and bandages

As with all clinical procedures, well-fitting gloves must be worn when drawing blood. Gloves with powder are contraindicated because the powder can contaminate the collected blood.

Gauze pads (2 × 2 inch) are needed for applying pressure to the venipuncture site after the needle has been removed. They can also be folded into quarters and placed under the bandage for additional protection of the puncture site. Cotton balls are not recommended for this purpose because they can adhere to the puncture site. When the cotton ball is later removed by the patient, the platelet plug may be removed as well, which can reinitiate bleeding.

Latex-free bandages or adhesive tape should be placed over the puncture site once the bleeding has ceased. Patients who are allergic to adhesive bandages can be treated with self-adhesive gauze (Fig 10-11).

Organizing and transporting equipment

As with any dental procedure, the first step in a successful venipuncture is to assemble the equipment and supplies needed in one convenient location. Large practices that use PRP on a routine basis may wish to invest in a mobile phlebotomy workstation on wheels (Fig 10-12), which can be configured to include a laptop computer, label printer, hazardous waste container, sharps container, and storage drawers and shelves. The workstation is designed to be

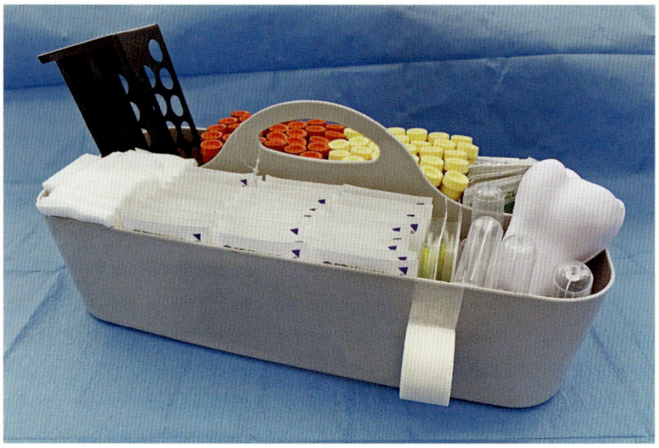

FIG 10-13 Blood collection tray.

wheeled from one operatory to another, eliminating the need to stock each operatory.

For most practices, the author recommends the use of a blood collection tray (Fig 10-13), which offers a convenient way to carry equipment between operatories. Once a venipuncture procedure has been started, it is critically important for safety reasons to be sure that all equipment is at hand. Furthermore, a well-organized tray instills confidence in the patient. Blood collection trays of various sizes and designs are available to meet individual needs.

Whether a mobile workstation or collection tray is used, it will need to be cleaned, disinfected, and restocked on at least a weekly basis.

Venipuncture Procedure

The practice of phlebotomy is not something that can be learned strictly by watching videos or reading a how-to guide; it must be practiced repeatedly in a controlled environment under expert guidance. However, the following step-by-step guide to best practices will enable the novice to be well prepared and get started on the right path.

1. Positioning the patient

Provide a comfortable and safe place for the patient to sit. Never draw blood from a patient who is in a standing position because of the risk of fainting. Seat the patient in a chair with armrests and have the patient place his or her arm on a hard surface. The patient's arm should be firmly supported and extended downward in a straight line from the shoulder to the wrist, allowing the collection tube to fill from the bottom up. Do not hyperextend the elbow, which can make vein palpation difficult. If necessary, ask the patient to uncross their legs. Be alert for any changes in the patient's condition while the procedure is being performed.

2. Sanitizing your hands and applying gloves

Always sanitize your hands in front of the patient prior to starting the procedure using either soap and water or an alcohol-based hand-cleansing solution. Afterward, apply a clean pair of well-fitting nonlatex gloves (Fig 10-14).

3. Locating and identifying veins

The area of the arm just below the bend in the elbow, known as the *antecubital fossa*, is the preferred venipuncture site because the veins located there are superficial and can be identified more easily than at other sites (Fig 10-15). There are three major veins in this area: the median cubital, the cephalic, and the basilic. The path and distribution of these veins varies from one person to another, but an "H-shaped" or "M-shaped" arrangement can be found in most people.

The H-shaped pattern (Fig 10-16) is found in about 70% of the population and includes the cephalic, median cubital, and basilic veins. In this pattern, the median cubital vein forms an anastomose between the cephalic and basilic veins, creating the appearance of a slanted letter H.

In the M-shaped pattern (Fig 10-17), found in approximately 20% of people, the median antebrachial vein divides into median basilic and median cephalic veins, adjoining the basilic and cephalic veins, respectively.

These veins continue down the forearm to the wrist area, where they become smaller and less stable. Never use the veins on the underside of the wrist to avoid the risk of accidentally puncturing an artery, nerve, or tendon. The back of the hand contains several small but prominent veins that can be used for venipuncture if necessary, but they may require a smaller-gauge needle or a butterfly needle.

Median cubital vein

Of the three superficial veins located in the antecubital area, the median cubital vein should be the first choice, for several reasons. It is a relatively large vessel that is close

FIG 10-14 *(a to c)* Health care professionals understand the importance of sanitizing their hands using an antiseptic solution or soap and hot water before putting on well-fitting gloves prior to any procedure.

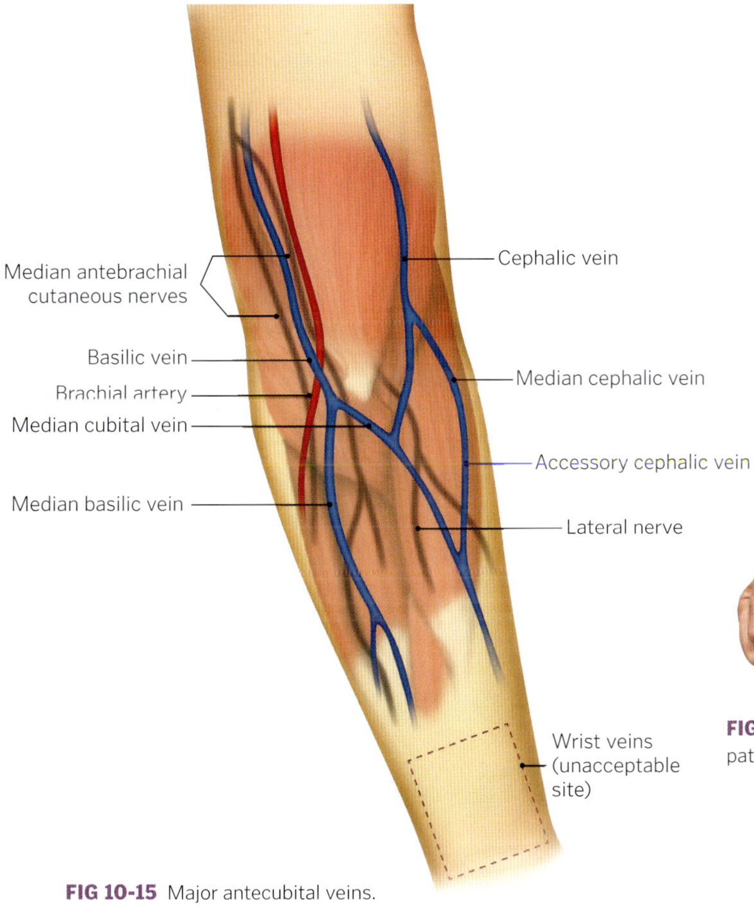

FIG 10-15 Major antecubital veins.

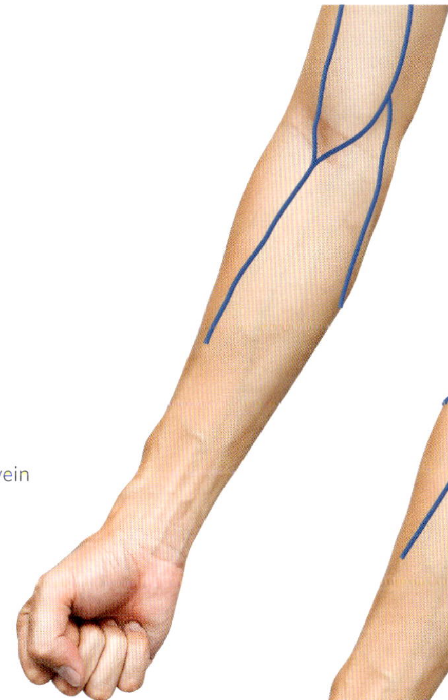

FIG 10-16 H-shaped vein pattern.

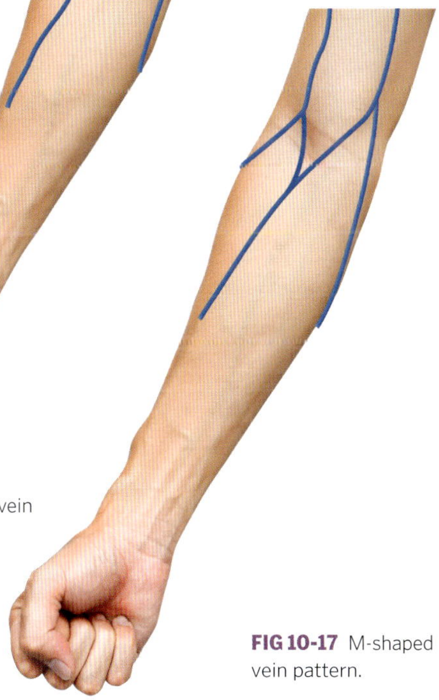

FIG 10-17 M-shaped vein pattern.

to the skin's surface, and it does not have a tendency to roll when a needle is inserted into it. Because it is located in the center of the cubital fossa (see Fig 10-15), where there are fewer nerve endings than in other areas, the median cubital vein is also the least painful to puncture. Finally, there is a lower risk of injuring a nerve or accidentally puncturing an artery because the median cubital vein lies over the bicipital aponeurosis, a fibrous membrane that serves to protect the underlying nerves and veins.

Cephalic vein

The cephalic vein on the lateral or thumb side of the arm (see Fig 10-15) is the vein of second choice when the median cubital vein is inaccessible on both arms. The

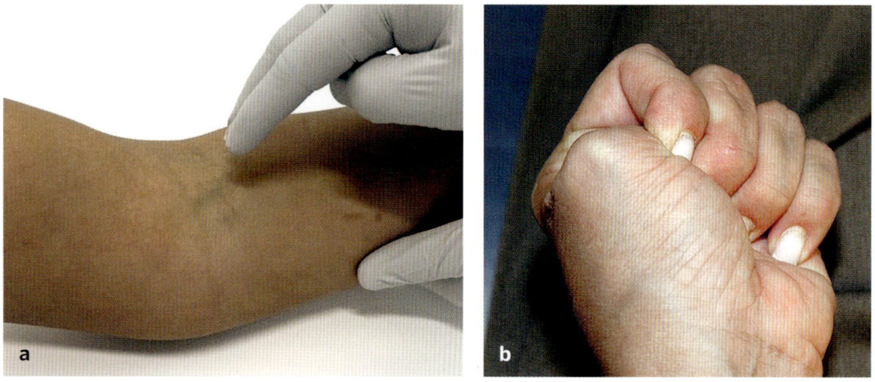

FIG 10-18 *(a)* When palpating the vein, do not use the thumb. *(b)* Ask the patient to make a loose fist just before you insert the needle. Pumping the fist, however, can lead to hemoconcentration and should be avoided.

FIG 10-19 ➤ Veinlite and other vein-detecting devices can be of great assistance in locating very small and/or very deep veins.

cephalic vein is usually more difficult to locate compared to the median cubital vein, and it has a tendency to roll. It is important to note that the most lateral aspect of the cephalic vein should be avoided because of the proximity of the lateral cutaneous nerve in that area.

Basilic vein

The basilic vein should be the vein of last resort in the antecubital area. It is located on the medial periphery of the antecubital area (see Fig 10-15), in the vicinity of the median nerve and the brachial artery. The basilic vein has a much greater tendency to roll and is associated with a greater likelihood of forming a hematoma than the other two veins. Far more permanent nerve injuries, arterial nicks, and hence legal actions are the result of collecting blood from the basilic vein than from any other vein. Therefore, when it is not visible or palpable, or when the first puncture attempt fails, continued probing in the area is highly risky and is strongly discouraged.

4. Performing palpation

Along with visual inspection of the puncture site, palpation is an essential step in locating and identifying veins because when it comes time to insert the needle, the ability to feel a vein is much more useful than the ability to see it. Palpation also can help prevent arterial punctures because arteries have a pulse that can be felt.

Palpation is performed using the tip of the index finger of the nondominant hand (Fig 10-18). Carefully probe the antecubital area both vertically and horizontally using a pushing motion. Apply gentle pressure to feel for spongy, resilient, tubelike structures, which are veins, and try to distinguish them from more rigid structures, which are tendons. Tilting the patient's arm can sometimes help. Tactile sense improves with practice, so do not become discouraged if you have a hard time identifying these differences when you first get started.

If your efforts to palpate a vein are unsuccessful on both arms, there are several techniques that may be helpful. Placing the patient's arm below the level of the heart for several minutes will decrease the return of blood to the heart and allow the veins to fill. Using a warm compress is another method to dilate the veins, which may help you to see or palpate them. Most importantly, take your time and try to release any tension you may feel, which can only increase the anxiety of the patient and make matters worse.

Some patients have very small and/or very deep veins that will elude even your best efforts to find them. Fortunately there are several different types of vein-detecting devices available to help you. The Veinlite (TransLite) uses high-intensity LED technology to transilluminate hard-to-find veins through the patient's subcutaneous tissue by absorbing light rather than reflecting it (Fig 10-19). The Vein Entry Indicator Device (Vascular Technologies) uses sensor technology and emits a continuous beeping sound that indicates a change in pressure when the needle penetrates a vein.

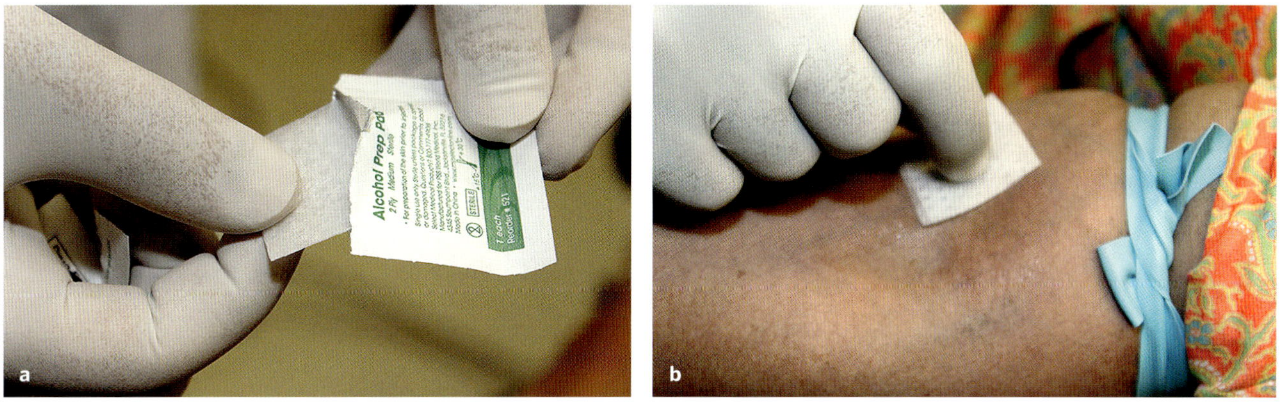

FIG 10-20 *(a and b)* Clean the venipuncture site with a 70% isopropyl alcohol pad using a back-and-forth (not a circular) motion and allow it to dry fully before proceeding.

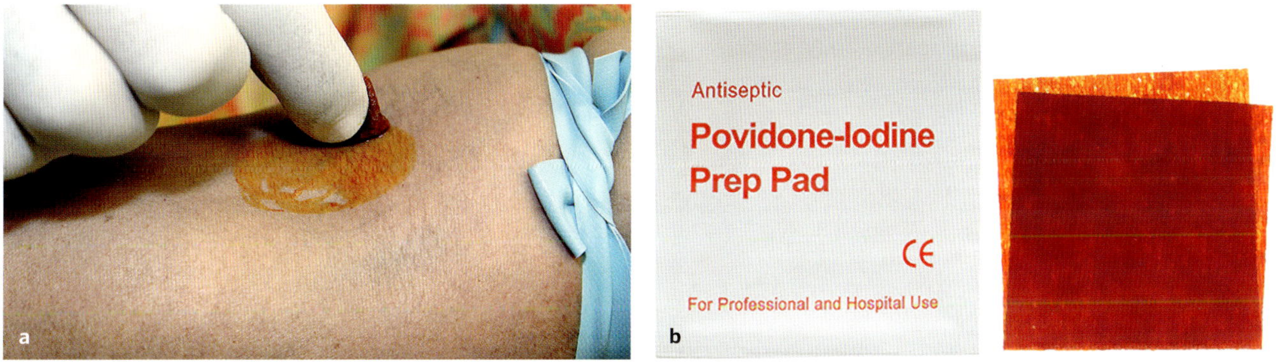

FIG 10-21 *(a and b)* A povidone-iodine prep pad is also acceptable for cleaning the site, but ask the patient if they have an iodine allergy beforehand, and be sure to remove the iodine from the patient's skin with alcohol after the phlebotomy procedure.

5. Cleaning the puncture site

Venipuncture provides an entry point for microorganisms and puts the patient at potential risk for the transmission of diseases. In most hospitals and clinics, the standard method for cleaning a venipuncture site is to scrub it with 70% isopropyl alcohol, which kills microorganisms by damaging their cell walls (Fig 10-20). It dries quickly, leaves no residue, almost never irritates or dries the skin, and does not obscure the venipuncture site. Povidone-iodine or chlorhexidine gluconate can also be used (Fig 10-21).

Once you identify a vein that is relatively large and easily palpated, open a single-use 70% isopropyl alcohol prep pad and apply it in a back-and-forth motion to clean a 2- to 3-inch diameter area over and around the vein. Wait approximately 30 seconds for the alcohol to dry, which helps kill any bacteria that are present. If the venipuncture is performed before the alcohol dries, the patient will experience a stinging sensation. Do not blow on, fan, or touch the site after it has been cleaned, as this will contaminate it.

While the alcohol is drying, do a quick visual survey of the equipment in your mobile cart or blood collection tray to make sure nothing is missing. After the alcohol has dried, apply the tourniquet.

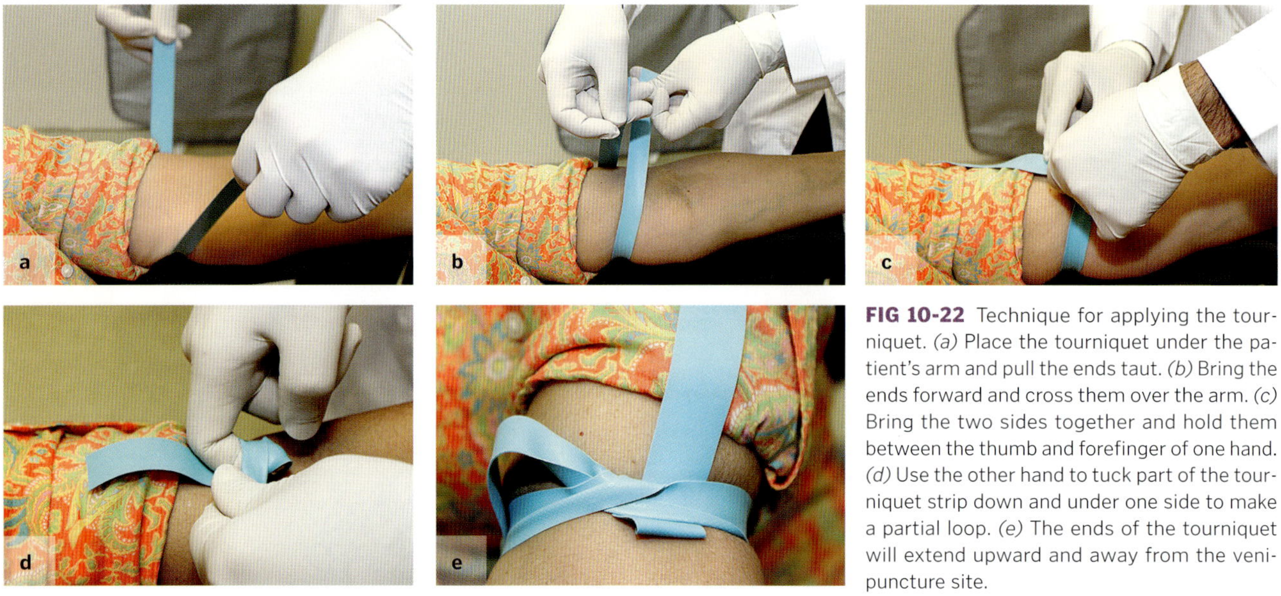

FIG 10-22 Technique for applying the tourniquet. *(a)* Place the tourniquet under the patient's arm and pull the ends taut. *(b)* Bring the ends forward and cross them over the arm. *(c)* Bring the two sides together and hold them between the thumb and forefinger of one hand. *(d)* Use the other hand to tuck part of the tourniquet strip down and under one side to make a partial loop. *(e)* The ends of the tourniquet will extend upward and away from the venipuncture site.

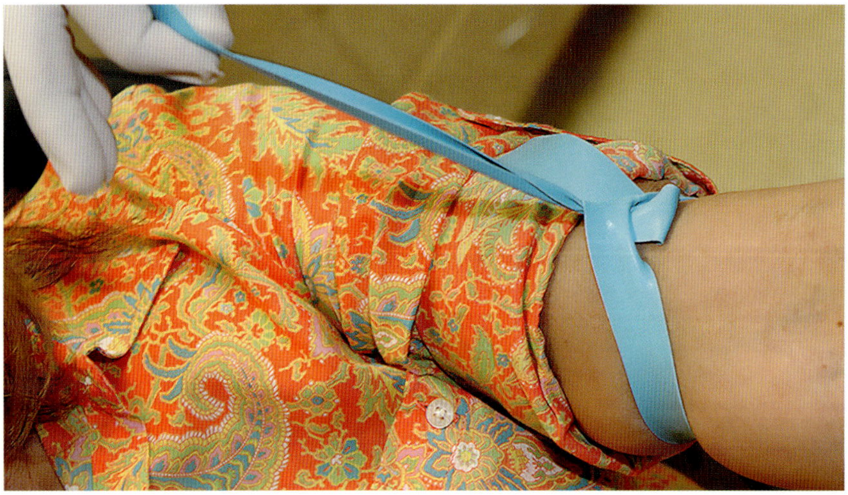

FIG 10-23 Remove the tourniquet within 1 minute of applying it using one hand and one motion.

6. Applying the tourniquet

Place the tourniquet around the patient's arm about 3 to 4 inches above the venipuncture site. To tie the latex strip so that adequate pressure is achieved without causing discomfort to the patient, hold one end in each hand and cross them over each other without tying a knot (Fig 10-22). Allow the friction of the latex to help hold the tourniquet in place. When tied properly, the ends of the tourniquet will be pointing up, away from the venipuncture site. Afterward, it can be quickly and easily released with one hand (Fig 10-23).

The tourniquet causes the inner pressure of the vein to increase artificially. When left in place for too long, it can result in a decrease in plasma volume, causing a simultaneous increase in the concentration of red blood cells. This will compromise the quality of your PRP. Therefore, never leave the tourniquet in place for more than 1 minute.

7. Performing the venipuncture

If the needle and holder that you use does not come preassembled by the manufacturer, assemble it now. Screw the stopper-puncturing end of the double-ended needle

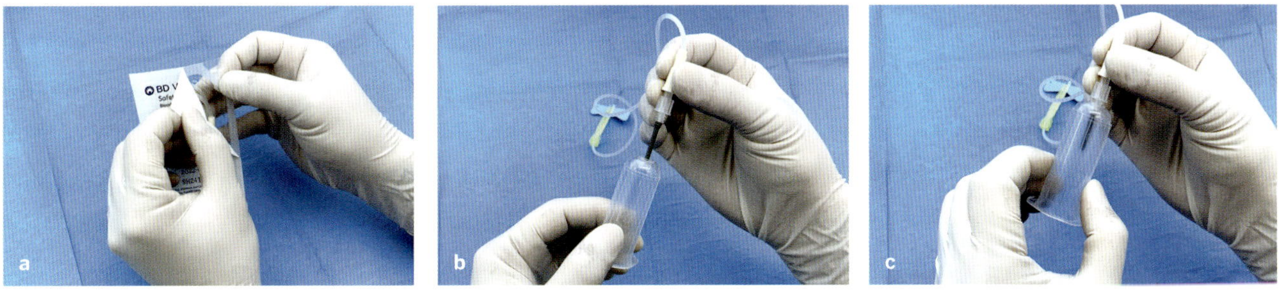

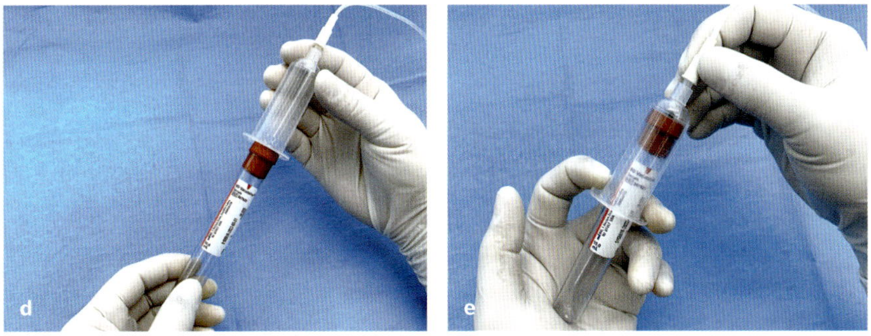

FIG 10-24 To assemble the winged needle system, open the package containing the single-use needle and tubing unit with gloved hands *(a)* and attach it to the tube holder *(b and c)*. Stretch out the coiled tubing, insert the first evacuated tube into the tube holder *(d)*, and advance the tube to the line indicated on the holder *(e)*.

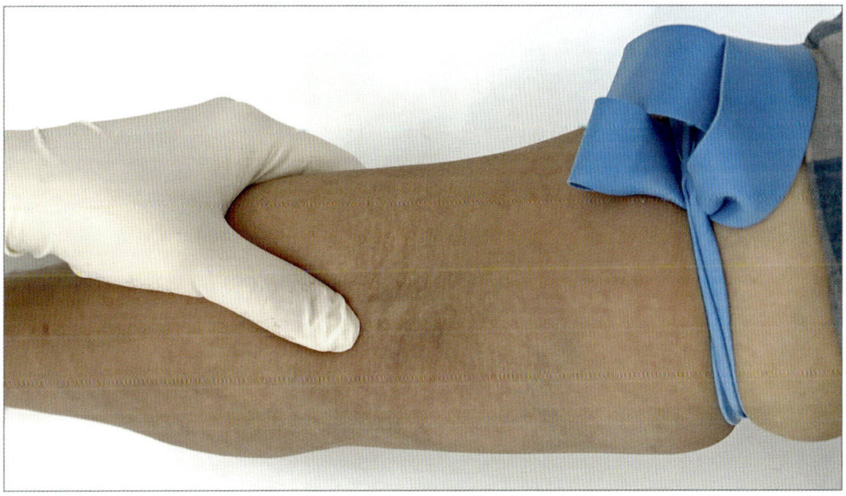

FIG 10-25 Proper placement of the nondominant hand to stabilize the patient's arm and anchor the vein.

into the evacuated tube holder; however, do not remove the colored cap from the puncturing end of the needle at this time. Insert the first evacuated tube into the holder and advance it only to the designated mark. Examine the tube to ensure that the vacuum has not been lost, which is a fairly common occurrence (Fig 10-24). For this reason, it is important to have more tubes than you actually need in your cart or tray at all times.

Stabilizing the vein

Use the thumb of your nondominant hand to stabilize the selected vein, taking care not to touch the area that was cleaned. Place it about 2 inches below and slightly to the left of the site of entry. Your four fingers should be placed on the back of the patient's arm (Fig 10-25). Using these fingers, pull the skin taut. This will stabilize the vein, prevent it from rolling, and minimize the pain of the venipuncture. When a needle punctures skin that is loose and flaccid, it is very painful and the venipuncture procedure is more likely to fail.

Inspecting the needle

Immediately before entering the vein, use your dominant hand to pick up the holder and remove the plastic cap from

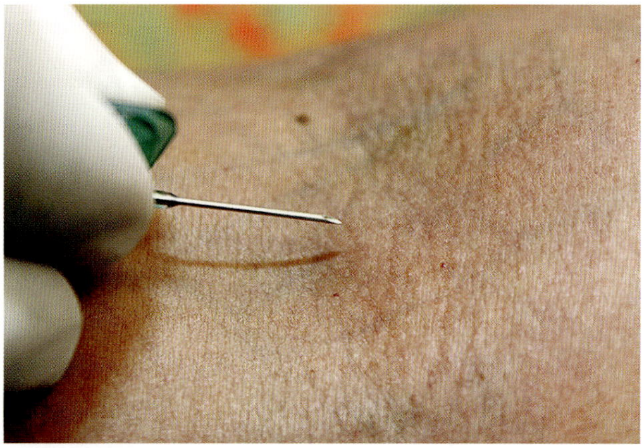

FIG 10-26 When using a winged needle, insert it at an angle of about 15 degrees.

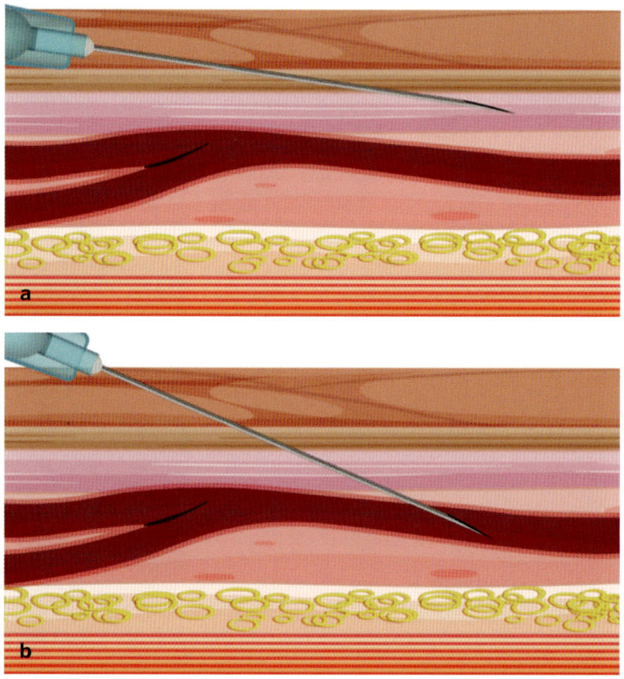

FIG 10-27 *(a and b)* For venipuncture, the angle of needle insertion should be no more than 30 degrees depending on the depth of the vein.

the needle. Inspect the point of the needle for defects, making sure the end is not bent or blunt or barbed. Discard any defective needles and assemble a new one.

Inserting the needle

With the vein securely stabilized, align the needle with the vein and insert it, bevel facing up, at an angle of 15 to 30 degrees based on the depth of the vein (Figs 10-26 and 10-27). Inserting a needle with the bevel facing down makes the venipuncture procedure more painful for the patient and also introduces the slight possibility that the back wall of the vein will occlude the needle opening.

The insertion of the needle should be done in one smooth, quick motion so that the pain it causes will be as brief as possible. Always tell the patient when you are going to insert the needle, and check with the patient afterward to make sure they are okay.

Knowing how much force to use when inserting the needle requires practice. Once the needle has entered the vein, the resistance to the needle movement that you encountered in the beginning will palpably give way. That is your cue to push the evacuated tube fully into the holder.

Technique for back of hand venipuncture

Provide support for the patient's hand and ask him or her to make a loose fist. Apply a tourniquet 3 to 4 inches above the wrist bone. Use the techniques described above to palpate the back of the patient's hand, and locate a vein that is large and straight (Fig 10-28). Stabilize the vein by placing the thumb of your nondominant hand below the patient's knuckles and pull the skin taut. With your dominant hand, insert the needle into the vein at an angle of 10 to 15 degrees, with the bevel facing up. Release the tourniquet and ask the patient to relax their fist.

8. Filling the tubes

Use the thumb of your nondominant hand to push the evacuated tube into the needle so that the needle penetrates the rubber stopper, and grasp the flanges of the

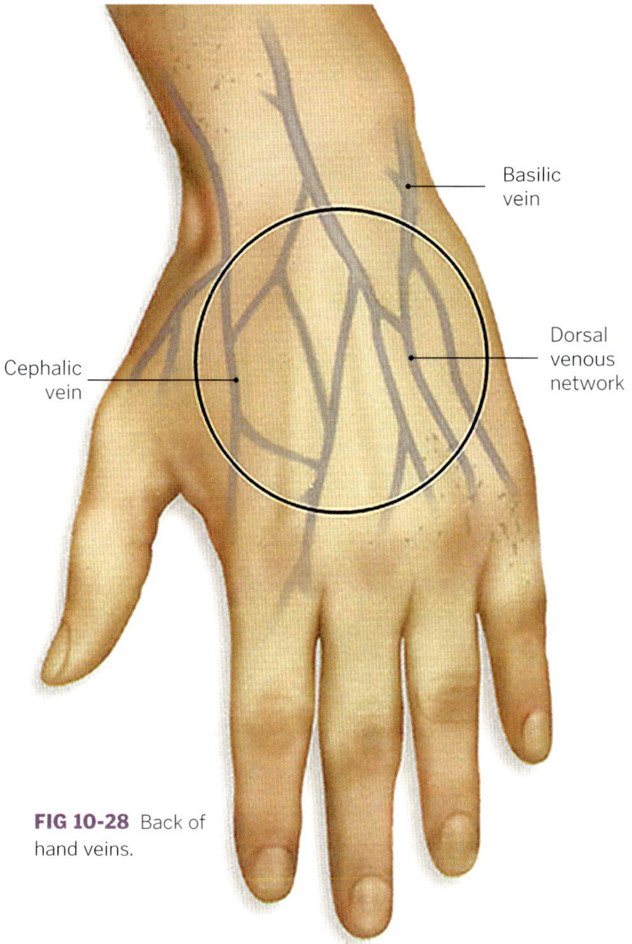

FIG 10-28 Back of hand veins.

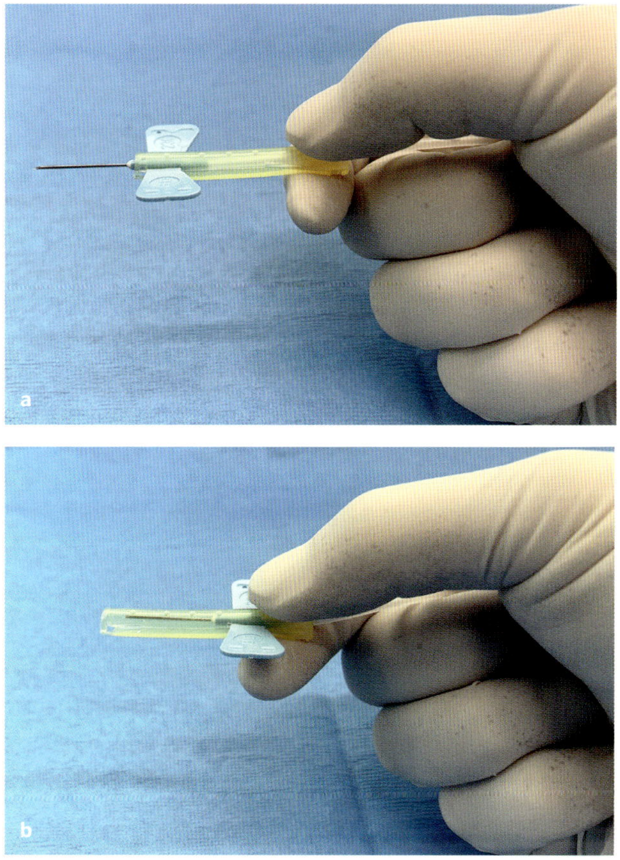

FIG 10-29 *(a and b)* After the needle is removed from the vein, the shield is automatically pushed forward and locked in place over the needle to protect the operator.

holder with the middle and index fingers to prevent it from moving. Once blood begins to flow into the tube, release the tourniquet from the patient's arm. Be sure to hold the tube at a downward angle so that the blood cannot reflux back into the needle. Meanwhile, your dominant hand should continue to hold the needle braced against the patient's arm, so that it remains immobile.

Blood will flow into the first tube until the vacuum is exhausted, and then it will stop. Do not remove the tube from the holder prematurely because the ratio of blood to anticoagulant in the yellow-stopper tubes will then be incorrect. The tube will not be completely full when it is finished filling. Once the first tube has been filled and the blood has stopped flowing, gently twist it off the puncturing needle using the flanges of the holder as a brace. Blood samples that are collected in yellow-top (ACD-A) tubes should be inverted 5 to 10 times for proper mixing of the anticoagulant and blood. If the tube is not mixed, small fibrin clots may form, causing a falsely decreased platelet count. Place the tube upright in a metal collection rack.

When the final tube has been filled, remove it from the holder and invert it the appropriate number of times before beginning the process of removing the needle. If the needle is removed before the evacuated tube, blood will drip from the needle and potentially result in contamination.

9. Removing the needle

Gently place folded gauze over the venipuncture site and then withdraw the needle in one smooth motion (Fig 10-29). Immediately apply firm pressure to the venipuncture site. After a few seconds, ask the patient to continue to apply firm pressure for approximately 30 seconds and then place a latex-free bandage or adhesive tape over the gauze (Fig 10-30). Instruct the patient to leave the bandage in place for at least 15 minutes.

Be sure to remove the tourniquet before removing the needle to avoid the formation of a hematoma.

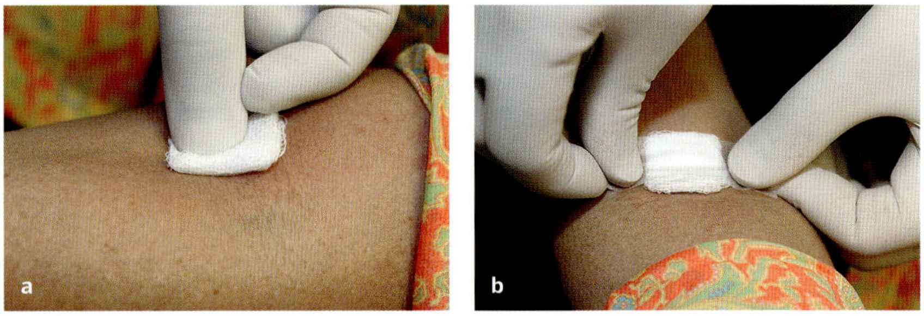

FIG 10-30 *(a)* Apply firm pressure to the venipuncture site using a gauze pad. *(b)* Then apply adhesive tape or a bandage over it to keep it in place once the bleeding has stopped.

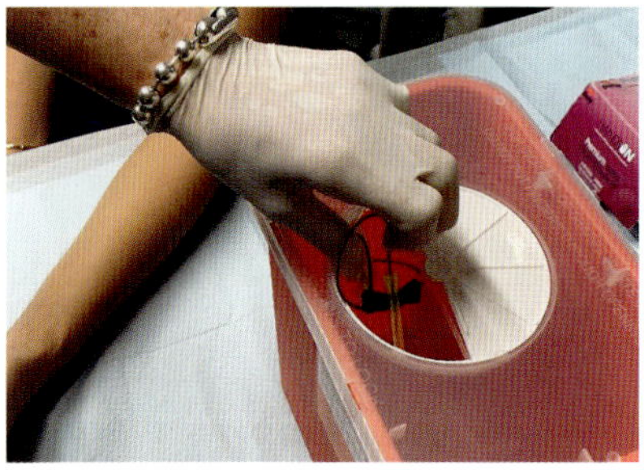

FIG 10-31 Dispose of the needle set immediately after the procedure.

10. Disposing of the needle

Immediately after the venipuncture has been completed, dispose of the contaminated needle and holder in a designated sharps container (Fig 10-31), which should be located within arm's reach of the patient.

Technical Complications

Lack of blood flow

Needle position

If you have inserted the needle but there is no appearance of blood, do not remove the needle and do not assume that the needle is not in the vein. There are a number of possible explanations and techniques for addressing them.

Bevel of needle obstructed by wall of vein. If the bevel of the needle is resting against the wall of the vein, blood will be blocked from entering the needle. Usually, the reason for this is an incorrect angle of entry of the needle. If the angle of entry was too shallow, the bevel will be obstructed by the upper wall of the vein (Fig 10-32a). Conversely, if the angle of entry was too steep, it will be obstructed by the lower wall of the vein (Fig 10-32b). Remember that the needle should be inserted at an angle of 15 to 30 degrees (depending on the size of the vein) and that the bevel of the needle should be facing up (Fig 10-32c).

The first step in addressing this problem is to remove the evacuated tube from the holder. (If the evacuated tube is engaged when the obstruction is removed, the vein can be damaged and the patient will experience pain.) Next, rotate the needle approximately 25 degrees. Finally, insert a new evacuated tube into the holder. The bevel should no longer be obstructed, and blood should start to flow.

Needle inserted too deep. When the needle is inserted at an angle greater than 30 degrees (Fig 10-32d) or is not adequately braced when a new evacuated tube is being inserted, there is a strong risk that it will pierce the back wall of the vein and project into the soft tissues. When this happens, blood can leak from the vein into the tissues and form a hematoma. To correct this, gently pull back on the needle to restore the flow of blood.

Needle inserted too shallow. When the needle is inserted at an angle of less than 15 degrees (Fig 10-32e), there is a possibility that it is only partially in the lumen of the vein. If this is the case, you may see just a small amount of blood flow very slowly into the tube. When this happens, it can usually be resolved by slowly and gently advancing the needle a bit further into the vein.

Be aware that, because the bevel of the needle was partly outside of the vein, it is possible that blood has leaked into the soft tissues and a hematoma will form under the skin. If you see this, you should remove the tourniquet and the needle immediately and apply pressure to the puncture site to minimize the bleeding and prevent injury to the patient.

Collapsed vein

A vein that closes up and constricts the flow of blood is known as a *collapsed vein* (Fig 10-32f). In venipuncture, the three main causes of a collapsed vein are *(1)* the use

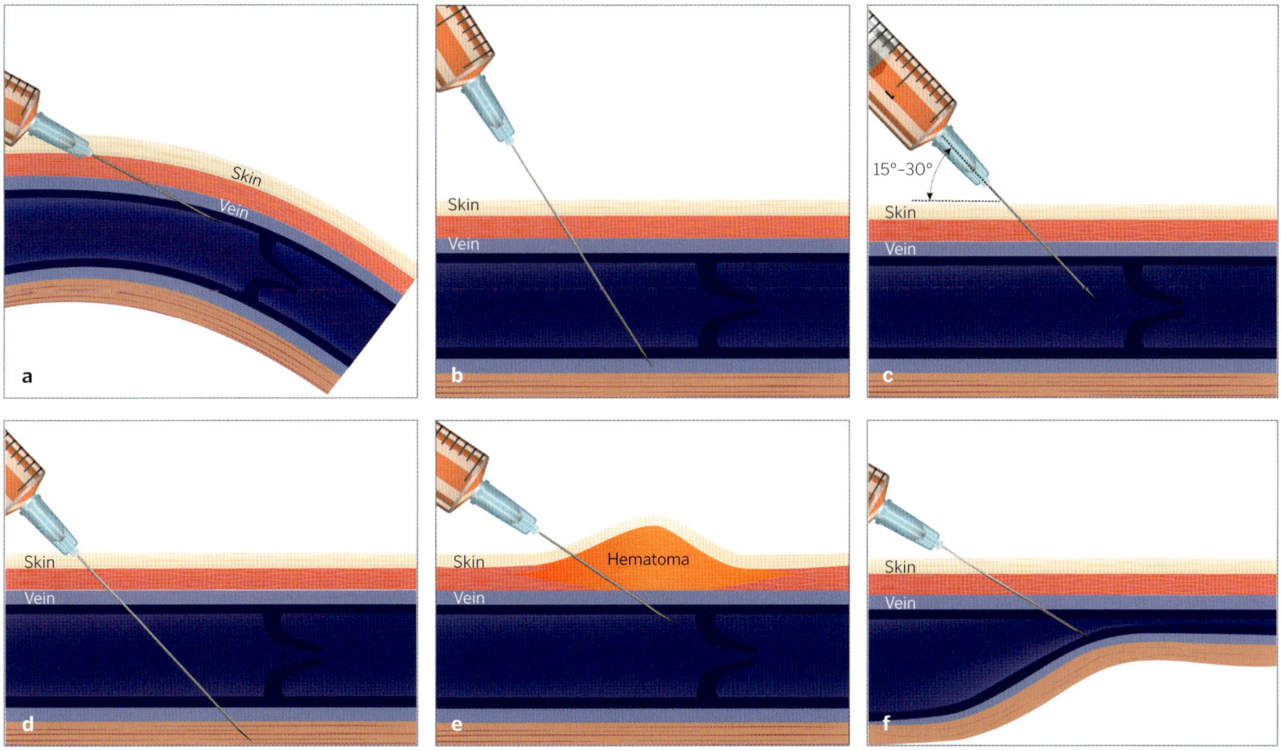

FIG 10-32 Common causes of lack of blood flow. *(a)* When the angle of needle entry is too shallow, the needle bevel will be obstructed by the upper wall of the vein, and there will be no blood flow. *(b)* When the angle of entry is too steep, the bevel will be obstructed by the lower wall of the vein. *(c)* The proper angle of entry is 15 to 30 degrees, with the bevel facing up. *(d)* When the needle is inserted too deep, it may pierce the back wall of the vein and project into the soft tissues. *(e)* When the needle is inserted too shallow, a hematoma may form under the skin. *(f)* Collapsed vein.

of an evacuated tube that is too large for the vein; *(2)* the improper use of a syringe blood collection device involving rapid withdrawal of the plunger, which places excessive pressure on the vein; and *(3)* placing the tourniquet too close to the venipuncture site or tying it too tight.

An evacuated tube that is too large for the vein should be replaced with a smaller-volume tube that will exert less pressure on the vein. Reapply a tourniquet that has been positioned improperly or tied too tight. These actions should allow the vein to refill within a few seconds.

Bad evacuated tube

If you ruled out the above causes of an obstructed blood flow, the problem may be an old and/or faulty evacuated tube that has been punctured, cracked, or otherwise lost its vacuum. In this case, replacing the evacuated tube will resolve the problem.

Regardless of the cause, when your first puncture does not yield adequate blood flow, you should withdraw the needle and repeat the procedure from the beginning on the other arm using a new needle and new evacuated tube. Never use a needle to puncture the skin more than once.

When your first attempt at venipuncture fails, it is extremely important for you to remain calm and focused on ensuring the comfort of the patient. Never insert a needle into a patient unless you have located and stabilized a vein first. Blindly probing for a vein is inhumane and can result in nerve damage, arterial puncture, or hematoma formation. If your second attempt at venipuncture also fails, do not make a third attempt. Instead, ask someone else to take over, providing the patient will consent to it.

Nerve injury

Nerve injury is an inherent risk of venipuncture. Although most anatomy textbooks tell us that the antecubital nerves lie on a plane below the antecubital veins, a 2000 study of the dissected arms of seven randomly chosen cadavers found that the nerves often overlaid the veins or were intertwined with them.[3] Therefore, although it is impossible to eliminate the risk of nerve injury entirely, it is imperative that we reduce this risk to the lowest degree possible. Box 10-2 lists the technical errors that can lead to nerve injury.

Box 10-2 Errors in technique that can cause nerve injury

- Blind probing
- Excessive manipulation of the needle
- Inserting the needle too far (angle greater than 30 degrees)
- Lateral redirection of the needle
- Movement by the patient while the needle is in the vein
- Selecting high-risk venipuncture sites (underside of the wrist, basilic vein)
- Using jerky movements

Adapted from Strassinger and Di Lorenzo.[1]

Types of nerve damage injuries

A nicked nerve can occur during venipuncture when you insert the needle without first locating the vein or when you realize that you missed the vein and engage in blind probing to try to locate it. The patient will experience an immediate sensation of an electric shock, a shooting pain, a "pins and needles" feeling, or numbness along the area of the stick in the arm or in the fingers of the arm being used for the venipuncture. If a patient informs you or demonstrates through body language that they have experienced any of these sensations, you must terminate the venipuncture procedure immediately. Treat the injured area with a cold ice pack initially followed by warm compresses, and document the details of the injury in the patient's file.

A nerve compression injury can be the result of a hematoma in subcutaneous tissue or of a tourniquet that is left on for too long or tied too tight. The swelling and numbness from nerve compression injuries usually become apparent 24 to 96 hours after the venipuncture. A diagnostic workup is warranted in the patient who complains of weakness, burning, or any degree of paresthesia during, immediately after, or within the first week following routine venipuncture.

The most frequently injured nerve is the median nerve, which lies in close proximity to the basilic vein and is the largest nerve in the arm. It runs inside the antecubital fossa and passes through the forearm into the palm of the hand. In some individuals, the median nerve passes over the basilic vein; therefore, selecting this vein before ruling out the other two less risky veins represents practice below the standard of care. Injury to the median nerve can have major implications for the patient, including loss of extension, flexion, and sensation in the forearm and hand.

Among blood donors, the incidence of nerve injury was recently found to be between 1 in 21,000 people and 1 in 26,000 venipunctures. The majority of these injuries are self-limiting and resolve spontaneously.[4]

Arterial puncture

Arterial puncture during venipuncture is a rare occurrence: The American Red Cross reported that just 12 cases of arterial puncture occurred in a 170-person blood donor center over a period of 26 months, an incidence of 0.0029%.[5] Compared with the thin walls and low pressure of a cutaneous vein, a brachial artery is firm, stiff, and under high pressure. In addition, it transports well-oxygenated blood, resulting in a faster flow and a brighter hue of red when an artery is punctured. Other signs include a needle or tubing that pulsates and the development of a hematoma. In the even rarer event that the hole created by an arterial puncture does not heal, it can develop into a brachial artery pseudoaneurysm, which can be repaired only with surgery.

If you detect any signs of an arterial puncture, you should stop the venipuncture procedure immediately and apply pressure to the site for 5 minutes. (Do not place this responsibility on the patient.) The event should be documented in the patient's record. Note, however, that if the arterial puncture yielded a sufficient quantity of blood, there is no reason why it cannot be used to prepare PRP.

Hematoma formation

A hematoma can form during or after a venipuncture procedure for a variety of reasons, although the vast

Box 10-3 Errors in technique that can cause hematoma formation

- Applying inadequate pressure to the site after needle removal
- Bending the patient's arm while applying pressure
- Excessive probing to obtain blood
- Failure to insert the needle far enough into the vein
- Inserting the needle through the vein
- Using a needle that is too large for the vein
- Selecting veins that are small and fragile
- Accidentally puncturing a brachial artery

Adapted from Strassinger and Di Lorenzo.[1]

majority are associated with an error in technique (Box 10-3). In some older patients, decreased elasticity of the vein walls can also lead to a hematoma. The use of a smaller-gauge needle and careful attention to stabilizing the vein prior to needle insertion can help prevent hematoma formation in these patients.

If you see a hematoma begin to form during venipuncture, remove the tourniquet and the needle immediately and apply firm pressure to the puncture site for 2 minutes. A cold compress should then be applied to the site to minimize swelling, and the patient should apply a warm compress upon returning home. If the patient consents to a second attempt, use the other arm.

Patient-Related Complications

Anxious or nervous patients

For some patients, the prospect of having blood drawn causes them extreme anxiety. It is helpful to identify these patients during the consultation so that arrangements can be made ahead of time to have a friend or relative accompany the patient to their appointment, or you can enlist the help of another staff member to provide support. The presence of another person can be a source of physical comfort and can help distract the patient's attention from the procedure.

Needle phobia

Although it is unlikely that an individual with a true needle phobia would consent to having an elective procedure such as PRP therapy, it is critically important to know the signs and symptoms of this condition because the consequences can be serious. Be aware that the pain these patients experience is much greater than the small pinprick that the average person feels when the needle punctures their skin. A patient with an intense fear of needles may turn pale, become light-headed, sweat profusely, and even faint. In extreme cases, these patients can exhibit symptoms consistent with being in a state of shock, potentially putting the patient in arrhythmia or cardiac arrest during or after venipuncture.

Syncope

Syncope, or fainting, is not necessarily the result of fear or apprehension but rather a malfunction of the involuntary nervous system that causes a vasovagal reaction. In many people, the condition is triggered by the sight of blood, and it leads to a sudden drop in heart rate, dilation of the veins in the legs, and reduced blood pressure. Always be alert to symptoms such as skin pallor, hyperventilation, dizziness, nausea, and/or cold clammy skin, which often immediately precede a fainting episode.

If a patient begins to faint during the procedure, remove the tourniquet and needle immediately and apply pressure to the venipuncture site. Instruct the patient to lower their head, take deep breaths, and alternately point and flex their toes, and do not allow them to leave for at least 15 minutes to be sure they are fully recovered and steady on their feet.

References

1. Strassinger SK, Di Lorenzo MS. The Phlebotomy Textbook. Philadelphia: FA Davis, 2019.
2. Ialongo C, Bernardini S. Phlebotomy, a bridge between laboratory and patient. Biochem Med (Zagreb) 2016;26:17–33.
3. Horowitz SH. Venipuncture-induced causalgia: Anatomic relations of upper extremity superficial veins and nerves, and clinical considerations. Transfusion 2000;40:1036–1040.
4. Oven SD, Johnson JD. Radial nerve injury after venipuncture. J Hand Microsurg 2017;9:43–44.
5. Newman BH. Arterial puncture phlebotomy in whole-blood donors. Transfusion 2001;41:1390–1392.

Index

Page numbers followed by "t" denote tables; those followed by "f" denote figures; and those followed by "b" denote boxes.